ROBOT ARMS

Edited by **Satoru Goto**

Robot Arms
Edited by Satoru Goto

Published by InTech
Janeza Trdine 9, 51000 Rijeka, Croatia

Publishing Process Manager Sandra Bakic

Technical Editor Teodora Smiljanic

Cover Designer InTech Design Team

Additional hard copies can be obtained from orders@intechopen.com

Robot Arms, Edited by Satoru Goto
p. cm.
ISBN 978-953-307-160-2

For EU product safety concerns:
IN TECH d.o.o., Prolaz Marije Krucifikse Kozulić 3, 51000 Rijeka, Croatia,
info@intechopen.com or visit our website at intechopen.com.

Meet the editor

Satoru Goto received B.E. and M.E. degrees in Applied Physics from Osaka University, Osaka, Japan, in 1988 and 1990, respectively, and Ph.D. degree from Osaka University in 1995. Since 1990, he has been with the Faculty of Science and Engineering, Saga University, Saga, Japan, where he was a Research Associate from 1990 to 1996, a Lecturer from 1996 to 1998, an Associate Professor from 1998 to 2007 and a Professor since 2007. Robot arms have been developing since 1960's, and those are widely used in industrial factories such as welding, painting, assembly, transportation, etc. Nowadays, the robot arms are indispensable for automation of factories. Moreover, applications of the robot arms are not limited to the industrial factory but expanded to living space or outer space. The robot arm is an integrated technology, and its technological elements are actuators, sensors, mechanism, control and system, etc.

Contents

Preface

Robot arms have been developing since 1960's, and those are widely used in industrial factories such as welding, painting, assembly, transportation, etc. Nowadays, the robot arms are indispensable for automation of factories. Moreover, applications of the robot arms are not limited to the industrial factory but expanded to living space or outer space. The robot arm is an integrated technology, and its technological elements are actuators, sensors, mechanism, control and system, etc.

Hot topics related to the robot arms are widely treated in this book such as model construction and control strategy of robot arms, robotic grasping and object handling, applications to sensing system and tele-sonography and human-robot interaction in a social setting.

I hope that the reader will be able to strengthen his/her research interests in robot arms by reading this book.

I would like to thank all the authors for their contribution and I am also grateful to the InTech staff for their support to complete this book.

Satoru Goto
Saga University
Japan

Part 1

Model and Control

1

Modeling Identification of the Nonlinear Robot Arm System Using MISO NARX Fuzzy Model and Genetic Algorithm

Ho Pham Huy Anh[1], Kyoung Kwan Ahn[2] and Nguyen Thanh Nam[3]
[1]Ho Chi Minh City University of Technology, Ho Chi Minh City
[2]FPMI Lab, Ulsan University, S. Korea
[3]DCSELAB, Viet Nam National University
Ho Chi Minh City (VNU-HCM)
Viet Nam

1. Introduction

The PAM robot arm is belonged to highly nonlinear systems where perfect knowledge of their parameters is unattainable by conventional modeling techniques because of the time-varying inertia, hysteresis and other joint friction model uncertainties. To guarantee a good tracking performance, robust-adaptive control approaches combining conventional methods with new learning techniques are required. Thanks to their universal approximation capabilities, neural networks provide the implementation tool for modeling the complex input-output relations of the multiple n DOF PAM robot arm dynamics being able to solve problems like variable-coupling complexity and state-dependency. During the last decade several neural network models and learning schemes have been applied to on-line learning of manipulator dynamics (Karakasoglu *et al.*, 1993), (Katic *et al.*, 1995). (Ahn and Anh, 2006a) have optimized successfully a pseudo-linear ARX model of the PAM robot arm using genetic algorithm. These authors in (Ahn and Anh, 2007) have identified the PAM manipulator based on recurrent neural networks. The drawback of all these results is considered the n-DOF robot arm as n independent decoupling joints. Consequently, all intrinsic coupling features of the n-DOF robot arm have not represented in its recurrent NN model respectively.

To overcome this disadvantage, in this study, a new approach of intelligent dynamic model, namely MISO NARX Fuzzy model, firstly utilized in simultaneous modeling and identification both joints of the prototype 2-axes pneumatic artificial muscle (PAM) robot arm system. This novel model concept is also applied to (Ahn and Anh, 2009) by authors.

The rest of chapter is organized as follows. Section 2 describes concisely the genetic algorithm for identifying the nonlinear NARX Fuzzy model. Section 3 is dedicated to the modeling and identification of the 2-axes PAM robot arm based on the MISO NAR Fuzzy model. Section 4 presents the experimental set-up configuration for MISO NARX Fuzzy model-based identification. The results from the MISO NARX Fuzzy model-based identification of the 2-axes PAM robot arm are presented in Section 5. Finally, in Section 6 a conclusion remark is made for this paper.

2. Genetic algorithm for NARX Fuzzy Model identification

The classic GA involves three basic operations: reproduction, crossover and mutation. As to derive a solution to a near optimal problem, GA creates a sequence of populations which corresponds to numerical values of a particular variable. Each population represents a potential solution of the problem in question. Selection is the process by which chromosomes in population containing better fitness value having greater probability of reproducing. In this paper, the roulette-wheel selection scheme is used. Through selection, chromosomes encoded with better fitness are chosen for recombination to yield off-springs for successive generations. Then natural evolution (including Crossover and Mutation) of the population will be continued until a desired termination or error criterion achieved. Resulting in a final generation contained of highly fitted chromosomes represent the optimal solution to the searching problems. Fig. 1 shows the procedure of conventional GA optimization.

It needs to tune following parameters before running the GA algorithm:

D: number of chromosomes chosen for mating as parents

N : number of chromosomes in each generation

L_t: number of generations tolerated for no improvement on the value of the fitness before MGA terminated

L_e: number of generations tolerated for no improvement on the value of the fitness before the extinction operator is applied. It need to pay attention that $L_e \langle\langle L_t$.

ρ : portion of chosen parents permitted to be survived into the next generation

q: percentage of chromosomes are survived according to their fitness values in the extinction strategy

The steps of MGA-based NARX Fuzzy model identification procedure are summarized as:

Step 1. Implement tuning parameters described as above. Encode estimated parameters into genes and chromosomes as a string of binary digits. Considering that parameters lie in several bounded region η_k

$$|w_k| \leq \eta_k \text{ for } k=1,\ldots,h. \tag{1}$$

The length of chromosome needed to encode w_k is based on η_k and the desired accuracy δ_k. Set $i=k=m=0$.

Step 2. Generate randomly the initial generation of N chromosomes. Set $i=i+1$.

Step 3. Decode the chromosomes then calculate the fitness value for every chromosome of population in the generation. Consider $F^i_{\max}$ the maximum fitness value in the ith generation.

Step 4. Apply the Elitist strategies to guarantee the survival of the best chromosome in each generation. Then apply the *G-bit strategy* to this chromosome for improving the efficiency of MGA in local search.

Step 5.

1. *Reproduction:* In this paper, reproduction is set as a linear search through roulette wheel values weighted proportional to the fitness value of the individual chromosome. Each chromosome is reproduced with the probability of

$$\frac{F_j}{\sum_{j=1}^{N} F_j}$$

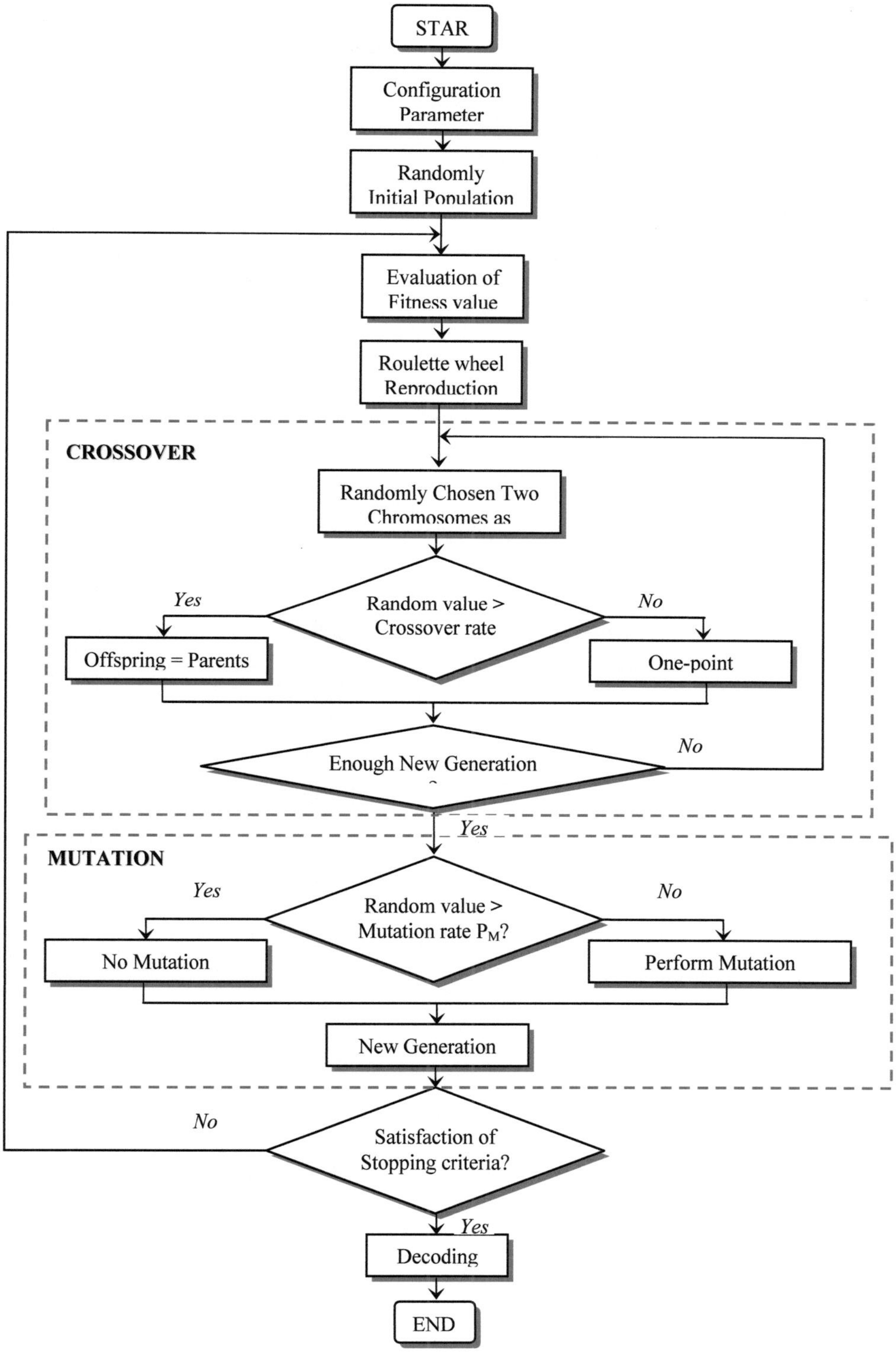

Fig. 1. The flow chart of conventional GA optimization procedure.

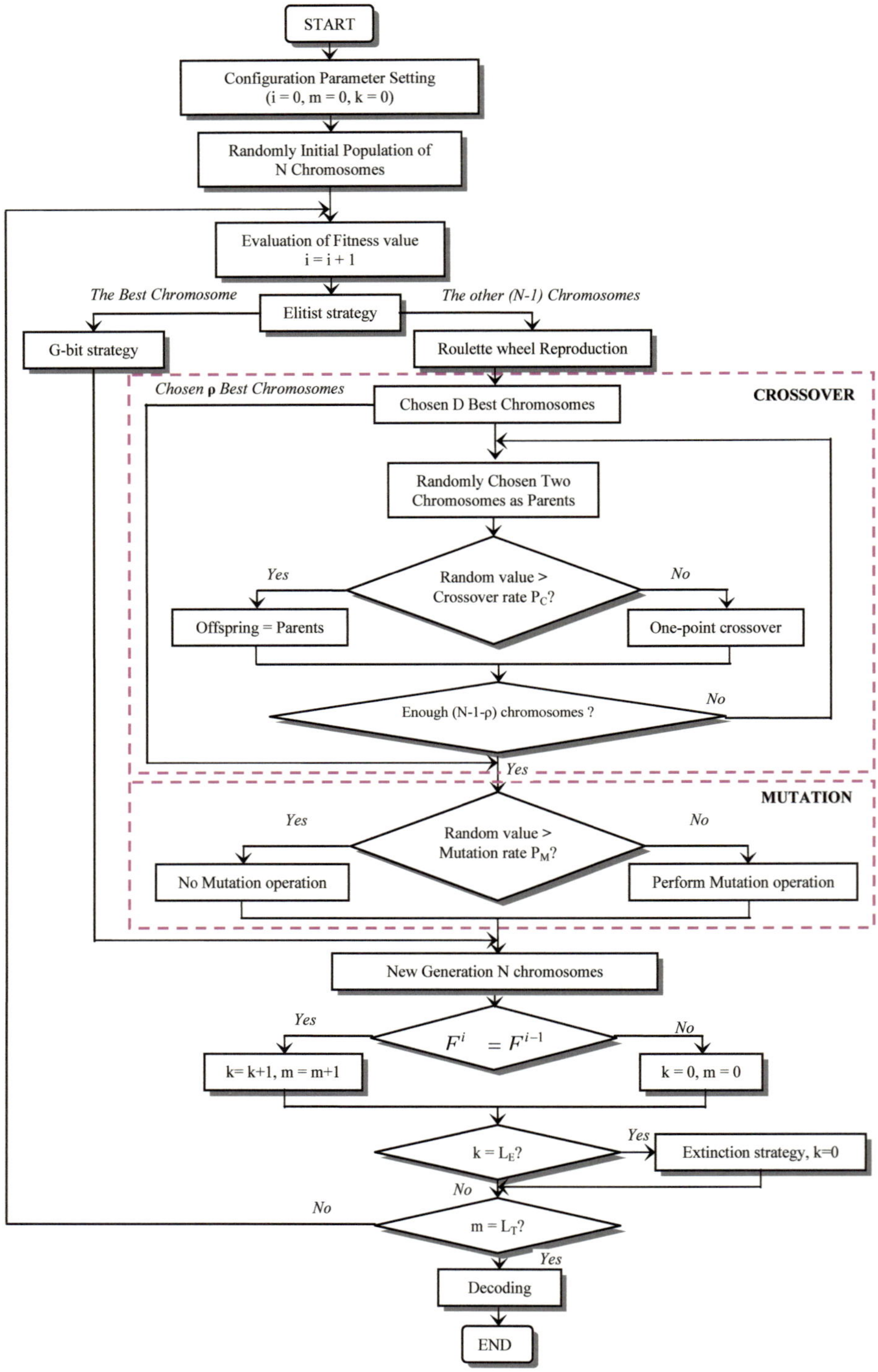

Fig. 2. The flow chart of the modified GA optimization procedure.

with j being the index of the chromosome ($j=1,\ldots,N$). Furthermore, in order to prevent some strings possess relatively high fitness values which would lead to premature parameter convergence, in practice, linear fitness scaling will be applied.

2. *Crossover:* Choose D chromosomes possessing maximum fitness value among N chromosomes of the present gene pool for mating and then some of them, called ρ best chromosomes, are allowed to survive into the next generation. The process of mating D parents with the crossover rate p_c will generate $(N-\rho)$ children. Pay attention that, in the identification process, it is focused the mating on parameter level rather than on chromosome level.
3. *Mutation:* Mutate a bit of string ($0 \leftrightarrow 1$) with the mutation rate P_m.

Step 6. Compare if $F^{i}_{\max} = F^{i-1}_{\max}$, then $k=k+1$, $m=m+1$; otherwise, $k=0$ and $m=0$.

Step 7. Compare if $k=L_e$, then apply the extinction strategy with $k=0$.

Step 8. Compare if $m=L_t$, then terminate the MGA algorithm; otherwise go to Step 3.

Fig. 2 shows the procedure of modified genetic algorithm (MGA) optimization.

3. Identification of the 2-Axes PAM robot arm based on MISO NARX fuzzy model

3.1 Assumptions and constraints

Firstly, it is assumed that symmetrical membership functions about the y-axis will provide a valid fuzzy model. A symmetrical rule-base is also assumed. Other constraints are also introduced to the design of the MISO NARX Fuzzy Model (MNFM).

- All universes of discourses are normalized to lie between -1 and 1 with scaling factors external to the DNFM used to give appropriate values to the input and output variables.
- It is assumed that the first and last membership functions have their apexes at -1 and 1 respectively. This can be justified by the fact that changing the external scaling would have similar effect to changing these positions.
- Only triangular membership functions are to be used.
- The number of fuzzy sets is constrained to be an odd integer greater than unity. In combination with the symmetry requirement, this means that the central membership function for all variables will have its apex at zero.
- The base vertices of membership functions are coincident with the apex of the adjacent membership functions. This ensures the value of any input variable is a member of at most two fuzzy sets, which is an intuitively sensible situation. It also ensures that when a variable's membership of any set is certain, i.e. unity, it is a member of no other sets.

Using these constraints the design of the DNFM input and output membership functions can be described using two parameters which include the number of membership functions and the positioning of the triangle apexes.

3.2 Spacing parameter

The second parameter specifies how the centers are spaced out across the universe of discourse. A value of one indicates even spacing, while a value larger than unity indicates that the membership functions are closer together in the center of the range and more spaced out at the extremes as shown in Fig.3. The position of each center is calculated by taking the position the centre would be if the spacing were even and by raising this to the power of the spacing parameter. For example, in the case where there are five sets, with even spacing (*p* =1) the center of one set would be at 0.5. If p is modified to two, the position of this center moves to 0.25. If the spacing parameter is set to 0.5, this center moves to $(0.5)^{0.5}$ = 0.707 in the normalized universe of discourse. Fig. 3 presents Triangle input membership function with spacing factor = 0.5.

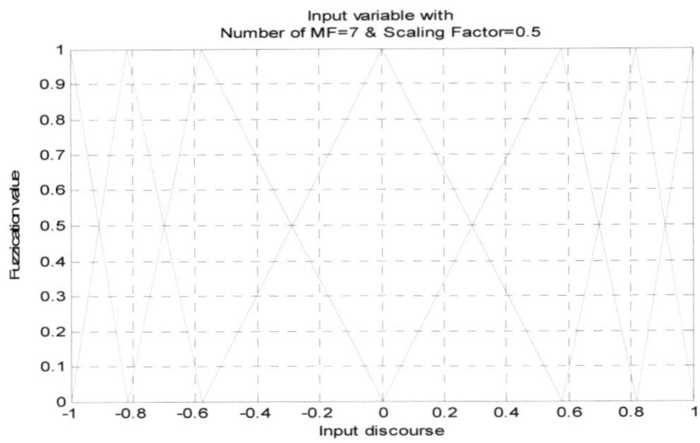

Fig. 3. Triangle input membership function with spacing factor = 0.5.

3.3 Designing the rule base

As well as specifying the membership functions, the rule-base also needs to be designed. Again idea presented by Cheong in was applied. In specifying a rule base, characteristic spacing parameters for each variable and characteristic angle for each output variable are used to construct the rules.

Certain characteristics of the rule-base are assumed in using the proposed construction method:

- Extreme outputs more usually occur when the inputs have extreme values while mid-range outputs generally are generated when the input values are mid-range.
- Similar combinations of input linguistic values lead to similar output values.

Using these assumptions the output space is partitioned into different regions corresponding to different output linguistic values. How the space is partitioned is determined by the characteristic spacing parameters and the characteristic angle. The angle determines the slope of a line through the origin on which seed points are placed. The positioning of the seed points is determined by a similar spacing method as was used to determine the center of the membership function.

Grid points are also placed in the output space representing each possible combination of input linguistic values. These are spaced in the same way as before. The rule-base is determined by calculating which seed-point is closest to each grid point. The output linguistic value representing the seed-point is set as the consequent of the antecedent represented by the grid point. This is illustrated in Fig. 4a, which is a graph showing seed points (blue circles) and grid-points (red circles). Fig. 4b shows the derived rule base. The lines on the graph delineate the different regions corresponding to different consequents. The parameters for this example are 0.9 for both input spacing parameters, 1 for the output spacing parameter and 45° for the angle theta parameter.

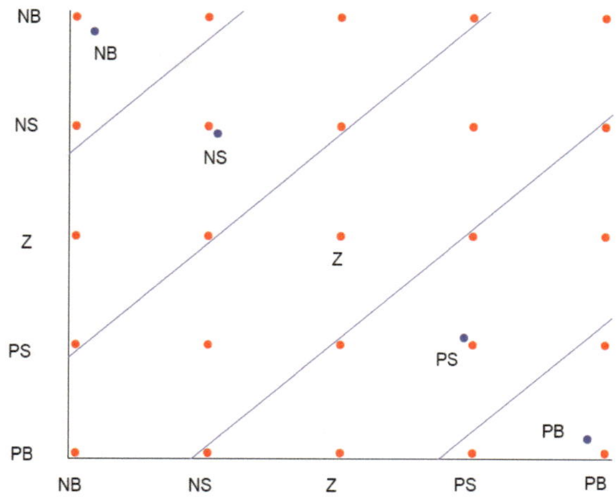

Fig. 4a. The Seed Points and the Grid Points for Rule-Base Construction

de \ e	NB	NS	Z	PS	PB
NB	NB	NB	NS	NS	Z
NS	NB	NS	NS	Z	PS
Z	NS	NS	Z	PS	PS
PS	NS	Z	PS	PS	PB
PB	Z	PS	PS	PB	PB

Fig. 4b. Derived Rule Base.

3.4 Parameter encoding

To run a MGA, a suitable encoding for each of the parameters and bounds for each of them needs to be carefully decided. For this task the parameters given in Table 1 are used with the shown ranges and precisions. Binary encoding is used as it is felt that this allows the MGA more flexible to search the solution space more thoroughly. The numbers of membership functions are limited to the odd integers inclusive between (3 - 9) in case MGA-based PAM robot arm Inverse and Forward TS fuzzy model and between (3-5) in case MGA-based PAM robot arm Inverse and Forward NARX Fuzzy model identification. Experimentally, this was considered to be a reasonable constraint to apply. The advantage of doing this is that this parameter can be captured in just one to two bits per variable.
For the spacing parameters, two separate parameters are used. The first, within the range [0.1- 1.0], determines the magnitude and the second, which takes only the values -1 or 1, is the power by which the magnitude is to be raised. This determines whether the membership functions compress in the center or at the extremes. Consequently, each spacing parameter obtains the range [0.1 - 10]. The precision required for the magnitude is 0.01, meaning that 8 bits are used in total for each spacing parameter. The scaling for the

input variables is allowed to vary in the range [0 – 100], while that of the output variable is given the range [0 – 1000].

Parameter	Range	Precision	No. of Bits
Number of Membership Functions	3-9	2	2
Membership Function Spacing	0.1 – 1.0	0.1	7
Membership Function	-1 - 1	2	1
Rule-Base Scaling	0.1 – 1.0	0.01	7
Rule-Base Spacing	-1 - 1	2	1
Input Scaling	0 - 100	0.1	10
Output Scaling	0 - 1000	0.1	17
Rule-Base Angle	0 - 2π	π/512	11

Table 1. MGA-based Inverse and Forward NARX Fuzzy Model Parameters used for encoding.

3.5 Inverse and forward MISO NARX fuzzy models of the 2-Axes PAM robot arm

The newly proposed Inverse and Forward MISO NARX Fuzzy model of the PAM robot arm presented in this paper is improved by combining the extraordinary predictive and adaptive features of the Nonlinear Auto-Regressive with eXogenous input (NARX) model structure. The resulting model established a nonlinear relation between the past inputs and outputs and the predicted output, the system prediction output is combination of system output produced by real inputs and system historical behaviors. It can be expressed as:

$$\hat{y}(k) = f\left(y(k-1),\dots,y(k-n_a),u(k-n_d),\dots,u(k-n_b-n_d)\right) \tag{2}$$

Here, n_a and n_b are the maximum lag considered for the output, and input terms, respectively, n_d is the discrete dead time, and f represents the mapping of fuzzy model.
The structure of the newly proposed MISO NARX TS fuzzy model is that this MISO NARX TS fuzzy model interpolates between local linear, time-invariant (LTI) ARX models as follows:
Rule j: if $z_1(k)$ *is* $A_{1,j}$ *and ... and* $z_n(k)$ *is* $A_{n,j}$ then

$$\hat{y}(k) = \sum_{i=1}^{na} a_i^j y(k-i) + \sum_{i=1}^{nb} b_i^j u(k-i-n_d) + c^j \tag{3}$$

where the element of $z(k)$ *"scheduling vector"* are usually a subset of the $x(k)$ regressors that contains the variables relevant to the nonlinear behaviors of the system,

$$Z(k) \in \left\{y(k-1),\dots,y(k-n_a),u(k-n_d),\dots,u(k-n_b-n_d)\right\} \tag{4}$$

while the $f_j(q(k))$ consequent function contains all the regressor $q(k)=[X(k)\ 1]$,

$$f_j(q(k)) = \sum_{i=1}^{na} a_i^j y(k-i) + \sum_{i=1}^{nb} b_i^j u(k-i-n_d) + c^j \tag{5}$$

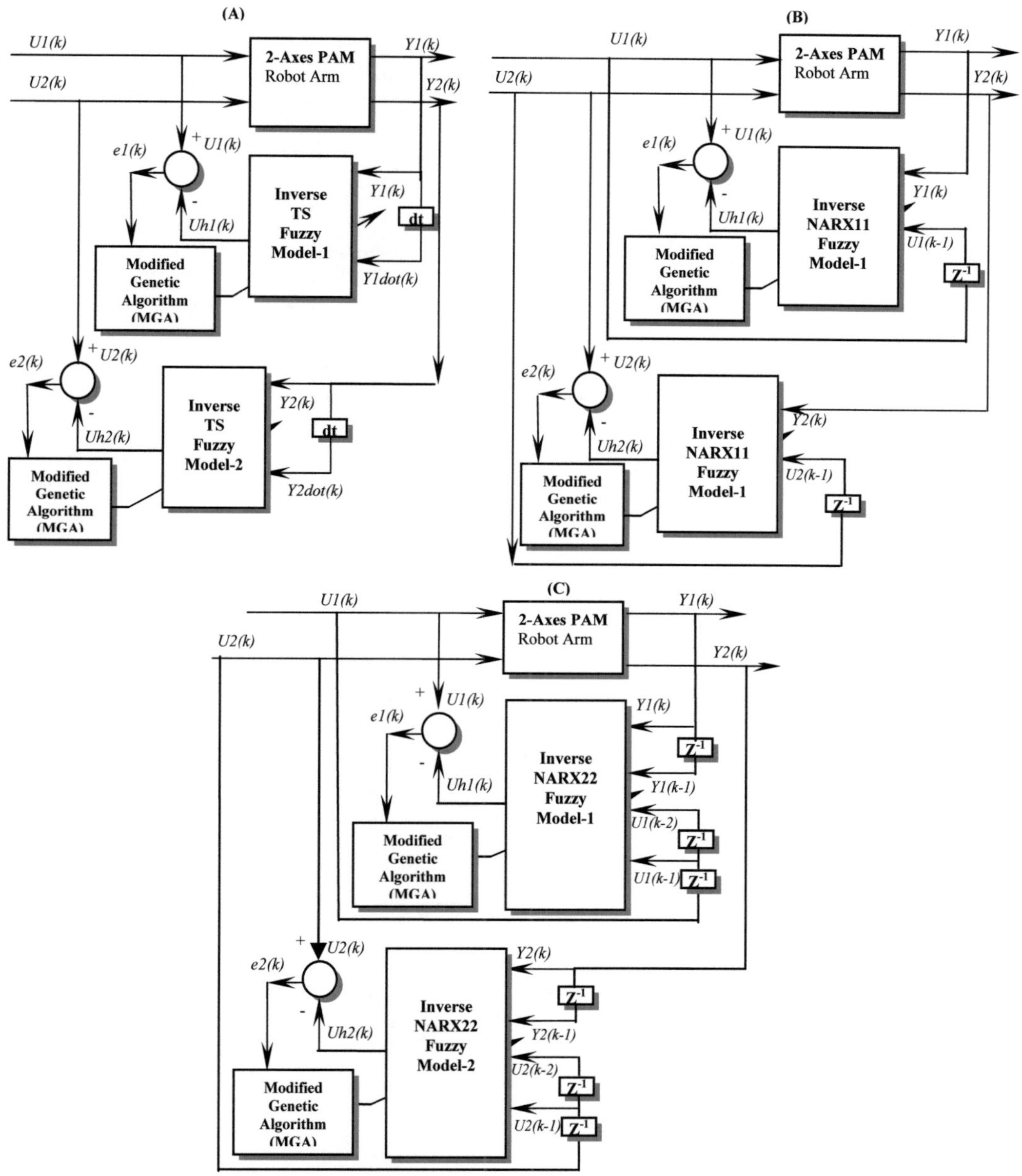

Fig. 5. Block diagrams of The MGA-based 2-Axes PAM robot arm Inverse MISO Fuzzy Model Identification.

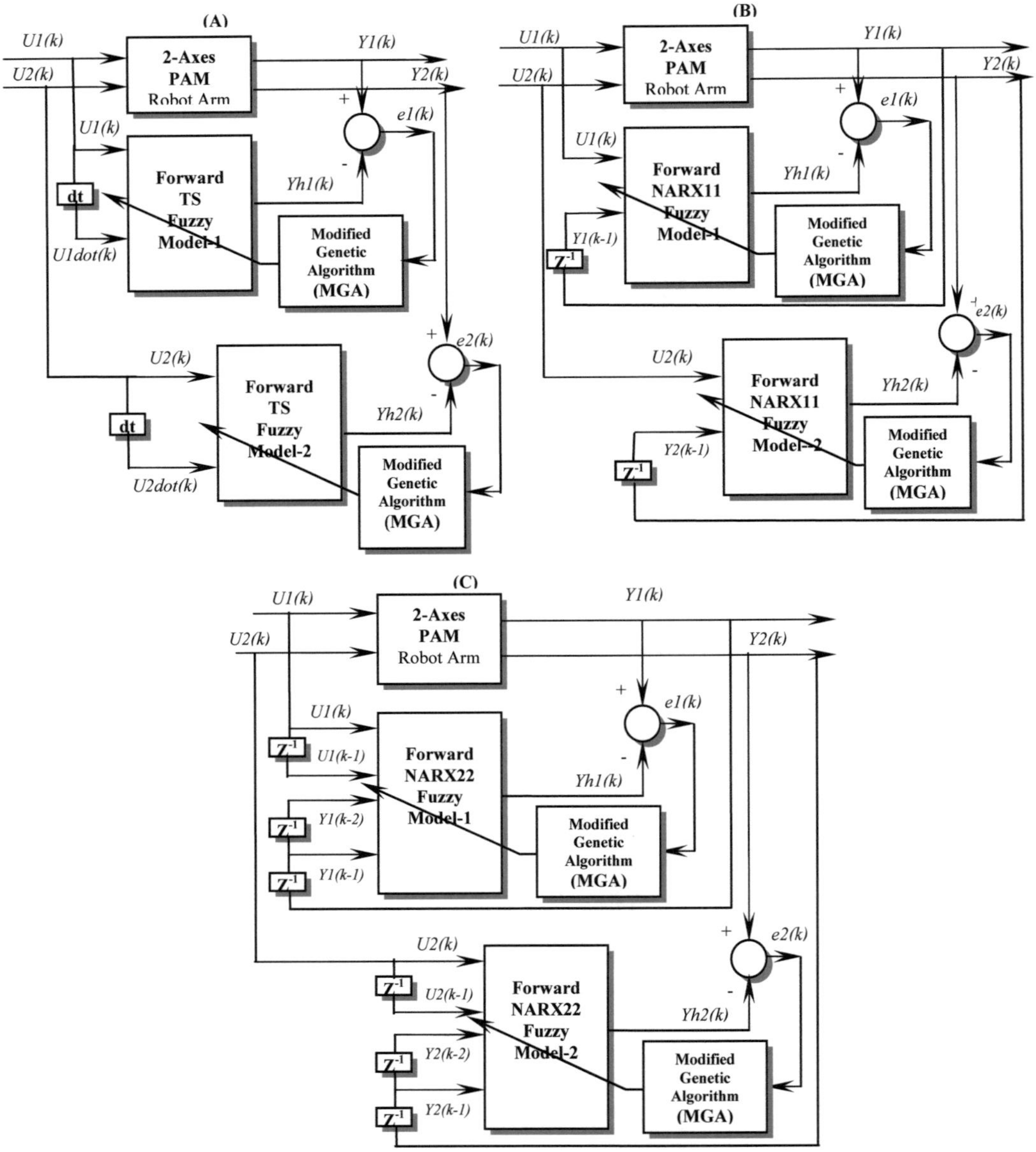

Fig. 6. Block diagrams of The MGA-based 2-Axes PAM robot arm Forward MISO Fuzzy Model Identification.

In the simplest case, the NARX type zero-order TS fuzzy model (singleton or Sugeno fuzzy model which is not applied in this paper) is formulated by simple rules consequents as:

Rule j: if $z_1(k)$ *is* $A_{1,j}$ *and … and* $z_n(k)$ *is* $A_{n,j}$ then

$$\hat{y}(k) = c^j \tag{6}$$

with the *z(k)* contains all inputs of the NARX model:

$$Z(k) = X(k) = \{y(k-1),\ldots,y(k-n_a),u(k-n_d),\ldots,u(k-n_b-n_d)\} \tag{7}$$

Thus the difference between NARX fuzzy model and Fuzzy TS model method is that the output from Inverse TS fuzzy model is linear and constant, and the output from Inverse NARX fuzzy model is NARX function. But they have same fuzzy inference structure (FIS).
The block diagrams presented in Fig. 5a and Fig. 5b illustrate the difference between the MGA-based PAM robot arm Inverse MISO TS Fuzzy model and the MGA-based PAM robot arm Inverse MISO NARX Fuzzy model identification. Forwardly, the block diagrams presented in Fig. 5b and Fig.5c illustrate the difference between the MGA-based PAM robot arm Inverse MISO NARX11 Fuzzy model identification and Inverse MISO NARX22 Fuzzy model identification.
Likewise, the block diagrams presented in Fig. 6a and Fig. 6b illustrate the difference between the MGA-based PAM robot arm Forward MISO TS Fuzzy model and the MGA-based PAM robot arm Forward MISO NARX Fuzzy model identification. Forwardly, the block diagrams presented in Fig. 6b and Fig.6c illustrate the difference between the MGA-based PAM robot arm Forward MISO NARX11 Fuzzy model identification and Forward MISO NARX22 Fuzzy model identification.

4. Identification of inverse and forward MISO NARX fuzzy models

The schematic diagram of the prototype 2-Axes PAM robot arm and the block diagram of the experimental apparatus are shown in Fig.7 and Fig. 8.

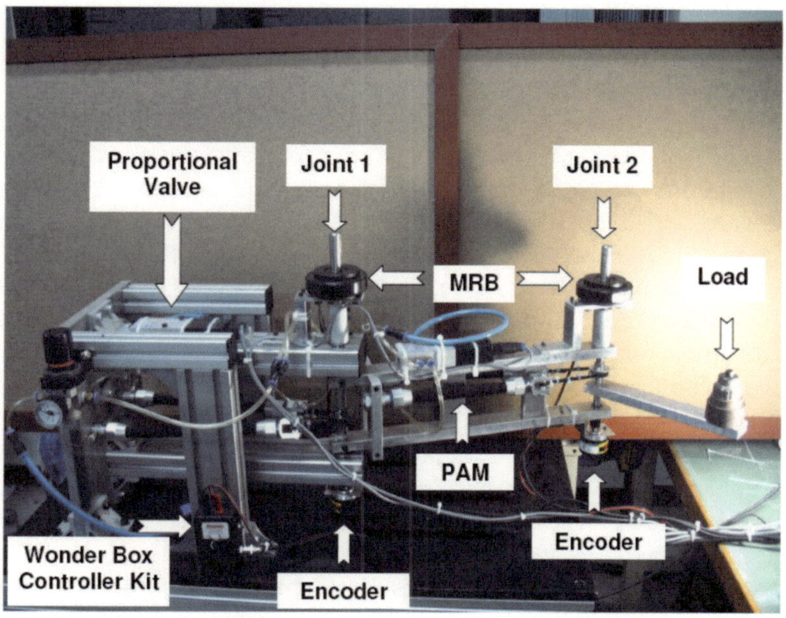

Fig. 7. General configuration of 2- axes PAM robot arm

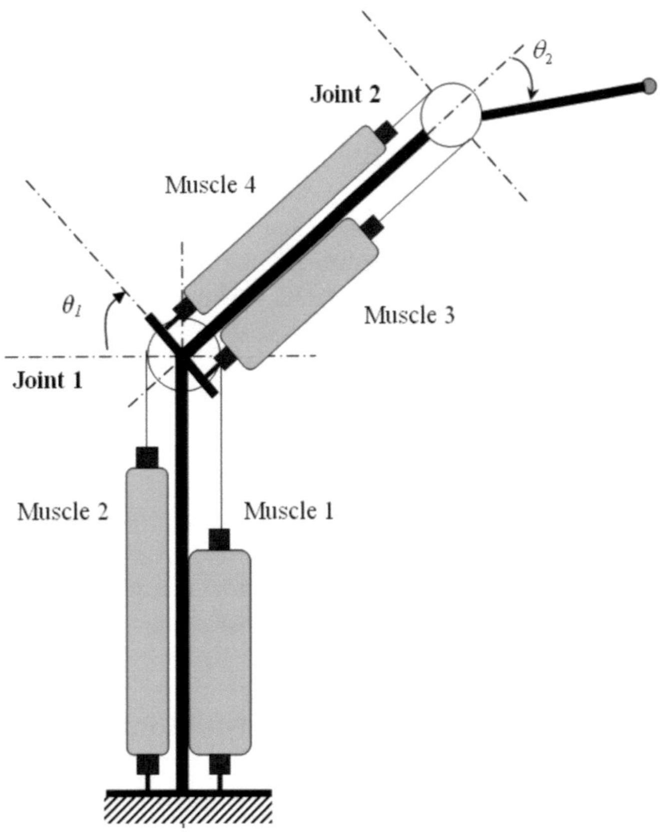

Fig. 8. Working principle of the 2-axes PAM robot arm.

In general, the procedure which must be executed when attempting to identify a dynamical system consists of four basic steps (Fig. 9).

To realize Step 1, Fig. 10 presents the PRBS input applied simultaneously to the 2 joints of the tested 2-axes PAM robot arm and the responding joint angle outputs collected from both of them. This experimental PRBS input-output data is used for training and validating not only the Forward MISO NARX Fuzzy model (see Fig. 10a) but also for training and validating the Inverse MISO NARX Fuzzy model (see Fig. 10b) of the whole dynamic two-joint structure of the 2-axes PAM robot arm.

PRBS input and Joint Angle output from (40–80)[s] will be used for training, while PRBS input and Joint Angle output from (0–40)[s] will be used for validation purpose. The range (4.3 - 5.7) [V] and the shape of PRBS voltage input applied to the 1st joint as well as the range (4.5 - 5.5) [V] and the shape of PRBS voltage input applied to rotate the 2nd joint of the 2-axes PAM robot arm is chosen carefully from practical experience based on the hardware set-up using proportional valve to control rotating joint angle of both of PAM antagonistic pair. The experiment results of 2-axes PAM robot arm position control prove that experimental control voltages $u_1(t)$ and $u_2(t)$ applied to both of PAM antagonistic pairs of the 2-axes PAM robot arm is to function well in these ranges. Furthermore, the chosen frequency of PRBS signal is also chosen carefully based on the working frequency of the 2-axes PAM robot arm will be used as an elbow and wrist rehabilitation device in the range of (0.025 - 0.2) [Hz].

5. Experiment results

Three different identification models were carried out, which include MGA-based 2-axes PAM robot arm's MISO *UUdot* fuzzy model identification, MGA-based 2-axes PAM robot arm's MISO NARX11 fuzzy model identification, and MGA-based PAM 2-axes robot arm's MISO NARX22 fuzzy model identification, respectively.

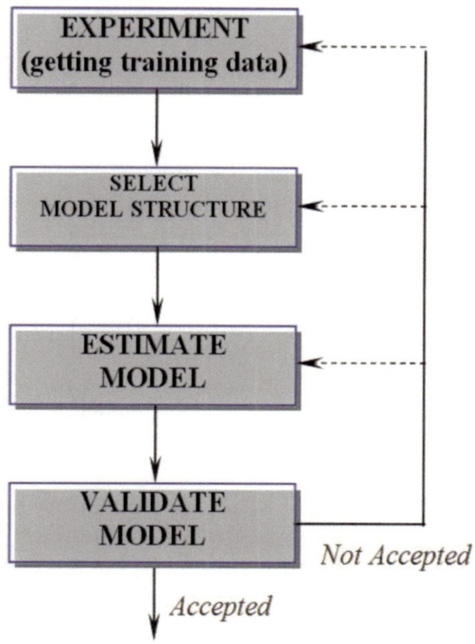

Fig. 9. MISO NARX Fuzzy Model Identification procedure

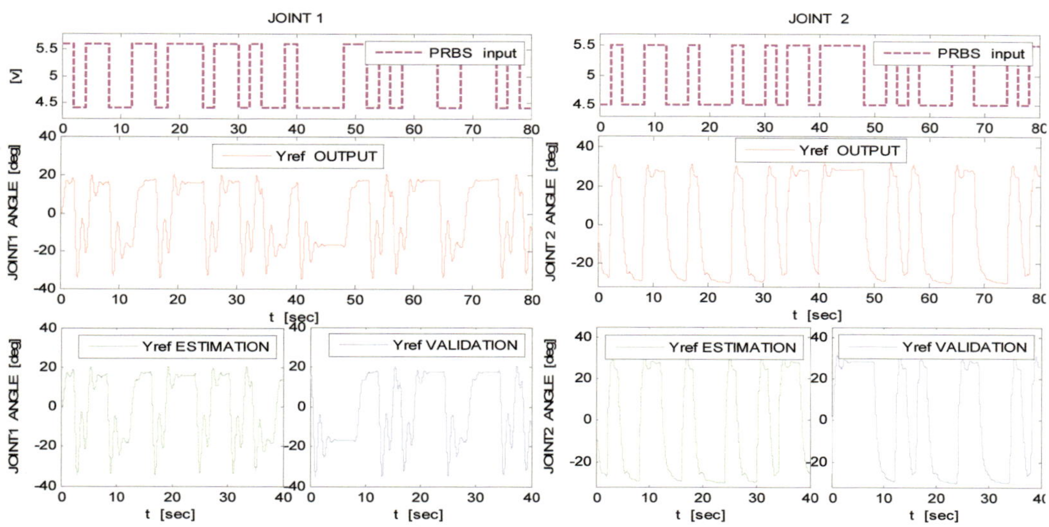

Fig. 10a. Forward MISO NARX Fuzzy Model Training data obtained by experiment.

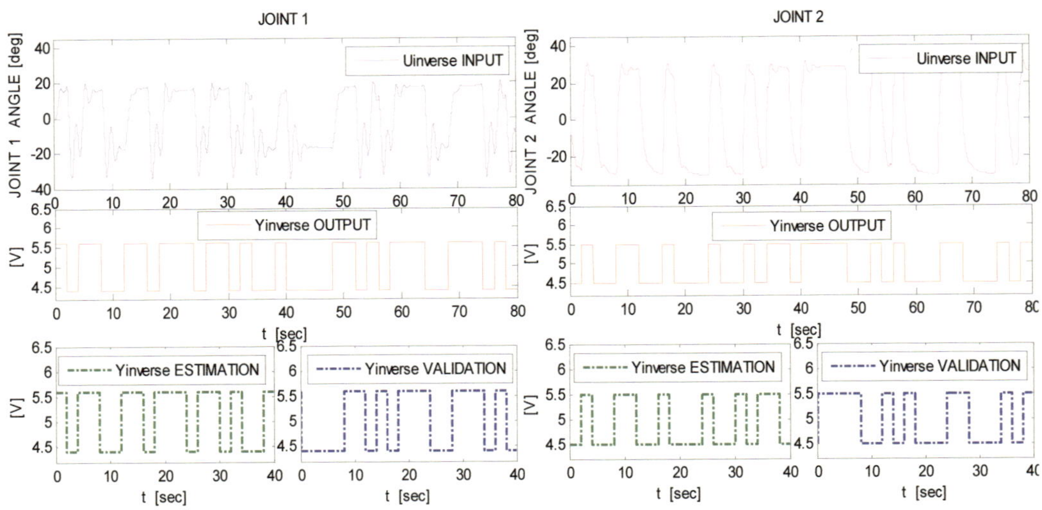

Fig. 10b. Inverse MISO NARX Fuzzy Model Training data obtained by experiment.

5.1 MGA-based 2-axes PAM robot arm forward MISO NARX fuzzy model identification

The identification procedure bases on the experimental input-output data values measured from the 2-axes PAM robot arm. Table 1 tabulates fuzzy model parameters used for encoding as optimized input values of MGA optimization algorithm. The range (3–5) permits the variable of number of membership functions obtaining 2 different odd values would be chosen by MGA (3 and 5). Block diagrams in Fig.5a, Fig.5b and Fig.5c illustrate the MGA-Based 2-axes PAM robot arm's forward MISO Fuzzy model identification.

The fitness value of MGA-based optimization calculated based on Eq. (8) is presented in Fig. 11 (with population = 40 and generation = 150).

$$F_j = 10^4.(\frac{1}{M}\sum_{k=1}^{M}(y(k)-\hat{y}_j(k))^2)^{-1} \tag{8}$$

This Figure represents the fitness convergence values of both Forward Fuzzy models of both joints of the 2-axes PAM robot arm corresponding to three identification methods. This Figure shows that the fitness value of Forward MISO UUdot fuzzy model falls early at 10th generation into a local optimal trap equal 1050 with joint 1 and 1250 with joint 2. The reason is that UUdot fuzzy model can't cover nonlinear features of the 2-axes PAM robot arm implied in input signals *U* [v] and *Udot* [v/s]. On the contrary, the fitness value of Forward MISO NARX fuzzy model obtains excellently the global optimal value (equal 2350 with joint 1 and 12600 with joint 2 in case of Forward MISO NARX11 fuzzy model and equal 9350 with joint 1 and 10400 with joint 2 in case of Forward MISO NARX22 fuzzy model). The cause is due to novel Forward MISO NARX fuzzy model combines the extraordinary approximating capacity of fuzzy system with powerful predictive and adaptive potentiality of the nonlinear NARX structure implied in Forward NARX Fuzzy Model. Consequently, resulting Forward MISO NARX11 and Forward MISO NARX22 fuzzy model as well cover excellently most of nonlinear features of the 2-axes PAM robot arm implied in input signals *U(z)*[v] and *Y(z-1)* [deg].

Consequently, the validating result of the MGA-based identified 2-axes PAM robot arm's Forward MISO NARX fuzzy model presented in Fig. 12 also shows a very good range of error (< [$\pm 5^o$] with joint 1 and < [$\pm 1^o$] with joint 2 in case of Forward MISO NARX11 fuzzy

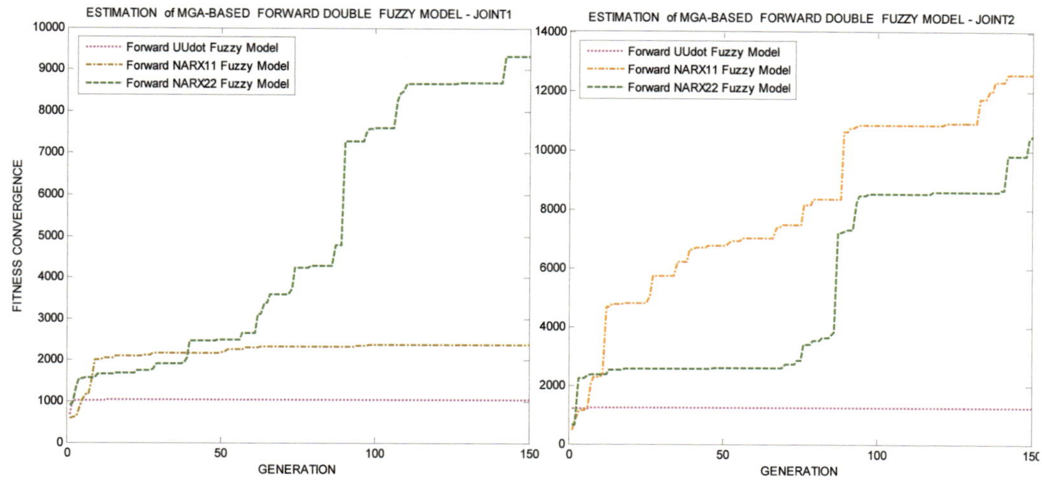

Fig. 11. Fitness Convergence of MGA-based Forward MISO Fuzzy Model optimization of the 2-axes PAM robot arm

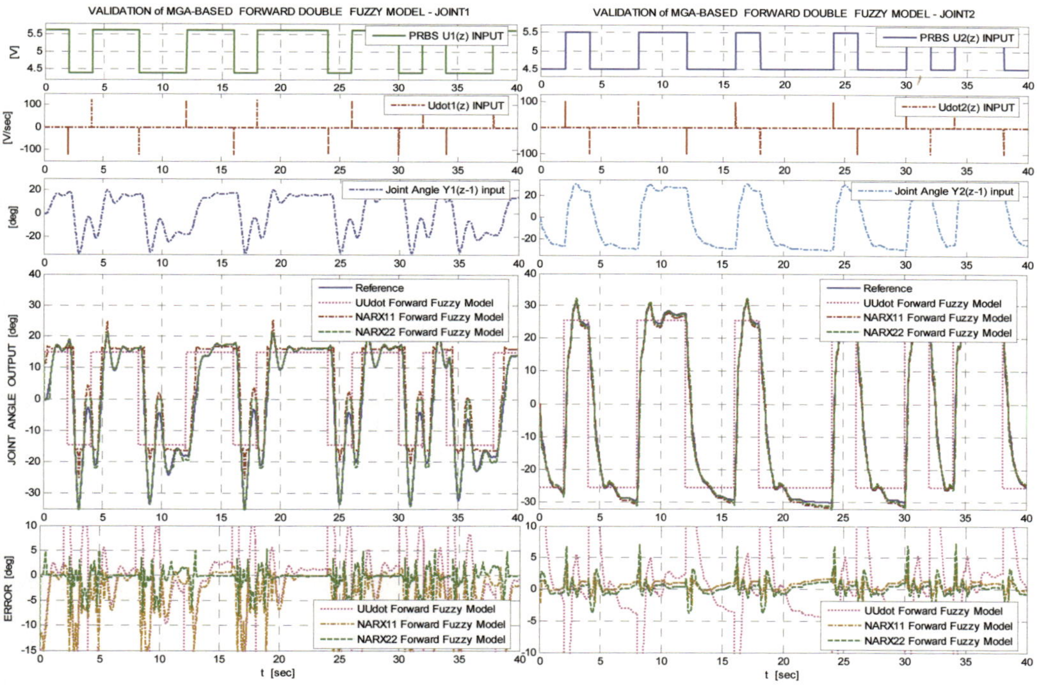

Fig. 12. Validation of MGA-based Forward MISO Fuzzy Model of the 2-axes PAM robot arm

model and <[$\pm 3^{\circ}$] with joint 1 and <[$\pm 2.5^{\circ}$] with joint 2 in case of Forward MISO NARX22 fuzzy model). These results are very impressive in comparison with Forward MISO UUdot fuzzy model (error > [$\pm 10^{\circ}$] for both joints.

These results assert the outstanding potentiality of the novel proposed MISO NARX fuzzy model not only in modeling and identification but also in advanced control application as well.

5.2 MGA-based 2-axes PAM robot arm Inverse MISO NARX fuzzy model identification

The identification procedure bases on the experimental input-output data values measured from the 2-axes PAM robot arm. Table 1 tabulates fuzzy model parameters used for encoding as optimized input values of MGA optimization algorithm. The range (3–5) permits the variable of number of membership functions obtaining 2 different odd values would be chosen by MGA (3 and 5). Block diagrams in Fig.6a, Fig.6b and Fig.6c illustrate the MGA-Based 2-axes PAM robot arm's Inverse MISO Fuzzy model identification.

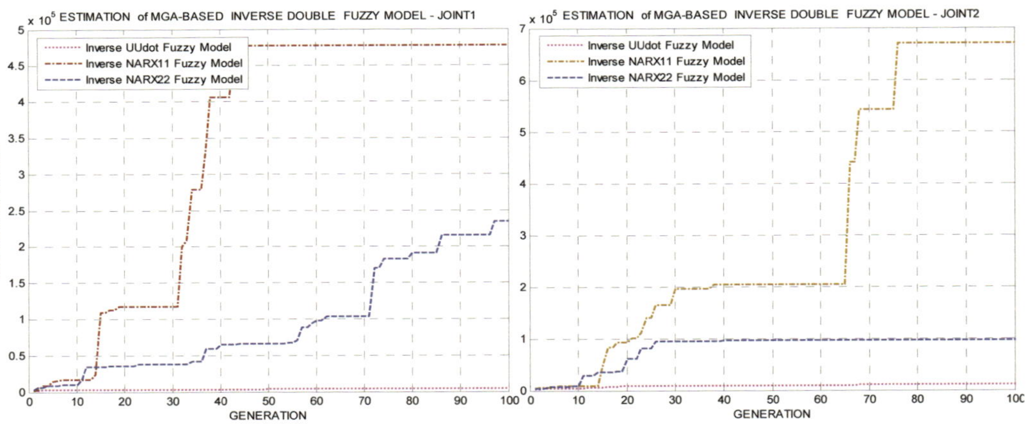

Fig. 13. Fitness Convergence of MGA-based Inverse MISO Fuzzy Model optimization of the 2-axes PAM robot arm

The fitness value of MGA-based optimization calculated based on equation (8) is presented in Fig. 13 (with population = 40 and generation = 100). This Figure represents the fitness convergence values of both Inverse Fuzzy models of both joints of the 2-axes PAM robot arm corresponding to three different identification methods. This Figure shows that the fitness value of Inverse MISO *UUdot* fuzzy model falls early (at 8th generation with joint-1 and 48th generation with joint-2) into a local optimal trap equal 5250 with joint 1 and 7480 with joint 2. The reason is that *UUdot* fuzzy model seems impossible to learn nonlinear features of the 2-axes PAM robot arm implied in input signals *U* [deg] and *Udot* [deg/s]. On the contrary, the fitness value of Inverse MISO NARX fuzzy model obtains excellently the global optimal value (equal 485000 with joint 1 and 676000 with joint 2 in case of Inverse MISO NARX11 fuzzy model and equal 235000 with joint 1 and 98400 with joint 2 in case of Inverse MISO NARX22 fuzzy model). The cause is due to proposed Inverse MISO NARX fuzzy model combines the extraordinary approximating capacity of fuzzy system with powerful predictive and adaptive potentiality of the nonlinear NARX structure implied in Inverse NARX Fuzzy Model. Consequently, MGA-based Inverse MISO NARX11 and Inverse MISO NARX22 fuzzy model as well cover excellently all of nonlinear features of the 2-axes PAM robot arm implied in input signals *U(z)*[deg] and *Y(z-1)* [V].
Consequently, the validating result of the MGA-based identified 2-axes PAM robot arm's Inverse MISO NARX fuzzy model presented in Fig. 14 also shows a very good range of error (< [±0.1[*V*]] with joint 1 and < [±0.05[*V*]] with joint 2 in case of Inverse MISO NARX11 fuzzy model and < [±0.15[*V*]] with joint 1 and < [±0.3[*V*]] with joint 2 in case of Inverse MISO NARX22 fuzzy model). These results are very impressive in comparison with Inverse MISO *UUdot* fuzzy model (error > [±1[*V*]] with joint 1 and > [±0.5[*V*]] with joint 2 respectively).

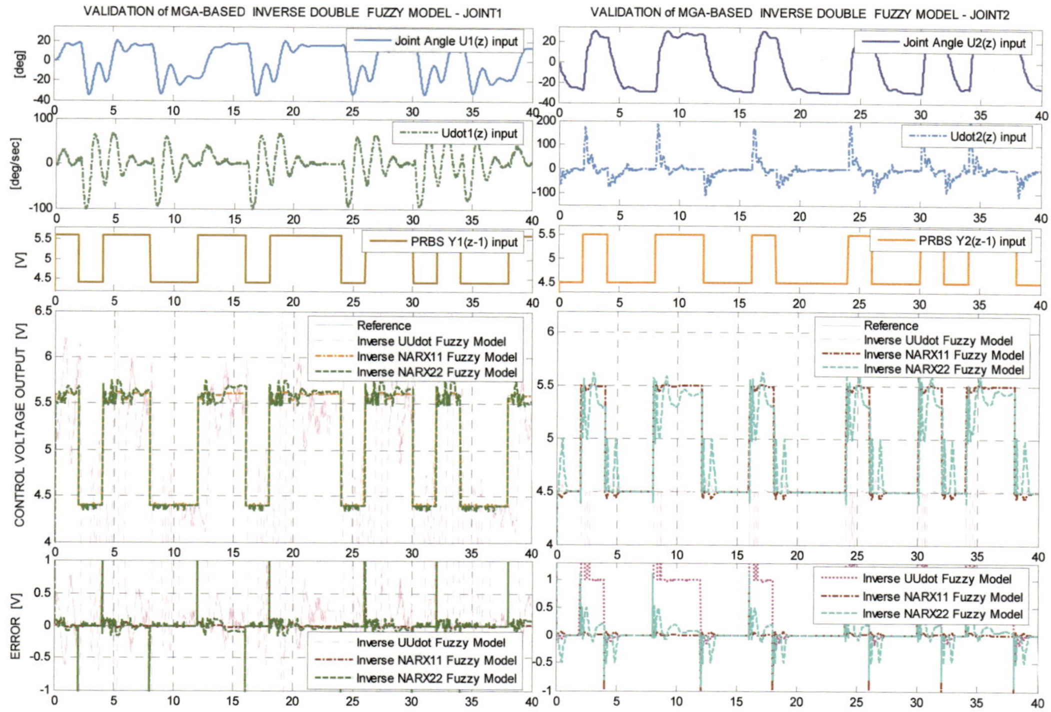

Fig. 14. Validation of MGA-based Inverse MISO Fuzzy Model of the 2-axes PAM robot arm

These results assert the outstanding potentiality of the novel proposed Forward and Inverse MISO NARX fuzzy model not only in modeling and identification of the 2-axes PAM robot arm but also in advanced control application of nonlinear MIMO systems as well.

6. Conclusion

In this study, a new approach of MISO NARX Fuzzy model firstly utilized in modeling and identification of the prototype 2-axes pneumatic artificial muscle (PAM) robot arm system which has overcome successfully the nonlinear characteristic of the prototype 2-axes PAM robot arm and resulting Forward and Inverse MISO NARX Fuzzy model surely enhance the control performance of the 2-axes PAM robot arm, due to the extraordinary capacity in learning nonlinear characteristics and coupled effects as well of MISO NARX Fuzzy model. Results of training and testing on the complex dynamic systems such as PAM robot arm show that the newly proposed MISO NARX Fuzzy model which is trained and optimized by modified genetic algorithm presented in this study can be used in online control with better dynamic property and strong robustness. This resulting MISO NARX Fuzzy model is quite suitable to be applied for the modeling, identification and control of various plants, including linear and nonlinear process without regard greatly changing external environments.

7. Acknowledgements

This research was supported by the DCSELAB - Viet Nam National University Ho Chi Minh City (VNU-HCM) and the NAFOSTED, Viet Nam.

8. References

Ahn K.K., Anh H.P.H., 2006. System modeling and identification of the two-link pneumatic artificial muscle (PAM) manipulator optimized with genetic algorithm. In: Proceedings of the 2006 IEEE-ICASE Int. Conf., Busan, Korea, pp. 356–61.

Ahn K.K., Anh H.P.H., 2007. A new approach of modeling and identification of the pneumatic artificial muscle (PAM) manipulator based on recurrent neural network. In Proc. IMechE, Part I: Journal of Systems and Control Engineering, 2007, 221(I8), 1101-1122.

Ahn K.K., Anh H.P.H., 2009. Identification of the pneumatic artificial muscle manipulators by MGA-based nonlinear NARX fuzzy model. In MECHATRONICS, 2009, Volume 19, Issue 1, pp. 106-133.

Ahn K.K., Thanh T.D.C., 2005. Nonlinear PID control to improve the control performance of PAM manipulators using neural network. In KSME, Int., Jour., 19(1):pp.106~15.

Balasubramanian K, Rattan K.S., 2003a. Fuzzy logic control of a pneumatic muscle system using a linearizing control scheme. In: Proceedings of Int. Conf., North American Fuzzy Information Processing Society, pp. 432-6.

Chan S.W., Lilly J., Berlin J.E., May 2003. Fuzzy PD+I learning control for a pneumatic muscle. In: Proceedings of IEEE Int. Conf. Fuzzy Systems, St. Louis, MO, pp. 278–83.

Hesselroth T, Sarkar K, Van der Smagt P, Schulten K., 1994. Neural network control of a pneumatic robot arm. IEEE Trans. System Man Cybernetics 24(1): pp.28–38.

Karakasoglu,A., Sudharsanan, S.I. and Sundareshan,M.K., "Identification and decentralized adaptive control using dynamical neural networks with application to robotic manipulators," *IEEE Trans. on neural networks*, Vol. 4, No. 6, (Nov. 1993), pp. 919-930.

Katic, D.M. and Vukobratovic, M.K., "Highly efficient robot dynamics learning by decomposed connectionist feed-forward control structure," *IEEE Trans. on syst. man and cybern.*, Vol. 25, No. 1, (1995), pp. 145-158.

Lilly J., Sep. 2003.Adaptive tracking for pneumatic muscle actuators in bicep and tricep configurations. IEEE Trans. Neural Syst. Rehabil. Eng. 11(3):pp.333–9.

Lilly J.H., Chang X., Sep.2003. Tracking control of a pneumatic muscle by an evolutionary fuzzy controller. In IEEE Intell. Automat. Soft Comput., 9(3):pp. 227–44.

Medrano-Cerda G.A., Bowler C.J., Caldwell D.G., Aug. 1995. Adaptive position control of antagonistic pneumatic muscle actuators. In: Proceedings of IEEE Int. Conf. Intelligent Robots and Systems, Pittsburgh, PA, pp. 378–83.

Nelles O., Nonlinear system identification, Springer, 2000.

Repperger D.W., Johnson K.R., Phillips C.A., 1998. VSC position tracking system involving a large scale pneumatic muscle actuator. In: Proceedings of IEEE Conf. Decision Control, Tampa, FL, Dec. pp. 4302–7.

Repperger D.W., Phillips C.A., Krier M., Aug. 1999. Controller design involving gain scheduling for a large scale pneumatic muscle actuator," In: Proceedings of IEEE Conf. Control Applications, Kohala Coast, HI, pp. 285–90.

Reynolds D.B., Repperger D.W., Phillips C.A., Bandry G., 2003. Dynamic characteristics of pneumatic muscle. In IEEE Ann. Biomed. Eng., 31(3):pp.310–7.

Tsagarakis N, Darwin G.C., 2000. Improved modeling and assessment of pneumatic muscle actuators. In: Proceedings of IEEE Int. Conf. Robotics and Automation, San Francisco, CA, pp. 3641–6.

2

Kinematics of AdeptThree Robot Arm

Adelhard Beni Rehiara
University of Papua
Indonesia

1. Introduction

Robots are very powerful elements of today's industry. They are capable of performing many different tasks and operations precisely and do not require common safety and comfort elements humans need. However, it takes much effort and many resources to make a robot function properly. Most companies that made industrial robots can be found in the market such as Adept Robotics, Staubli Robotics and Fanuc Robotics. As a result, there are many thousands of robots in industry.

An AdeptThree robot arm is a selectively compliant assembly robot arm (SCARA) manufactured by the Adept Company. In general, traditional SCARA's are 4-axis robot arms within their work envelope. They have the jointed two-link arm layout similar to our human arms and commonly used in pick-and-place, assembly, and packaging applications. As a SCARA robot, an AdeptThree robot has 4 joints which denote that it has 4 degree of freedom (DOF). The robot has been designed with completed components including operating system and programming language namely V+ (Rehiara and Smit, 2010).

In robotic, there are two important studies which are kinematics and dynamics studies. Robot kinematics is the study of robot motion without regards to the forces that result it. On the other hand, the relationship between motion, and the associated forces and torques is studied in robot dynamics. In this chapter, kinematics problem for an AdeptThree robot will be explained in detail.

2. AdeptThree robot system

An AdeptThree robot is a 4-axis SCARA robot which is designed for assembling and part-handling tasks. The body of the robot is too big compared to the most SCARAs but it has strength and rigidity to carry a load about 25 kg (55 lb) as its maximum payload. For the working envelope, it has a 1067 mm maximum radial that can make more than two meters in diameter and also it has 305 mm Z-axis stroke. Fig. 1 shows the physical system of an AdepthThree robot arm. All of the figures in this section are provided by Adept company (1991).

As manufactured by Adept Company, AdeptThree robot is designed to be compatible with the other Adept products either the Adept MC or the Adept CC controller interface. All of the control and operation of the AdeptThree robot are programmed through the selected controller. In this case, the robot is using the Adept MC controller.

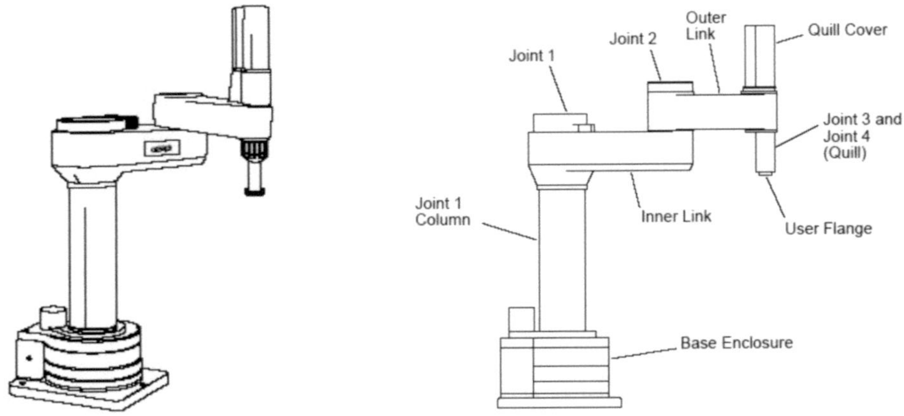

(a) (b)

Fig. 1. (a) Physics of an AdeptThree Robot Arm and (b) Joints and links names

2.1 Joints motion

An AdeptThree robot has 4 joints which are linked to the robot. Joint 3 is a translational joint which can move along Z-axis while joint 1, 2, and 4 are rotational joints. Working envelope of the robot is shown in fig.2 (b).

First joint is the base joint and it is also called"the shoulder" as its function looks like a human shoulder. In this joint, the rotational movement of the inner link and the column will be provided. The joint has a maximum movement of about 300^0 that can be separated in 150^0 to the left and 150^0 to the right as in fig.2 (a).

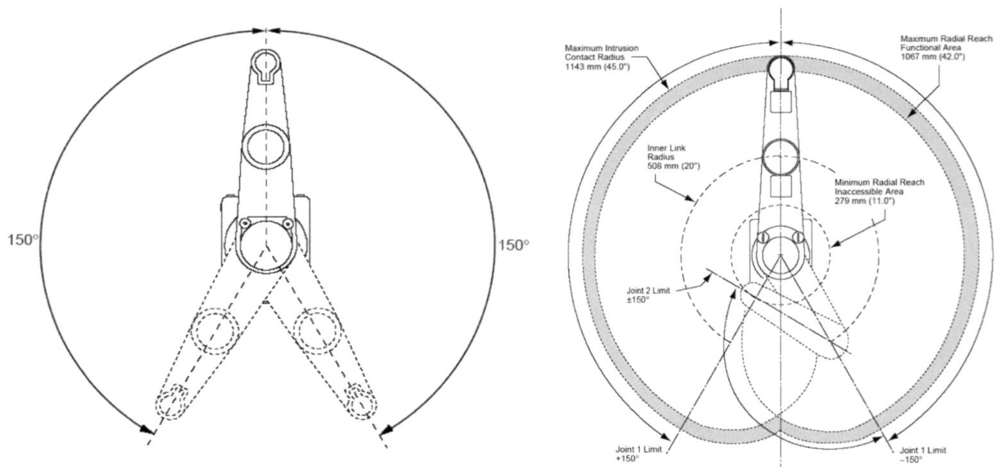

(a) (b)

Fig. 2. (a) 1st joint motion and (b) working envelope

Second joint is called "the elbow" as its function looks like a human elbow. In this joint, both the outer link and inner link are linked. Furthermore this joint is similar to 1st joint, the maximum movement of the joint is also about 300^0.

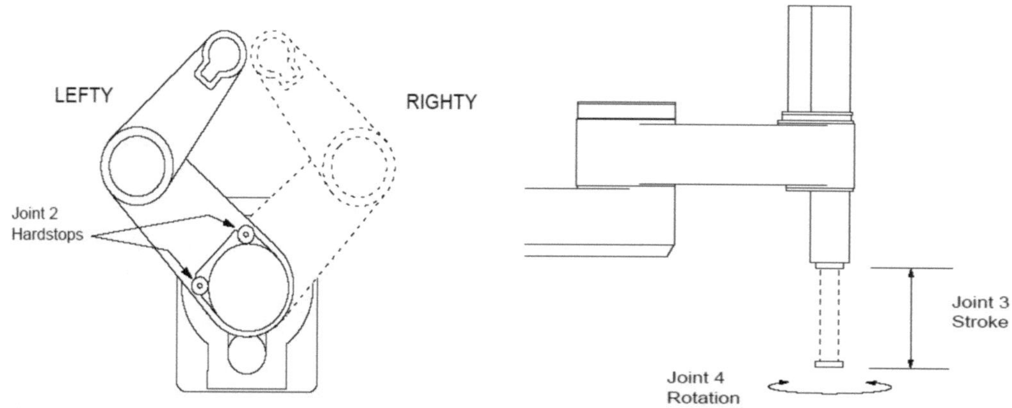

(a) (b)

Fig. 3. (a) 2nd joint, (b) 3rd and 4th joint movements

Figure 3(a) shows the movement of the 2nd joint. In order to avoid any ambiguity to program the robot, the robot can be programmed to move like a human left or right arm by using the syntax "LEFTY" or "RIGHTY".

Third joint is placed at the end of the outer link. It has a maximum stroke of about 12 inches or 30.5 cm. Fig. 3(b) shows the 3rd joint and also 4th joint.

Fourth joint is also called the "the wrist". The joint can be moved over a range of 540^0. Its function is similar to a human wrist and it can be rotated as a human hand to tighten a bolt or unscrew a screw.

Although the AdeptThree robot has the widest working envelope, it still has a limitation. The limitation is about the travelling of each joint and it was built to avoid the damage of the robot. The maximum joint travel is confined by soft-stop and hard-stop. Soft-stop and hard stop occur when the joint is expected to pass the limit angle. While both stops happen, robot power will be turned off.

Soft-stop can be a programmed cancellation and it requires the robot arm to be moved manually into its working envelope. After the arm into the working envelope, the robot arm can be used directly without any other setting. On the other hand, the action of the hard-stop is to cancel all of the robot operations and it requires to move the robot manually by using the manual control pendant (MCP) to its working area.

2.2 Operating system

The AdeptThree robot has its own operating system called V+ that also can recognize some syntaxes in programming the robot. As an operating system, the V+ can handle all of the system operations. The programming language in the robot operating system is a high level programming language. It is similar to C or Pascal programming and it can transfer syntaxes to machine language.

The V+ real-time and multi-tasking operating system manages all system level operations, such as input/output (I/O), program execution, task management, memory management and disk file operations. As a programming language, V+ has a rich history and has evolved into the most powerful, safe and predictable, robot programming language available today. V+ is the only language to provide an integrated solution to all of the programming needs in a robotic work cell, including safety, robot motion, vision operations, force sensing and I/O.

In general, the syntaxes using by V+ can be categorized into 4 parts:

- Monitor command, it can be used directly by typing it one by one.
- Program command, it will be run if it is used in a program lines.
- Real-time command, it only can be run in a program.
- String command, it is used to handle all operations with string variable. It can be used in monitor and program command.

2.3 Robot setup

Before using the robot, it is needed to be booted by using its operation system V+. The booting screen of the Adept + is placed in fig. 4. Dot (.) command in the last line means that the robot is ready to be commanded by applying the V+ syntaxes.

```
Adept V+

Copyright (c) 1984, 1985, 1986, 1987, 1988, 1989, 1990, 1991, 1992, 1993, 1994
By Adept Technology, Inc. All rights reserved

Adept External Encoder Module.
Adept SCARA Robot Control Module.

Software   : 10.4 1917-301 2960
Controller : 310-2960-3 2048 Kb
Robot 1    : 640-319-10 1
.
```

Fig. 4. Adept V+ booting screen

As shown in fig.4, the robot consists of some modules which are software (V+ version 10.4), controller module and a robot arm. Unlike most computers, the controller does not have BIOS (basic input output system) memory; therefore the robot time needs to be changed with the actual time every time after tuned on.

3. Kinematics

Kinematics in robotics is a statement form about geometrical description of a robot structure. From the geometrical equation we can get relationship among joints spatial geometry concept on a robot with ordinary co-ordinate concept which is used to determine the position of an object. In other word, kinematics is the relationships between the positions, velocities, and accelerations of the links of a robot arm. The aim of kinematics is to define position relative of a frame to its original coordinates.

Using kinematics model, a programmer can determine the configuration of input reference that should be fed to every actuator so that the robot can do coincide movements of all joint to reach the desired position. On the other hand, with information of position that is shown by every joint while robot is doing a movement, the programmer by means of kinematics analysis can determine where is arm tip position or which parts of the robot should be moved in spatial coordinate. Kinematics problem consists of forward and inverse kinematics and each type of the kinematics has its own function as illustrated in fig.5.

From fig. 5, forward kinematics is used for transferring joint variable to get end-effector position. On the other hand, inverse kinematics will be applied to find joint variable from end-effector position.

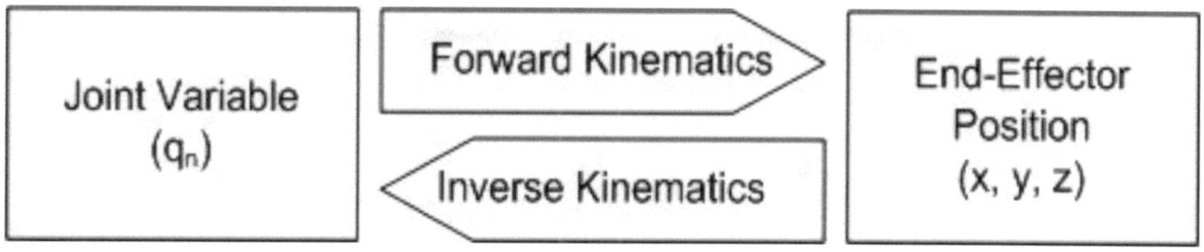

Fig. 5. Forward and inverse kinematics diagram

3.1 Forward kinematics

Forward kinematics problem is deal with finding the position and orientation of a robot end-effector as a function of its joint angles. Forward kinematics problem is relatively simple and it is easy to be implemented. There are two methods for building forward kinematics provided in this section.

3.1.1 Graphical method

A simple forward kinematics can be derived from its space using graphical solution. With a three link planar robot in fig.6, the graphical method for solving forward kinematics will be described in this section.

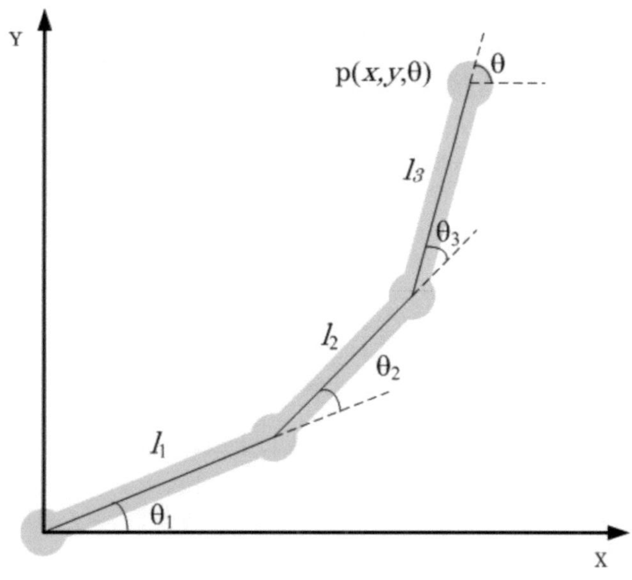

Fig. 6. Geometric of three link planar robot

Using the vector algebra solution to analyse the graph, the coordinate of the robot end-effector can be solved as follows.

$$
\begin{aligned}
x &= l_1\cos(\theta_1) + l_2\cos(\theta_1+\theta_2) + l_3\cos(\theta_1+\theta_2+\theta_3) \\
y &= l_1\sin(\theta_1) + l_2\sin(\theta_1+\theta_2) + l_3\sin(\theta_1+\theta_2+\theta_3) \\
\theta &= \theta_1+\theta_2+\theta_3
\end{aligned}
\tag{1}
$$

Maple is mathematical software which is widely used in computation, modelling and simulation. In each section of the kinematics and Jacobean, the script of the software is provided. The Maple script for building forward kinematics using the graphical method is listed as follows.

```
> restart:
> n:=3:y:=0:c:=0:
> for i from 1 to n do
> for j from i to n do
> c:=c+theta[j];
> end do;
> y:=y+l[n-i+1]*cos(c):c:=0:
> end do;
  ...
```

3.1.2 D-H convention

The steps to get the position in using D-H convention are finding the Denavid-Hartenberg (D-H) parameters, building A matrices, and calculating T matrix with the coordinate position which is desired.

D-H Parameters

D-H notation is a method of assigning coordinate frames to the different joints of a robotic manipulator. The method involves determining four parameters to build a complete homogeneous transformation matrix. These parameters are the twist angle ai , link length ai, link offset di , and joint angle θi (Jaydev, 2005). Based on the manipulator geometry, two of the parameters which are αi and ai have constant values, while the di and θi parameters can be variable depending on whether the joint is prismatic or revolute.

Jaydev (2005) has provided 10 steps to denote the systematic derivation of the D-H parameters as :

1. Label each axis in the manipulator with a number starting from 1 as the base to n as the end-effector. Every joint must have an axis assigned to it.
2. Set up a coordinate frame for each joint. Starting with the base joint, set up a right handed coordinate frame for each joint. For a rotational joint, the axis of rotation for axis i is always along Z_{i-1}. If the joint is a prismatic joint, Z_{i-1} should point in the direction of translation.
3. The X_i axis should always point away from the Z_{i-1}axis.
4. Y_i should be directed such that a right-handed orthonormal coordinate frame is created.
5. For the next joint, if it is not the end-effector frame, steps 2–4 should be repeated.
6. For the end-effector, the Z_n axis should point in the direction of the end-effector approach.
7. Joint angle θ_i is the rotation about Z_{i-1} to make X_{i-1} parallel to X_i.
8. Twist angle α_i is the rotation about Xi axis to make Z_{i-1}parallel to Z_i.
9. Link length a_i is the perpendicular distance between axis i and axis $i + 1$.
10. Link offset d_i is the offset along the Z_{i-1} axis.

A Matrix

The A matrix is a homogenous 4x4 transformation matrix which describe the position of a point on an object and the orientation of the object in a three dimensional space. The homogeneous transformation matrix from one frame to the next frame can be derived by the determining D-H parameters. The homogenous rotation matrix along an axis is given by

$$Rot = \left[\begin{array}{ccc|c} \cos\theta & -\cos\alpha\ \sin\theta & \sin\alpha\ \sin\theta & 0 \\ \sin\theta & \cos\alpha\ \cos\theta & -\sin\alpha\ \sin\theta & 0 \\ 0 & \sin\alpha & \cos\alpha & 0 \\ \hline 0 & 0 & 0 & 1 \end{array}\right] \tag{2}$$

and the homogeneous translation matrix transforming coordinates from a frame to the next frame is given by

$$Trans = \left[\begin{array}{ccc|c} 1 & 0 & 0 & a \\ 0 & 1 & 0 & 0 \\ 0 & 0 & 1 & d \\ \hline 0 & 0 & 0 & 1 \end{array}\right] \tag{3}$$

Where the four quantities θ_i, a_i, d_i, α_i are the names joint angle, link length, link offset, and twist angle respectively. These names derive from specific aspects of the geometric relationship between two coordinate frames. The four parameters are associated with link *i* and joint *i*.

In Denavit-Hartenberg convention, each homogeneous transformation matrix A_i is represented as a product of four basic transformations as follows.

$$A_i = Rot(z,\theta_i)\, Trans\,(z,d_i)\, Trans\,(x,a_i)\, Rot(x,\alpha_i) \tag{4}$$

or in completed form as

$$A_i = \begin{bmatrix} \cos(\theta_i) & -\sin(\theta_i) & 0 & 0 \\ \sin(\theta_i) & \cos(\theta_i) & 0 & 0 \\ 0 & 0 & 1 & 0 \\ 0 & 0 & 0 & 1 \end{bmatrix} \begin{bmatrix} 1 & 0 & 0 & 0 \\ 0 & 1 & 0 & 0 \\ 0 & 0 & 1 & d_i \\ 0 & 0 & 0 & 1 \end{bmatrix} \begin{bmatrix} 1 & 0 & 0 & a_i \\ 0 & 1 & 0 & 0 \\ 0 & 0 & 1 & 0 \\ 0 & 0 & 0 & 1 \end{bmatrix} \begin{bmatrix} 1 & 0 & 0 & 0 \\ 0 & \cos(\alpha_i) & -\sin(\alpha_i) & 0 \\ 0 & \sin(\alpha_i) & \cos(\alpha_i) & 0 \\ 0 & 0 & 0 & 1 \end{bmatrix} \tag{5}$$

By simplifying equation 5, the matrix A_i which is known as D-H convention matrix is given in equation 6.

$$A_i = \left[\begin{array}{ccc|c} \cos\theta_i & -\cos\alpha_i \sin\theta_i & \sin\alpha_i \sin\theta_i & a_i \cos\theta_i \\ \sin\theta_i & \cos\alpha_i \cos\theta_i & -\sin\alpha_i \sin\theta_i & a_i \sin\theta_i \\ 0 & \sin\alpha_i & \cos\alpha_i & d_i \\ \hline 0 & 0 & 0 & 1 \end{array}\right] \tag{6}$$

In the matrix A_i, about three of the four quantities are constant for a given link. While the other parameter which is θ_i for a revolute joint and d_i for a prismatic joint is variable for a joint. The A_i matrix contains a 3x3 rotation matrix, a 3x1 translation vector, a 1x3 perspective vector and a scaling factor. The A_i matrix can be simplified as follows.

$$A_i = \begin{bmatrix} R_{i3x3} & P_{i3x1} \\ 0_{i1x3} & 1 \end{bmatrix} \tag{7}$$

T Matrix

The T matrix is a kinematics chain of transformation. The matrix can be used to obtain coordinates of an end-effector in terms of the base link. The matrix can be built from 2 or more A matrices depending on the number of manipulator joint(s). The T matrix can be formulated as

$$T \equiv T_n = A_1 A_2, ..., A_n \tag{8}$$

Inside the T matrix, the direct kinematics can be found in the translation matrix P_i while the X, Y and Z positions are P_1, P_2 and P_3 respectively.

Solution for the robot

An AdeptThree robot arm with four joints is figured in fig. 7. The AdeptThree robot joint motions are revolution, revolution, prismatic and revolution (RRPR) respectively from joint 1 to 4. So the robot has four degrees of freedom.
From fig. 7, joints 1, 2, and 4 are revolute joints; then the values of θ_i are variable. Since there is no rotation about prismatic joint in joint 3, the θ_i values for joint 3 is zero while d_i is variable.

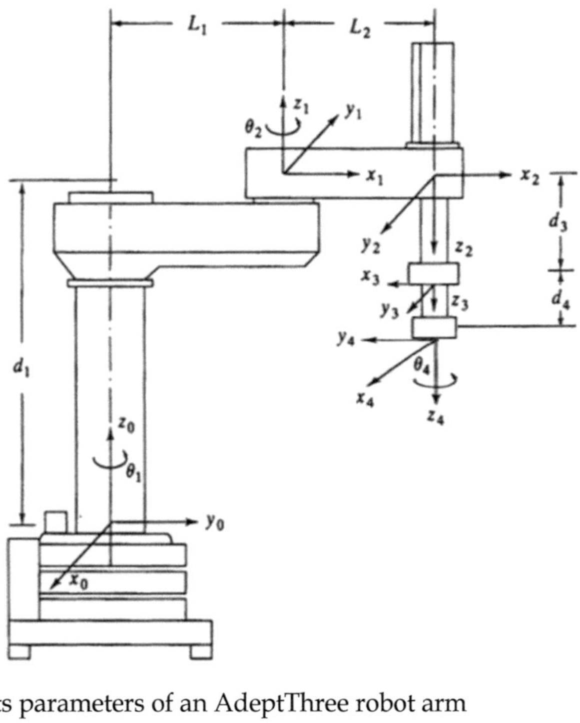

Fig. 7. Links and joints parameters of an AdeptThree robot arm

Each axis of the AdeptThree robot was numbered from 1 to 4 based on the algorithm explained before. After established coordinate frames, the next step is to determine the D-H parameters by first determining α_i. The α_i is the rotation about X_i to make Z_{i-1} parallel with Z_i. Starting from axis 1, α_1 is 0 because Z_0 and Z_1 are parallel. For axis 2, the α_2 is π or 180° because Z_2 is opposite of Z_0 which is pointing down along the translation of the prismatic joint. α_3 and α_4 values are zero because Z_3 is parallel with Z_2 and Z_4 is also parallel with Z_3.
The next step is to determine a_i and d_i. For axis 1, there is an offset d_1 between axes 1 and 2 in the Z_0 direction. There is also a distance a_1 between both axes. For axis 2, there is a distance a_2 between axes 2 and 3 away from the Z_1 axis. No offset is found in this axis so d_2 is zero. In axis 3, due to prismatic joint, the offset d_3 is variable. Between axes 3 and 4, there is an offset d_4 which is equal to this distance, while a_3 and a_4 are zero. The completed D-H parameters are listed in table 1.

Axis Number	*Joint Angle* θ_i	*Link Offset* d_i	*Link Length* a_i	*Twist Angle* α_i
1	θ_1	d_1	l_1	0
2	θ_2	0	l_2	π
3	0	d_3	0	0
4	θ_4	d_4	0	0

Table 1. D-H Parameters of an AdeptThree Robot

The transformation matrix A_i can now be computed. Using the expression in equation 6 the A matrices of each joint can be build as

$$A_1^0 = \begin{bmatrix} c_1 & -s_1 & 0 & l_1c_1 \\ s_1 & c_1 & 0 & l_1s_1 \\ 0 & 0 & 1 & d_1 \\ 0 & 0 & 0 & 1 \end{bmatrix} \tag{9}$$

$$A_2^1 = \begin{bmatrix} c_2 & s_2 & 0 & l_2c_2 \\ s_2 & -c_2 & 0 & l_2s_2 \\ 0 & 0 & -1 & 0 \\ 0 & 0 & 0 & 1 \end{bmatrix} \tag{10}$$

$$A_3^2 = \begin{bmatrix} 1 & 0 & 0 & 0 \\ 0 & 1 & 0 & 0 \\ 0 & 0 & 1 & d_3 \\ 0 & 0 & 0 & 1 \end{bmatrix} \tag{11}$$

$$A_4^3 = \begin{bmatrix} c_4 & -s_4 & 0 & 0 \\ s_4 & c_4 & 0 & 0 \\ 0 & 0 & 1 & d4 \\ 0 & 0 & 0 & 1 \end{bmatrix} \quad (12)$$

T matrix is created by multiplying each A matrix defined using equation 9 to 12 and the result is as follows.

$$T = \begin{bmatrix} s_4 s_{1+2} + c_4 c_{1+2} & s_4 c_{1+2} + c_4 s_{1+2} & 0 & l_2 c_{1+2} + l_1 c_1 \\ s_4 c_{1+2} + c_4 s_{1+2} & -s_4 s_{1+2} + c_4 c_{1+2} & 0 & l_2 s_{1+2} + l_1 s_1 \\ 0 & 0 & -1 & -d_4 - d_3 + d_1 \\ 0 & 0 & 0 & 1 \end{bmatrix} \quad (13)$$

Where c_i and s_i are the cosines and sinus of θ_i, c_{1+2} and $s_{1+2}$are $\cos(\theta_1+\theta_2)$ and $\sin(\theta_1+\theta_2)$, l_i is the length of link i and d_i is the offset of link i.
By using the T matrix, it is possible to calculate the values of (P_x, P_y, P_z) with respect to the fixed coordinate system. Then the P_x, P_y, P_z which are obtained with direct kinematics are equations which are listed below:

$$\begin{aligned} P_x &= l_2 c_{1+2} + l_1 c_1 \\ P_y &= l_2 s_{1+2} + l_1 s_1 \\ P_z &= -d_4 - d_3 + d_1 \end{aligned} \quad (14)$$

Where constant parameters l_1=559 mm, l_2=508 mm, and d_1=876.3 mm. The direct kinematics can be used to find the end-effector coordinate of the robot movement by substituting the constant parameter values to the above equation.
Maple script for the D-H convention of forward kinematics is listed as follows.

```
> restart:
> DH:=Matrix(<<theta[1],theta[2],0,theta[4]>|<d[1],0,d[3],d[4]>|<l[1],l[2],
  0, 0>|<0,pi,0,0>>):
> for i from 1 to 4 do
> A[i]:=Matrix(<<cos(DH[i,1]),sin(DH[i,1]),0,0>|<-cos(DH[i,4])*sin(DH[i,1]),
  cos(DH[i,4])*cos(DH[i,1]),sin(DH[i,4]),0>|<sin(DH[i,4])*sin(DH[i,1]),-
  sin(DH[i,4])*sin(DH[i,1]),cos(DH[i,4]),0>|<DH[i,3]*cos(DH[i,1]),DH[i,3]*si
  n(DH[i,1]),DH[i,2],1>>);
> end do:
> T:=simplify(A1.A2.A3.A4);
  ...
```

3.2 Inverse kinematics

Inverse kinematics deals with the problem of finding the appropriated joint angles to get a certain desired position and orientation of the end-effector. Finding the inverse kinematics solution for a general manipulator can be a very tricky task. In general, inverse kinematics solutions are non linear. To find those equations can be complicated and sometimes there is no solution for the problem. Geometric and algebraic methods are provided in this section for solving inverse kinematics of a robot arm.

3.2.1 Geometric method

One of the simple ways to solve the inverse kinematics problem is by using geometric solution. With this method, cosines law can be used. A two planar manipulator will be used to review this kinematics problem as in following figure.

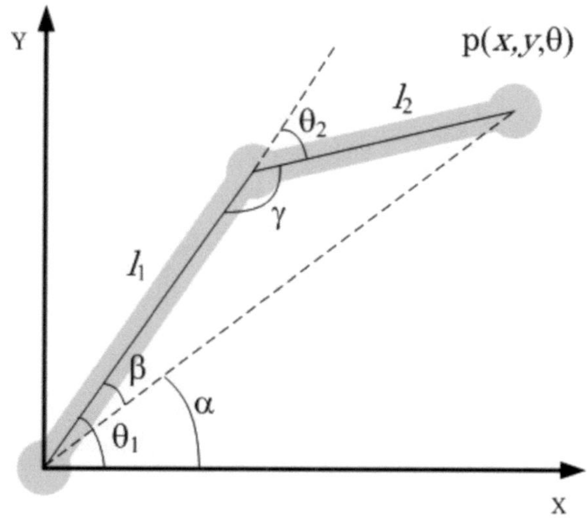

Fig. 8. Geometric of two link planar robot

With cosines law, we get

$$\left(x^2+y^2\right)=l_1^2+l_2^2-2l_1l_2\cos(180-\theta_2) \tag{15}$$

Since $\cos(180\text{-}\theta_2)$ = $-\cos(\theta_2)$ then the equation 15 will become

$$\left(x^2+y^2\right)=l_1^2+l_2^2+2l_1l_2\cos(\theta_2) \tag{16}$$

By solving the equation 16 for getting the $\cos(\theta_2)$,

$$\cos\theta_2=\frac{x^2+y^2-l_1^2-l_2^2}{2l_1l_2} \tag{17}$$

Therefore the θ_2 will be determined by taking inverse cosines as

$$\theta_2=\arccos\left(\frac{x^2+y^2-l_1^2-l_2^2}{2l_1l_2}\right) \tag{18}$$

Again looking the fig. 8, we get

$$\frac{\sin(\beta)}{l_2}=\frac{\sin(\gamma)}{\sqrt{x^2+y^2}};\quad \alpha=\arctan\left(\frac{y}{x}\right) \tag{19}$$

Where $\sin(\gamma)$ = $\sin(180\text{-}\theta_2)$ = $\sin(\theta_2)$. By replacing $\sin(\gamma)$ with $\sin(\theta_2)$, the equation 19 will become

$$\beta = \arcsin\left(\frac{l_2 \sin(\theta_2)}{\sqrt{x^2+y^2}}\right) \tag{20}$$

Since $\theta_1 = \beta + \alpha$, the θ_1 can be solved as

$$\theta_1 = \arcsin\left(\frac{l_2 \sin(\theta_2)}{\sqrt{x^2+y^2}}\right) + \arctan\left(\frac{y}{x}\right) \tag{21}$$

Maple script for the geometric method of inverse kinematics is listed as follows.

```
> restart:
> beta:=solve(sin(beta)/l2=sin(theta2)/sqrt(x^2+y^2), beta):
> alpha:=arctan(y,x):
> theta1:=beta+alpha;
  ...
> theta2:=solve(y^2+x^2=l1^2+l2^2+(2*l1*l2*cos(theta)),theta);
  ...
```

3.2.2 Algebraic method

The other simple ways to solve the inverse kinematics problem is by using algebraic solution. This method is used to make an invert of forward kinematics. Rewriting the end-effector coordinate from forward kinematics:

$$\begin{aligned} x &= l_1 c_1 + l_2 c_{1+2} \\ y &= l_1 s_1 + l_2 s_{1+2} \end{aligned} \tag{22}$$

Using the square of the coordinate, we get

$$\begin{aligned} x^2 + y^2 = l_1^2 c_1^2 + l_2^2 (c_{1+2})^2 + 2 l_1 l_2 c_1 (c_{1+2}) + \\ l_1^2 s_1^2 + l_2^2 (s_{1+2})^2 + 2 l_1 l_2 s_1 (s_{1+2}) \end{aligned} \tag{23}$$

Since $\cos(a)^2+\sin(a)^2 = 1$ and also $\cos(a+b)^2+\sin(a+b)^2 = 1$, the equation 23 can be simplify as

$$x^2 + y^2 = l_1^2 + l_2^2 + 2 l_1 l_2 \left[c_1 (c_{1+2}) + s_1 (s_{1+2}) \right] \tag{24}$$

Note that

$$\begin{aligned} \cos(a \pm b) &= \cos(a)\cos(b) \mp \sin(a)\sin(b) \\ \sin(a \pm b) &= \cos(a)\sin(b) \pm \sin(a)\cos(b) \end{aligned} \tag{25}$$

By simplifying the formulation inside the parenthesis in equation 24 with the rule in equation 25, the only left parameter is $\cos(\theta_2)$; so the equation 24 will become

$$x^2 + y^2 = l_1^2 + l_2^2 + 2 l_1 l_2 c_2 \tag{26}$$

Now the θ_2 can be formulated as the function of inverse cosines

$$\theta_2 = \arccos\left(\frac{x^2 + y^2 - l_1^2 - l_2^2}{2l_1l_2}\right) \tag{27}$$

Using the rule of sinus and cosines in equation 25, the end-effector coordinate can be rewritten as

$$\begin{aligned} x &= l_1c_1 + l_2c_1c_2 - l_2s_1s_2 \\ y &= l_1s_1 + l_2s_1c_2 + l_2c_1s_2 \end{aligned} \tag{28}$$

There are two unknown parameters inside the equation which are $\cos(\theta_1)$ and $\sin(\theta_1)$. The $\cos(\theta_1)$ can be defined from the rewritten x as

$$c_1 = \frac{x + l_2s_1s_2}{l_1 + l_2c_2} \tag{29}$$

The $\sin(\theta_1)$ is still a missing parameter and it is need to be solved. Substituting c_1 to y in equation 28, we get

$$y = \frac{x + l_2s_1s_2}{l_1 + l_2c_2}(l_2s_2) + l_1s_1 + l_2s_1c_2 \tag{30}$$

The equation 28 will become,

$$\begin{aligned} y = {} & \frac{xl_2s_2 + l_2^2s_1s_2^2}{l_1 + l_2c_2} + \frac{l_1^2s_1 + l_1l_2s_1c_2}{l_1 + l_2c_2} + \\ & \frac{l_1l_2s_1c_2 + l_2^2s_1c_2^2}{l_1 + l_2c_2} \end{aligned} \tag{31}$$

Simplifying the equation 31 we get

$$\mathrm{y} = \frac{\mathrm{xl_2s_2 + s_1\left(l_1^2 + l_2^2 + 2l_1l_2c_2\right)}}{\mathrm{l_1 + l_2c_2}} \tag{32}$$

The parenthesis in equation 32 can be replaced using cosines law with $x^2 + y^2$. Therefore the sinus of θ_1 can derived from the above equation as

$$y = \frac{xl_2s_2 + s_1\left(x^2 + y^2\right)}{l_1 + l_2c_2} \tag{33}$$

Now the θ_1 will be got as the function of inverse sinus as

$$\theta_1 = \arcsin\left(\frac{\mathrm{y(l_1 + l_2c_2) - xl_2s_2}}{\mathrm{x^2 + y^2}}\right) \tag{34}$$

Until now we had defined both θ_1 and θ_2 of a two planar robot that is similar to the AdeptThree robot. The joint angles can be used by applying link length of the robot to the equation of those angles.

Maple script for the algebraic method of inverse kinematics is listed below.

```
> restart:
> theta2:=solve(x^2+y^2=l1^2+l2^2+(2*l1*l2*cos(theta2)),theta2);
  ...
> restart:
> cos(theta1):=solve(x=l1*cos(theta1)+l2*cos(theta1)*cos(theta2)-
  l2*sin(theta1)*sin(theta2),cos(theta1)):
> theta1:=simplify(solve(y=l1*sin(theta1)+l2*sin(theta1)*cos(theta2)+
  l2*cos(theta1)*sin(theta2),theta1));
  ...
```

4. Jacobean

The Jacobean defines the transformation between the robot hand velocity and the joint velocity. Knowing the joint velocity, the joint angles and the parameters of the arm, the Jacobean can be computed and the hand velocity calculated in terms of the hand Cartesian coordinates. The Jacobean is an important component in many robot control algorithms. Normally, a control system receives sensory information about the robot's environment, most naturally implemented using Cartesian coordinates, yet robots operate in the joint or world coordinates. Transforms are needed between Cartesian coordinates and joint coordinates and vice versa. The transformation between the velocity of the arm, in terms of its joint speeds, and the velocity of the arm in Cartesian coordinates, in a particular frame of reference, is very important. Solving the inverse kinematics can provide a transform, but this would be a difficult task to perform in real-time and in most cases no unique solutions exist for the inverse kinematics. An alternative is to use the Jacobean (Zomaya et al. , 1999).
Many ways to design a Jacobean matrix of a robot arm were provided. Zomaya et al. (1999) had presented three kinds of algorithms to perform a Jacobean matrix. First algorithm is the simple way. Without using matrix calculation, the Jacobean can be built from T matrix. Second algorithm was found to perform very well using a sequential processing method. Third algorithm is also provided to sequential machine, but it would be interesting to study how well it maps onto the mesh with multiple buses. The other algorithm was provided by Manjunath (2007) and Frank (2006). It uses tool configuration vector to perform the Jacobean. The last algorithm will be used and explained in this paper (Rehiara, 2011).
Given joint variable coordinate of the end effectors:

$$q = [q_1 \; q_2 \; \dots \; q_n]^T \tag{35}$$

Where $q_i = \theta_i$ for a rotary joint and $q_i = d_i$ for a prismatic joint. Nonlinear transformation from joint variable $q(t)$ to $y(t)$ is defined as $y=h(q)$, then the velocities of joint axes is given by

$$\dot{y} = \frac{\partial h}{\partial q}\dot{q} = J\dot{q} \tag{36}$$

Where J is the Jacobean of manipulator. Inverse of the Jacobean J^{-1} relates the change in the end-effector to the change in axis displacements,

$$\dot{q} = J^{-1}\dot{y} \tag{37}$$

The Jacobean is not always invertible, in certain positions it will happen. These positions are called geometric singularities of the mechanism.
A rotation matrix in a T matrix is formed by three 3x1 vector. In simple, the T matrix can be rewriting as

$$T = \begin{bmatrix} n & o & a & p \\ 0 & 0 & 0 & 1 \end{bmatrix} \quad (38)$$

Where a is the approach vector of the end-effector, o is the orientation vector which is the direction specifying the orientation of the hand, from fingertip to fingertip while n is the normal vector which is chosen to complete the definition of a right-handed coordinate system (Frank, 2006).
The T matrix can be used to design the Jacobean by first defining the tool configuration vector w as follows.

$$\omega(q) = \begin{bmatrix} p_i \\ \hline a_i e^{(q n/\pi)} \end{bmatrix} \quad (39)$$

Rewriting p and a vector from equation 13, we get the tool configuration vector as

$$\omega(q) = \begin{bmatrix} l_2 c_{1+2} + l_1 c_1 \\ l_2 s_{1+2} + l_1 s_1 \\ -d_4 - d_3 + d_1 \\ \hline 0 \\ 0 \\ e^{\left(\frac{\theta 4}{\pi}\right)} \end{bmatrix} \quad (40)$$

Then the Jacobean matrix is the differential of the tool configuration vector ω as

$$J(q) = \frac{\partial w}{\partial q_i} \quad (41)$$

By taking a differentiation of the eq. 40, the Jacobean for the AdeptThree robot is defines as

$$J(q) = \begin{bmatrix} -l_1 s_1 - l_2 s_{1+2} & -l_2 s_{1+2} & 0 & 0 \\ l_1 c_1 + l_2 c_{1+2} & l_2 c_{1+2} & 0 & 0 \\ 0 & 0 & -1 & 0 \\ 0 & 0 & 0 & 0 \\ 0 & 0 & 0 & 0 \\ 0 & 0 & 0 & \frac{e^{\left(\theta 4/\pi\right)}}{\pi} \end{bmatrix} \quad (42)$$

The first 3x3 matrix in the Jacobean is also called direct Jacobean. Because the Jacobean in eq. 42 is not a square matrix, it is not invertible. In this condition, the direct Jacobean can be useful since it is a square and invertible matrix.

Maple script for forming the Jacobean is listed below.

```
> restart: with(LinearAlgebra):
> q:=vector(4,[phi1,phi2,d3,phi4]):
> J:=matrix(6,4):
> w[1]:=l1*cos(phi1)+l2*cos(phi1+phi2):
> w[2]:=l1*sin(phi1)+l2*sin(phi1+phi2):
> w[3]:=-d4-d3+d1;
> w[4]:=0;
> w[5]:=0;
> w[6]:=exp(q[4]/pi);
> for i from 1 to 4 do
> for j from 1 to 6 do
> J[j,i]:=diff(w[j],q[i]);
> end do;
> end do;
> print(J);
  ...
```

5. Kinematics simulation

A Virtual Instrumentation (VI) was built to the section of kinematics simulation for supporting the manual calculation of a four DOF SCARA robot. The VI is a product of

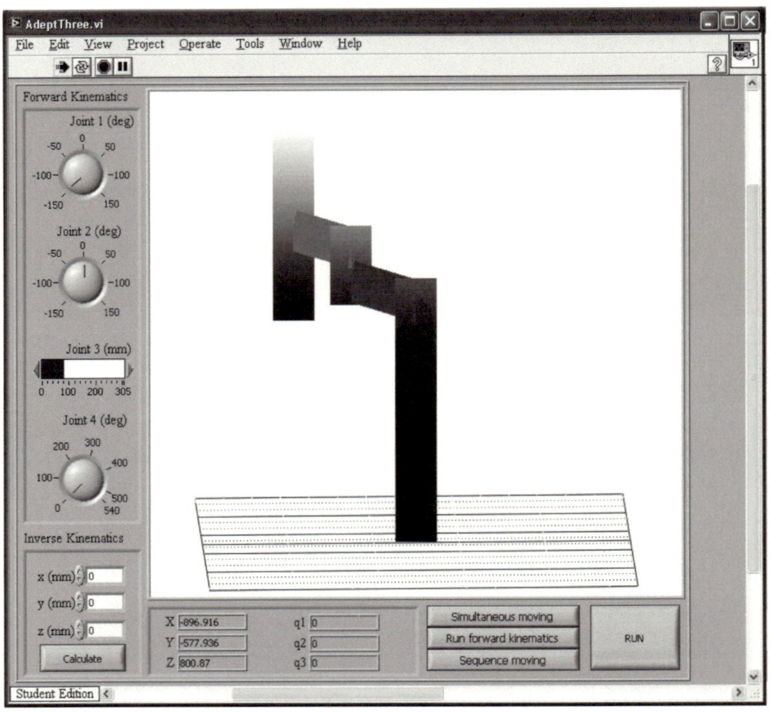

Fig. 9. SCARA robot simulation

graphical programming in LabView which is produced by National Instrumentation. The designed VI can simulate visual movement of the SCARA robot. The advantage of utilizing LabView is that the graphical programming language is easy and simple to be used. A user only needs to set each property to program the VI.
As shown in fig.9, the VI can be used to move the robot by applying the method of forward and inverse kinematics. To support the visual joint tracking, the VI is provided with simultaneous moving and sequence moving buttons. In simultaneous moving mode, each joint move together in same time. On the other hand sequence moving mode provides the motion of each joint one by one. Started from 1st joint to 4th joint, each joint will move after the other finished its task. The position of the end-effector is given in X, Y and Z boxes, while the joint variables are shown in q_1, q_2 and q_3 boxes.

6. Conclusion

This paper formulates and solves the kinematics problem for an AdeptThree robot arm. The forward kinematics of an AdeptThree robot was explained utilizing D-H convention while inverse kinematics of the robot was design using the principal cosines. Jacobean for the robot was design by using tool configuration vectors and direct Jacobean. Some script to design forward and inverse kinematics and also Jacobean matrix were provided using Maple. A graphical solution for simulating and calculating the robot kinematics was implemented in a virtual instrumentation (VI) of LabView. Using the VI, forward kinematics for a four dof SCARA robot can be simulated. Inverse kinematics for the robot can also be calculated with this VI.

7. References

[1] Jaydev P. Desai (2005). *D-H Convention*, Robot and Automation Handbook, CRC Press, USA, ISBN 0-8493-1804-1.

[2]Zomaya A.Y., Smitha H., Olariub S., Computing robot Jacobians on meshes with multiple buses, *Microprocessors and Microsystems*, no. 23, (1999), pp 309–324.

[3] Frank L.Lewis, Darren M.Dawson, Chaouki T.Abdallah (2006), *Robot Manipulators Control*, Marcel Dekker, Inc., New York.

[4] Bulent Ozkan, Kemal Ozgoren, Invalid Joint Arrangements and Actuator Related Singular Configuration of a System of two Cooperating SCARA Manipulator, *Journal of Mechatronics*, Vol.11, (2001), pp 491-507.

[5] Taylan Das M., L. Canan Dulger, Mathematical Modeling, Simulation and Experimental Verification of a SCARA Robot, *Journal of Simulation Modelling Practice and Theory*, Vol.13, (2005), pp 257-271.

[6] Mete Kalyoncu, Mustafa Tinkir (2006), Mathematical Modeling for Simulation and Control of Nonlinear Vibration of a Single Flexible Link, *Procedings of Intelligent Manufacturing Systems Symposium*, Sakarya University Turkey, May 29-31, 2006.

[7] Mustafa Nil, Ugur Yuzgec, Murat Sonmez, Bekir Cakir (2006),, Fuzzy Neural Network Based Intelligent Controller for 3-DOF Robot Manipulator *Procedings of Intelligent Manufacturing Systems Symposium*, Sakarya University Turkey, May 29-31, 2006.

[8] Rasit Koker, Cemil Oz, Tarik Cakar, Huseyin Ekiz, A Study of Neural Network Based Inverse Kinematics Solution for a Three-Joint Robot, *Journal of Robotics and Autonomous System*, Vol.49, (2004), pp 227-234

[9] Adept, (1991), *AdeptThree Robot: User's Guide*, Adept Technology, USA.
[10] Manjunath T.C., Ardil C., Development of a Jacobean Model for 4-Axes indigenously developed SCARA System, *International Journal of Computer and Information Science and Engineering*, Vol. 1 No 3, (2007), pp 152-158.
[11] John Faber Archila Diaz, Max Suell Dutra, Claudia Johana Diaz (2007), Design and Construction of a Manipulator Type Scara, Implementing a Control System, *Proceedings of COBEM, 19th International Congress of Mechanical Engineering*, November 5-9, 2007, Brasília.
[12] Rehiara Adelhard Beni, Smit Wim (2010), Controller Design of a Modeled AdeptThree Robot Arm, *Proceedings of the 2010 International Conference on Modelling, Identification and Control*, Japan, July 17-19, 2010, pp 854-858.
[13] Rehiara Adelhard Beni, System Identification Solution for Developing an AdeptThree Robot Arm Model, *Journal of Selected Areas in Robotics and Control*, February Edition, (2011), pp. 1-5 available at http://www.cyberjournals.com/Papers/Feb2011/06.pdf.

3

Solution to a System of Second Order Robot Arm by Parallel Runge-Kutta Arithmetic Mean Algorithm

S. Senthilkumar and Abd Rahni Mt Piah
Universiti Sains Malaysia, School of Mathematical Sciences,
Pulau Pinang
Malaysia

1. Introduction

Enormous amount of real time robot arm research work is still being carried out in different aspects, especially on dynamics of robotic motion and their governing equations. Taha [5] discussed the dynamics of robot arm problems. Research in this field is still on-going and its applications are massive. This is due to its nature of extending accuracy in order to determine approximate solutions and its flexibility. Many studies [4-8] have reported different aspects of linear and non-linear systems. Robust control of a general class of uncertian non-linear systems are investigated by zhihua [10].

Most of the initial value problems (IVPs) are solved using Runge-Kutta (RK) methods which in turn are employed in order to calculate numerical solutions for different problems, which are modelled in terms of differential equations, as in Alexander and Coyle [11], Evans [12], Shampine and Watts [14], Shampine and Gordan [18] codes for the Runge-Kutta fourth order method. Runge-Kutta formula of fifth order has been developed by Butcher [15-17]. Numerical solution of robot arm control problem has been described in detail by Gopal et al.[19]. The applications of non-linear differential–algebraic control systems to constrained robot systems have been discussed by Krishnan and Mcclamroch [22]. Asymptotic observer design for constrained robot systems have been analyzed by Huang and Tseng [21]. Using fourth order Runge-Kutta method based on Heronian mean (RKHeM) an attempt has been made to study the parameters concerning the control of a robot arm modelled along with the single term Walsh series (STWS) method [24]. Hung [23] discussed on the dissipitivity of Runge-Kutta methods for dynamical systems with delays. Ponalagusamy and Senthilkumar [25,26] discussed on the implementations and investigations of higher order techniques and algorithms for the robot arm problem. Evans and Sanugi [9] developed parallel integration techniques of Runge-Kutta form for the step by step solution of ordinary differential equations.

This paper is organized as follows. Section 2 describes the basics of robot arm model problem with variable structure control and controller design. A brief outline on parallel Runge-Kutta integration techniques is given in section 3. Finally, the results and conclusion on the overall notion of parallel 2-stage 3-order arithmetic mean Runge-Kutta algorithm and obtains almost accurate solution for a given robot arm problem are given in section 4.

2. Statement of the robot arm model problem and essential variable structure

2.1 Model of a robot arm

It is well known that both non-linearity and coupled characteristics are involved in designing a robot control system and its dynamic behavior. A set of coupled non-linear second order differential equations in the form of gravitational torques, coriolis and centrifugal represents dynamics of the robot. It is inevitable that the significance of the above three forces are dependent on the two physical parameters of the robot namely the load it carries and the speed at which the robot operates. The design of the control system becomes more complex when the end user needs more accuracy based on the variations of the parameters mentioned above. Keeping the objective of solving the robot dynamic equations in real time calculation in view, an efficient parallel numerical method is needed. Taha [5] discussed dynamics of robot arm problem represented by as

$$T = A(Q)\ddot{Q} + B(Q,\dot{Q}) + C(Q) \tag{1}$$

where $A(Q)$ represents the coupled inertia matrix, $B(Q,\dot{Q})$ is the matrix of coriolis and centrifugal forces. $C(Q)$ is the gravity matrix, T denotes the input torques applied at various joints.

For a robot with two degrees of freedom, by considering lumped equivalent massless links, i.e. it means point load or in this case the mass is concentrated at the end of the links, the dynamics are represented by

$$\left.\begin{aligned} T_1 &= D_{11}\ddot{q}_1 + D_{12}\ddot{q}_{21} + D_{122}(\ddot{q}_2)^2 + D_{112}(\dot{q}_1\dot{q}_2) + D_1, \\ T_2 &= D_{21}\ddot{q}_1 + D_{22}\ddot{q}_2 + D_{211}(\dot{q}_1)^2 + D_2, \end{aligned}\right\} \tag{2}$$

where

$$D_{11} = (M_1 + M_2)d_2^2 + 2M_2 d_1 d_2 \cos(q_2),$$

$$D_{12} = D_{21} = M_2 d_2^2 + M_2 d_1 d_2 \cos(q_2),$$

$$D_{22} = M_2 d_2^2,$$

$$D_{112} = -2M_2 d_1 d_2 \sin(q_2),$$

$$D_{122} = D_{211} = -M_2 d_1 d_2 \sin(q_2),$$

$$D_1 = [(M_1 + M_2)d_1 \sin(q_1) + M_2 d_2 \sin(q_1 + q_2)]g$$

and

$$D_2 = [M_2 d_2 \sin(q_1 + q_2)]g.$$

The values of the robot parameters used are M_1= 2kg, M_2 = 5kg, $d_1 = d_2 = 1$. For problem of set point regulation, the state vectors are represented as

$$X = (X_1, X_2, X_3, X_4)^T = (q_1 - q_{1d}, \dot{q}_1, q_2 - q_{2d}, \dot{q}_2)^T, \tag{3}$$

where
q_1 and q_2 are the angles at joints 1 and 2 respectively, and q_{1d} and q_{2d} are constants. Hence, equation (2) may be expressed in state space representation as

$$\left.\begin{aligned}
\dot{e}_1 &= x_2 \\
\dot{x}_2 &= \frac{D_{22}}{d}(D_{122}X_2^2 + D_{112}X_2X_4 + D_1 + T_1) - \frac{D_{12}}{d}(D_{211}X_4^2 + D_2 + T_2) \\
\dot{e}_3 &= x_4 \\
\dot{x}_4 &= \frac{-D_{12}}{d}(D_{122}X_2^2 + D_{112}X_2X_4 + D_1 + T_1) - \frac{D_{12}}{d}(D_{211}X_4^2 + D_2 + T_2).
\end{aligned}\right\} \tag{4}$$

Here, the robot is simply a double inverted pendulum and the Lagrangian approach is used to develop the equations.
In [5] it is found that by selecting suitable parameters, the non-linear equation (3) of the two-link robot-arm model may be reduced to the following system of linear equations:

$$\left.\begin{aligned}
\dot{e}_1 &= x_2, \\
\dot{x}_2 &= B_{10}T_1 - A_{11}x_2 - A_{10}e_1, \\
\dot{e}_3 &= x_4, \\
\dot{x}_4 &= B_{20}^2T_2 - A_{21}^2x_4 - A_{20}^2e_3,
\end{aligned}\right\} \tag{5}$$

where one can attain the system of second order linear equations:

$$\ddot{x}_1 = -A_{11}\dot{x}_1 - A_{10}x_1 + B_{10}T_1,$$

$$\ddot{x}_3 = -\dot{A}_{21}^2x_3 - A_{20}^2x_1 + B_{210}^2T_2,$$

with the parameters concerning joint-1 are given by
A_{10} = 0.1730, A_{11} = -0.2140, B_{10} = 0.00265,
and the parameters of joint-2 are given by
A_{20}= 0.0438, A_{21} = 0.3610, B_{20} = 0.0967
If we choose T_1 = r (constant) and T_2 = λ (constant), it is now possible to find the complementary functions of equation (4) because the nature of the roots of auxiliary equations (A. Es) of (4) is unpredictable. Due to this reason and for the sake of simplicity, we take $T_1 = T_2 = 1$.

considering $q_1 = q_2 = 0$, $q_{1d} = q_{2d} = 1$ and $\dot{q}_1 = \dot{q}_2 = 0$, the initial conditions are given by $e_1(0) = e_3(0) = -1$ and $e_2(0) = e_4(0) = 0$ and the corresponding exact solutions are,

$$\left.\begin{aligned}
&e_1(t) = e^{0.107t}[-1.15317919\cos(0.401934074t) + 0.306991074\sin(0.401934074t)] + 0.15317919, \\
&e_2(t) = e^{0.107t}[0.463502009\sin(0.401934074t) + 0.123390173\cos(0.401934074t)] + e^{0.107t}[-1.15317919\cos(0.401934074t) + 0.306991074\sin(0.401934074t)], \\
&e_3(t) = 1.029908976e^{-0.1134044161t} - 6.904124484e^{-0.0169168839t} + 4.874215508, \\
&e_4(t) = -0.116795962e^{-0.11340416t} + 0.116795962e^{-0.0169163891t}.
\end{aligned}\right\} \tag{6}$$

3. A brief sketch on parallel Runge-Kutta numerical integration techniques

The system of second order linear differential equations originates from mathematical formulation of problems in mechanics, electronic circuits, chemical process and electrical networks, etc. Hence, the concept of solving a second order equation is extended using parallel Runge-Kutta numerical integration algorithm to find the numerical solution of the system of second order equations as given below. It is important to mention that one has to determine the upper limit of the step-size *(h)* in order to have a stable numerical solution of the given ordinary differential equation with IVP. We thus consider the system of second order initial value problems,

$$\ddot{y}_j = f_j(x, y_j, \dot{y}_j), j = 1,2,.....m \tag{7}$$

with $y_j(x_0) = y_{j0}$

$\dot{y}_j(x_0) = \dot{y}_{j0}$ for all $j = 1,2,.....m$.

3.1 Parallel Runge-Kutta 2-stage 3-order arithmetic mean algorithm

A parallel 2-stage 3-order arithmetic mean Runge-Kutta technique is one of the simplest technique to solve ordinary differential equations. It is an explicit formula which adapts the Taylor's series expansion in order to calculate the approximation. A parallel Runge-Kutta 2-stage 3-order arithmetic mean formula is of the form,

$$k_1 = hf(x_n, y_n)$$

$$k_2 = hf(x_n + \frac{1}{2}, y_n + \frac{k_1}{2}) = k_2^*$$

$$k_3 = hf(x_n + k_1, y_n + k_1) = k_3^* .$$

Hence, the final integration is a weighted sum of three calculated derivatives per time step is given by,

$$y_{n+1} = y_n + \frac{h}{6}[k_1 + 4k_2 + k_3].$$

Parallel 2-stage 3-order arithmetic mean Runge-Kutta algorithm to determine y_j and $\dot{y}_j, j=1,2,3,....m$ is given by,

$$y_{jn+1} = y_{jn} + \frac{h}{6}[k_{1j} + 4k_{2j} + k_{3j}] \tag{8}$$

and

$$\dot{y}_{jn+1} = \dot{y}_{jn} + \frac{h}{6}[u_{1j} + 4u_{2j} + u_{3j}]$$

$$k_{1j} = \dot{y}_{jn},$$

$$k_{2j} = \dot{y}_{jn} + \frac{hu_{1j}}{2} = k_{2j}^{*}$$

$$k_{3j} = \dot{y}_{jn} + hu_{1j} = k_{3j}^{*} \tag{9}$$

$$u_{1j} = f(x_n, y_{jn}, \dot{y}_{jn}), \forall j=1,2,3....,m \tag{10}$$

$$u_{2j} = f(x_n + \frac{h}{2}, y_{1n} + \frac{hk_{11}}{2}, y_{2n} + \frac{hk_{12}}{2}, ..., y_{mn} + \frac{hk_{1m}}{2}, \dot{y}_{1n} + \frac{hu_{11}}{2}, \dot{y}_{2n} + \frac{hu_{12}}{2}, ..., \dot{y}_{mn} + \frac{hu_{1m}}{2})$$

$$u_{3j} = f(x_n + h, y_{1n} + hk_{11}, y_{2n} + hk_{12}, ..., y_{mn} + hk_{1m}, \dot{y}_{1n} + hu_{11}, \dot{y}_{2n} + hu_{12}, ..., \dot{y}_{mn} + hu_{1m}).$$

The corresponding parallel 2-stage 3-order arithmetic mean Runge-Kutta algorithm array to represent equation (9) takes the form

0			
$\frac{1}{2}$	$\frac{1}{2}$		
1	1		
	1	4	1

Therefore, the final integration is a weighted sum of three calculated derivatives per time step given by,

$$y_{n+1} = y_n + \frac{h}{6}[k_1 + 4k_2 + k_3] \tag{11}$$

3.2 Parallel Runge-Kutta 2-stage 3-order geometric mean algorithm of type-I

The parallel 2-stage 3-order geometric mean Runge-Kutta formula of type–I is of the form,

$$k_1 = hf(x_n, y_n),$$

$$k_2 = hf(x_n + \frac{2}{3}, y_n + \frac{2k_1}{3}) = k_2^*,$$

$$k_3 = hf(x_n + k_1, y_n + k_1) = k_3^*,$$

Hence, the final integration is a weighted sum of three calculated derivates per time step which is given by,

$$y_{n+1} = y_n + hk_1^{1/4} k_2^{3/4}.$$

Parallel 2-stage 3-order geometric mean Runge-Kutta algorithm of type–I to determine y_j and $\dot{y}_j, j = 1,2,3,....m$ is given by,

$$y_{jn+1} = y_{jn} + hk_{1j}^{1/4} k_{2j}^{3/4}, \tag{12}$$

and

$$\dot{y}_{jn+1} = \dot{y}_{jn} + hk_{1j}^{1/4} k_{2j}^{3/4},$$

$$k_{1j} = \dot{y}_{jn},$$

$$k_{2j} = \dot{y}_{jn} + \frac{2hu_{1j}}{3} = k_{2j}^*,$$

$$k_{3j} = \dot{y}_{jn} + hu_{1j} = k_{3j}^*, \tag{13}$$

$$u_{1j} = f(x_n, y_{jn}, \dot{y}_{jn}), \ \forall j = 1,2,3....,m \tag{14}$$

$$u_{2j} = f(x_n + \frac{2h}{3}, y_{1n} + \frac{2hk_{11}}{3}, y_{2n} + \frac{2hk_{12}}{3}, ..., y_{mn} + \frac{2hk_{1m}}{3},$$
$$\dot{y}_{1n} + \frac{2hu_{11}}{3}, \dot{y}_{2n} + \frac{2hu_{12}}{3}, ..., \dot{y}_{mn} + \frac{2hu_{1m}}{3}),$$

$$u_{3j} = f(x_n + h, y_{1n} + hk_{11}, y_{2n} + hk_{12}, ..., y_{mn} + hk_{1m}, \dot{y}_{1n} + hu_{11}, \dot{y}_{2n} + hu_{12}, ..., \dot{y}_{mn} + hu_{1m}).$$

parallel Runge-Kutta 2-stage 3-order geometric mean of type–I array represent equation (13) takes the form

Hence, the final integration is a weighted sum of three calculated derivatives per time step and the parallel Runge-Kutta 2-stage 3-order geometric mean of type–I formula is given by,

$$y_{n+1} = y_n + hk_1^{1/4} k_2^{3/4}. \tag{15}$$

0		
$\frac{2}{3}$	$\frac{2}{3}$	
1	1	
	$1\frac{1}{4}$	$1\frac{1}{4}$

3.3 Parallel 2-stage 3-order geometric mean runge-kutta formula of type–II

The parallel 2-stage 3-order geometric mean Runge-Kutta formula of type–II is of the form,

$$k_1 = hf(x_n, y_n),$$

$$k_3 = hf(x_n - \frac{k_1}{6}, y_n - \frac{k_1}{6}).$$

Hence, the final integration is a weighted sum of three calculated derivates per time step given by,

$$y_{n+1} = y_n + hk_1^{\ 4}k_3^{\ -3}.$$

Parallel 2-stage 3-order geometric Mean Runge-Kutta algorithm of type–II to determine y_j and $\dot{y}_j, j=1,2,3,....m$ is given by,

$$y_{jn+1} = y_{jn} + h[k_{1j}^{4}\ k_{3j}^{-3}]. \tag{16}$$

and

$$\dot{y}_{jn+1} = \dot{y}_{jn} + h[u_{1j}^{4} + u_{3j}^{-3}].$$

$$k_{1j} = \dot{y}_{jn},$$

$$k_{3j} = \dot{y}_{jn} - \frac{hu_{ij}}{6}, \tag{17}$$

$$u_{1j} = f(x_n, y_{jn}, \dot{y}_{jn}),\ \forall j=1,2,3....,m \tag{18}$$

$$u_{3j} = f(x_n - \frac{h}{6}, y_{1n}\frac{hk_{11}}{6}, y_{2n} + \frac{hk_{12}}{6}, ..., y_{mn} + \frac{hk_{1m}}{6},$$
$$\dot{y}_{1n} + \frac{hu_{11}}{6}, \dot{y}_{2n} + \frac{hu_{12}}{6}, ..., \dot{y}_{mn} + \frac{hu_{1m}}{6}).$$

The corresponding parallel Runge-Kutta 2-stage 3-order geometric mean algorithm of type-II array to represent Equation (17) takes the form:

$$
\begin{array}{c|cc}
0 & & \\
-\frac{1}{6} & -\frac{1}{6} & \\
\hline
 & 1^{4} & 1^{-3}
\end{array}
$$

Therefore, the final integration is a weighted sum of three calculated derivatives and the parallel Runge-Kutta 2-stage 3-order geometric mean algorithm formula is given by

$$y_{n+1} = y_n + hk_1^{\,4}k_3^{\,-3}\,. \tag{19}$$

4. Results and conclusion

In this paper, the ultimate idea is focused on making use of parallel integration algorithms of Runge-Kutta form for the step by step solution of ordinary differential equations to solve system of second order robot arm problem. The discrete and exact solutions of the robot arm model problem have been computed for different time intervals using equation (5) and y_{n+1}. The values of e_1(t), e_2(t),e_3(t) and e_4(t) can be calculated for any time t ranging from 0.25 to 1 and so on.

To obtain better accuracy for $e_1(t)$, $e_2(t)$, $e_3(t)$ *and* $e_4(t)$ by solving the equations (5) and y_{n+1}.

Sol. No.	Time	Exact Solution	Parallel RKAM Solution	Parallel RKAM Error
1	0.00	-1.00000	-1.00000	0.00000
2	0.25	-0.99365	-0.99533	-0.00167
3	0.50	-0.97424	-0.97864	-0.00440
4	0.75	-0.94124	-0.94943	-0.00819
5	1.00	-0.89429	-0.90733	-0.01303

Table 1. Solutions of equation (5) for $e_1(t)$

Sol.	Time	Exact Solution	Parallel RKAM Solution	Parallel RKAM Error
1	0.00	0.00000	0.00000	0.00000
2	0.25	0.05114	0.04598	0.00515
3	0.50	0.10452	0.09412	0.01044
4	0.75	0.15968	0.14389	0.01578
5	1.00	0.21610	0.19499	0.02110

Table 2. Solutions of equation (5) for $e_2(t)$

Sol. No.	Time	Exact Solution	Parallel RKAM Solution	Parallel RKAM Error
1	0.00	-1.00000	-1.00000	0.00000
2	0.25	-0.99965	-0.99973	-0.00008
3	0.50	-0.99862	-0.99871	0.00009
4	0.75	-0.99693	-0.99700	0.00007
5	1.00	-0.99460	-0.99462	0.00001

Table 3. Solutions of equation (5) for $e_3(t)$

Sol. No.	Time	Exact Solution	Parallel RKAM Solution	Parallel RKAM Error
1	0.00	0.00000	0.00000	0.00000
2	0.25	0.00277	0.00285	-0.00007
3	0.50	0.00545	0.00560	-0.00015
4	0.75	0.00805	0.00879	-0.00074
5	1.00	0.01056	0.01084	-0.00028

Table 4. Solutions of equations (5) for $e_4(t)$

Similarly, by repeating the same computation process for parallel Runge-Kutta 2- stage 3-order geometric mean algorithm of type-I and type-II respectively, yield the required results. It is pertinent to pinpoint out that the obtained discrete solutions for robot arm model problem using the 2-parallel 2-processor 2-Stage 3-order arithmetic mean Runge-Kutta algorithm gives better results as compared to 2-parallel 2-procesor 2-stage 3-order geometric mean Runge-Kutta algorithm of type-I and 2-parallel 2-procesor 2-stage 3-order geometric mean Runge-Kutta algorithm of type-II. The calculated numerical solutions using 2-parallel 2-procesor 2-stage 3-order arithmetic mean Runge- Kutta algorithm is closer to the exact solutions of the robot arm model problem while 2-parallel 2-procesor 2-stage 3-order geometric mean Runge-Kutta algorithm of type-I and type-II gives rise to a considerable error. Hence, a parallel Runge-Kutta 2-stage 3-order arithmetic mean algorithm is suitable for studying the system of second order robot arm model problem in a real time environment. This algorithm can be implemented for any length of independent variable on a digital computer.

5. Acknowledgement

The first author would like to extend his sincere gratitude to Universiti Sains Malaysia for supporting this work under its post-doctoral fellowship scheme. Much of this work was carried out during his stay at Universiti Sains Malaysia in 2011. He wishes to acknowledge Universiti Sains Malaysia's financial support.

6. References

[1] K. Murugesan, N.P.Gopalan and D. Gopal,(2005),"Error free Butcher algorithms for linear electrical circuits", ETRI Journal, Vol.27, No.2, pp.195-205.

[2] J.Y. Park, D.J.Evans, K. Murugesan, S. Sekar, and V.Murugesh,(2004),"Optimal control of singular systems using the RK-Butcher algorithm", International Journal of Computer Mathematics, Vol.81, No.2, pp.239-249.

[3] J.Y. Park, K. Murugesan, D.J.Evans, S. Sekar, and V.Murugesh,(2005), "Observer design of singular systems (transistor circuits) using the RK-Butcher algorithm", International Journal of Computer Mathematics, Vol.82, No.1, pp.111-123.

[4] S.Sekar,V.Murugesh and K.Murugesan (2004), "Numerical strategies for the system of second order IVPs Using the RK-Butcher algorithms", International Journal of Computer Science and Applications, Vol.1, No.2, pp.96-117.

[5] Z.Taha, "Approach to variable structure control of industrial robots", pp.53-59, Compiled by K.Warwick and A.Pugh, (1988), Robot Control: Theory and applications, (I E E Control Engineering Series, Peter Peregrinus Ltd, North-Holland.

[6] S.Oucheriah, (1999), "Robust tracking and model following of uncertain dynamic delay systems by memory less linear controllers", IEEE Transactions on automatic control, Vol.44. No.7, pp. 1473-1481.

[7] D.Lim and H. Seraji, (1997), "Configuration control of a mobile dexterous Robot: Real time implementation and experimentation", International Journal of Robotics Research, Vol. 16,No.5. pp. 601-618.

[8] M.M.Polvcarpou and P.A.Loannou, (1996), "A Robust adaptive non-linear control Design", Automatica (Journal of IFAC), Vol.32. No. 3. pp. 423-427.
[9] D.J.Evans and B.B.Sanugi, (1989), "A parallel Runge-Kutta integration method", Parallel Computing,Vol.11,pp.245-251.
[10] Zhihua. Qu, (1992), "Robot control of a class of non-linear uncertain systems", IEEE Transactions on Automatic Control, Vol.37, No. 9, pp. 1437-1442.
[11] R.K.Alexander and J.J.Coyle,(1990), "Runge-Kutta methods for differential-algebraic systems", SIAM Journal of Numerical Analysis, Vol. 27, No.3, pp. 736-752.
[12] D.J.Evans, (1991), "A new 4th order Runge-Kutta method for initial value problems with error control", International Journal of Computer Mathematics, Vol.139, pp. 217-227.
[13] C.Hung,(2000), "Dissipativity of Runge-Kutta methods for dynamical systems with delays", IMA Journal of Numerical Analysis, Vol.20, pp. 153-166.
[14] L.F.Shampine and H.A.Watts, (1977), "The art of a Runge-Kutta code. Part-I", Mathematical Software, Vol.3. pp. 257-275.
[15] J.C. Butcher, (1964), "On Runge processes of higher order', Journal of Australian Mathematical Society", Vol.4. p.179.
[16] J.C. Butcher,(1987), "The numerical analysis of ordinary differential equations: Runge-Kutta and general linear methods", John Wiley & Sons, U.K.
[17] J.C.Butcher,(1990), "On order reduction for Runge-Kutta methods applied to differential-algebraic systems and to stiff systems of ODEs", SIAM Journal of Numerical Analysis, Vol.27, pp. 447-456.
[18] L.F. Shampine and M.K.Gordon,(1975), "Computer solutions of ordinary differential equations", W.H.Freeman. San Francisco. CA. p. 23.
[19] D. Gopal, V. Murugesh and K.Murugesan,(2006), "Numerical solution of second-order robot arm control problem using Runge-Kutta-Butcher algorithm", International Journal of Computer Mathematics, Vol. 83, No. 3, pp.345–356.
[20] Z.Taha, (1987), Dynamics and Control of Robots, Ph.D Thesis, University of Wales.
[21] H.P. Huang and Tseng,W.L,(1991), "Asymtotic observer design for constrained robot systems", IEE Proceedings Pt-D,138,3, pp.211-216,1991.
[22] H.Krishnan and N.Haris Mcclamroch,(1994), "Tracking in non-linear differential algebra control systems with applications to constrained robot systems", Automatica, Vol.30,No.12,pp.1885-1897.
[23] C.Hung, (2000), "Dissipitivity of Runge-Kutta methods for dynamical systems with delays", IMA Journal of Numerical Analysis,Vol.20,pp.153-166.
[24] D. P.Dhayabaran, E.C. H.Amirtharaj, K. Murugesan and D.J. Evans, (2006), "Numerical solution robot arm model using STWS RKHEM methods", LNCS, computational methods, pp.1695–1699.
[25] R.Ponalagusamy and S.Senthilkumar,(2010), "System of second order robot arm problem by an efficient numerical integration algorithm", International Journal of Computational Materials Science and Surface Engineering, Vol.3, pp. 237-250.

[26] R.Ponalagusamy and S.Senthilkumar,(2009), "Investigation on numerical solution for robot arm problem", Journal of Automation, Mobile Robotics and Intelligent Systems, Vol.3, No.3, pp.34-40.

4

Knowledge-Based Control for Robot Arm

Aboubekeur Hamdi-Cherif
[1]Qassim University, Computer Science Department, Buraydah
[2]Université Ferhat Abbas Setif (UFAS), Computer Science Department, Setif
[1]Saudi Arabia
[2]Algeria

1. Introduction

The present research work reports the usability of knowledge-based control (KBC) as an alternative control method with specific concentration robot arm (RA). This novel control approach is based on the combination of inferences and calculations. It is dictated by the advent of microprocessor technology which has been one of the sources of inspiration for techniques spanning the whole spectrum of controllers design. KBC can contribute to build simple proportional integral and derivative (PID) control schemes (Åström et al., 1992) to large classes of regulators such as self-tuning regulators and model-reference adaptive controllers, among others (Hanlei, 2010). Because knowledge base systems (KBSs) research has focused on implementing heuristic techniques, the corresponding knowledge-based controllers can justly be considered as the next logical step in control design and implementation (Handelman *et al.*, 1990). The main characteristics of knowledge-based controllers is that they incorporate years-long human expertise under the form of machine-understandable heuristic rules. In KBC, the knowledge elicited from human experts is codified and embodied within the KB in the form of IF-THEN rules. As a result, the KB technology takes into account the increase in system complexity. This sophistication is naturally encountered as efforts are made to stretch the limits of system performance and integrate more capabilities as a response to technological advances (Calangiu *et al.*, 2010). In addition, the inherent ability of KBSs to support incremental expansion of capabilities and provide justification for recommendations or actions is offered by conventional programming techniques. Serious considerations are being given to increasing system reliability by predicting algorithm failure in RAs control and reconfiguring control laws in response to algorithm failure due to instability/chattering, or large RAs parameter variations.

The knowledge-based control (KBC) benefits as applied to RA are to:

- Implement/incorporate heuristics within the RA control schemes.
- Diagnose or predict algorithm failure.
- Identify changes in RA parameters or structure.
- Recalculate control laws based upon knowledge of the current RA parameters.
- Select appropriate control laws based on the current RA responses.
- Execute supportive control logic which has been used for practical controllers in the past.
- Provide an explanation of the situation to the user as and when requested.

However KBC approach is not without issues. Indeed, KBC design problem requires elicitation / acquisition and coding of the "useful expertise", gained by humans over a lifetime. It is highly difficult to find proper ways of extracting this expertise from the human experts. Discerning "usefulness", avoiding unnecessary data and finding ways of optimizing this knowledge representation is not a straight-forward task. How far are we from common sense-based control? This paper extends the limits of RA control using KBC approach in order to reach this distant end.

The main issue of the present work is to answer positively our central question, *i.e.*, whether it is possible to integrate the diversified methods dealing with dynamical systems control exemplified by RA control, while concentrating on KBC as an alternative control method. We describe the epistemological characteristics of a framework that is believed to integrate two distinct methodological fields of research *i.e.*, artificial intelligence (AI)-based methods where KBC is partly rooted, on the one hand, and control theory, where RA control is formulated, on the other hand. Blending research from both fields results in the appearance of a richer research community. Emphasis is now made on RA control as a prelude to other classes of robotic systems; ultimately enhancing full programmable self-assembly compounds (Klavins, 2007). The chapter is organized as follows. In Section 2, the main KBC issues are discussed. Section 3 presents KBC within the general area of intelligent control and places KBC with respect to generalized hybrid control. Section 4 summarizes RA control in standard mathematical terms. Section 5 deals with an architecture for KBC for RA as an alternative control method followed by a conclusion and future developments.

2. Knowledge-based control issues

2.1 Our specific problem

The specific problem we want to tackle can broadly be expressed as follows:

Given:

- A plant configuration library describing the actual system to be controlled,
- A library of control algorithms with various degrees of complexity,

Find:

One (class of) algorithm (s) that control one plant configuration.

Application:

Address simulation of RAs dynamics under various control schemes.

For doing this, consider two complementary environments, *i.e.* a numeric environment responsible for making calculations (trajectory, control law,...) and a symbolic environment responsible for making logical inferences incorporating human experience. These two environments are the main components of any KBC architecture. Two modes of operation are therefore possible. In the numerical or exploitation mode, the program generates the outputs using imposed algorithms. In the inferential or exploration mode, the algorithm is not known before hand. Using the codified expertise in the KB, the program has to choose it from a library before firing the numeric mode. For the sequel, we first start by considering standard RA control and then KBC within the larger context of intelligent control.

2.2 From standard RA control to KBC

RA control is the process whereby a physical system, namely a set of robotic linked arms, is made compliant with some prescribed task such as following an imposed trajectory or keeping in pace with a given angular velocity (Siciliano, 2009). Welding and assembly-line robots are popular examples of RA industrial applications. RA control is a much diversified field. As a result, it makes concentrated research a difficult task. While RA control has been extensively studied from the pure control side (Lewis *et al.*, 2003), for the last four decades, or so, very little attention has been made with regard to KBC. Indeed, the symbolic approach efforts as applied to control at large remain quite isolated (Martins *et al.* 2006). Our fundamental aim is to contribute to the integration of RA control within KBC, considered within a larger intelligent control methodology; this latter being defined as a computational methodology that provides automatic means of improving tasks from heuristics (Hamdi-Cherif & Kara-Mohamed, 2009). As a subfield of intelligent control, KBC attempts to elaborate a control law on the basis of heuristics. KBC aim is therefore consistent with the overall goal of intelligent control and, as such, automatically generates a control law from heuristic rules and actual facts describing the actual RA status (control law, errors, trajectories).

2.3 Pending control issues

Although KBC is a promising applied research area, there remain many challenges to be addressed. The main pending issues are:

- the system under control can be very complex (*e.g.* nonlinearities in robot arm (RA)) ;
- our knowledge of the system is imprecise (*e.g.* unknown RA parameters, unknown conditions of operation) although gradually increasing during operation, in the optimistic case of successful identification process,
- the influence of the environment is strong (*e.g.* outside perturbation, modeling errors), may vary and may even influence the current task,
- the goal of the system is described symbolically and may have internal hierarchy to be further investigated and structured.

If the answer to these challenges can be obtained from human experts, then this knowledge is codified within the KB by knowledge engineers. If the answer is unknown, then offline experimentation is done by control engineers to gradually build an answer and codify it in the KB. In any case, the KBC designer has to constantly upgrade the KB with human expertise and/or manual experimentations.

2.4 Overview of related works

Few authors have addressed the issue of designing and developing systems that cater for general-purpose RA control. For example (Yae *et al.* 1994) have extended the EASY5 - the Boeing Engineering and Analysis SYstem - incorporating constrained dynamics. (Polyakov *et al.* 1994) have developed, in MATHEMATICA™, a symbolic computer algebra system toolbox for nonlinear and adaptive control synthesis and simulation which provides flexible simulation *via* C and MATLAB™ code generation. MATHEMATICA™ has also been used in a simulation program that generates animated graphics representing the motion of a simple planar mechanical manipulator with three revolute joints for teaching purposes (Etxebarria, 1994). A toolbox is available for RA control running on MATLAB™ (Corke, 1996). For supplementary and more general applications of computer algebra to CACSD (computer-aided control system design), we refer to (Eldeib and Tsai, 1989). Recent research directions aim at the development of operating systems for robots, not necessarily for the

RA class. An overview of ROS, an open source robot operating system has been recently reported. ROS is not an operating system in the traditional sense of process handling and scheduling. It provides a structured communications layer above the host operating systems of a heterogeneous cluster. ROS was designed to meet a specific set of challenges encountered when developing large-scale service robots as part of the so-called STAIR project [http://stair.stanford.edu/papers.php]. The way how ROS relates to existing robot software frameworks, and a brief overview of some of the available application software which uses ROS are reported in (Quigley, *et al.* 2009). However, none of these works addressed the issue of using the KBC approach to solve the RA control problem. Hence our solution.

3. Solution components

3.1 KBC within intelligent control

3.1.1 The area of intelligent control

One of the fundamental issues that concerns intelligent control is the extent to which it is possible to control the dynamic behavior of a given system independently of

- its complexity,
- our capability of separating it from the environment and localizing it,
- the context in which this system operates,
- the forms of knowledge available and the categories it manipulates,
- the methods of representation.

As formulated, this issue cannot be handled by either control theory or artificial intelligence (AI). Indeed, control theory has a very localized mostly numerical vision of the problem. This prevents it from looking beyond the localized constraints self-imposed by the designer and hidden within the mechanism of the mathematical representation. From the standpoint of AI, the available knowledge-related methods cannot easily handle dynamic systems and have very little consideration for numerical manipulation. Indeed, computations of margins of stability, controllability, observability are alien to AI. Moreover, both control theory and artificial intelligence (AI) cannot properly operate out of the operations research (OR) paradigm. Its queues, graphs and game-theoretic situations are typical of the variety of control applications. That is why an early proposal for the definition of intelligent control is to consider this field as the intersection of the three previously-cited disciplines namely control theory, AI and OR, (Saridis, 1987). Other fields such as soft computing represented by fuzzy, genetic, neural systems and their combinations, on the one hand, and cognitive science, on the other hand have been progressively integrated within the intelligent control discipline over the last three decades, or so (Lewis *et al*; 2003).

3.1.2 Landmarks of intelligent control

Intelligent control is a term that first appeared in the seventies and later developed in (Saridis , 1987). An early, but constantly refined definition of this field describes itself as that area beyond adaptive, learning and self-organizing systems which represents the meeting point between artificial intelligence (AI), automatic control (AC) and operations research (OR). A tremendous body of literature has been developed to account for the description / design within this novel paradigm. International intelligent control symposia have been held every year since 1985 and numerous contributions appear regularly in the specialized and thoroughly documented literature where novel original definitions and applications of the field are proposed *e.g.* (Rao, 1992), (Handelman *et al.*, 2010). Extensions of the field are reported by (Åström, 1989), (Åström and MacAvoy, 1992), and (Cellier *et al.*, 1992). Other approaches have also been considered by researchers like the cognition-oriented approach with applications, (Meystel, 1994). Among the several advanced theoretical and applied results are those due to (Saridis, 1987) who proposed the so-called an entropy-based theory for hierarchical controller design based on the so-called "principle of decreasing precision with increasing intelligence". More recently, methods concentrated on soft computing methods such as:

a. *Neural networks (NNs).* In (Kwan *et al.*, 2001), a desired compensation adaptive law-based neural network (NN) controller is proposed for the robust position control of rigid-link robots where the NN is used to approximate a highly nonlinear function. Global asymptotic stability is obtained with tracking errors and boundedness of NN weights. No offline learning phase is required as learning is done on-line. Compared with classic adaptive RA controllers, parameters linearity and determination of a regression matrix are not needed. However, time for converging to a solution might be prohibitive.

b. *Fuzzy-Genetic.* In (Merchán-Cruz and Morris, 2006), a simple genetic algorithm planner is used to produce an initial estimation of the movements of two RAs' articulations and collision free motion is obtained by the corrective action of the collision-avoidance fuzzy units.

3.1.3 Scope of intelligent control

Intelligent control as a discipline provides generalization of the existing control theories and methods on the basis of the following elements (Aström and MacAvoy, 1992):

- combined analysis of the plant and its control criteria,
- processes of multisensor operation with information (knowledge) integration and recognition in the loop,
- man-machine cooperative activities, including imitation and substitution of the human operator,
- computer structures representing these elements.

3.1.4 Specific issues in intelligent control

One of the main drawbacks of intelligent control is that, up to now, there is no established terminology identifiable with this discipline. There remains an inertia in following conventional views and recommendations. This attitude hinders the development of intelligent control ideas and methods. For the purpose of immediate applications, we will concentrate on a small area of intelligent control. On the one hand, we will focus on the use of numerical/exploitation (procedural) and inferential/exploration processing (declarative, rule-based) systems. The former describe the RA control algorithms while the latter represent the way in which the expertise is explored and used in firing the adequate algorithm according to the actual situation (plant, errors). In the multiresolutional control architectures for intelligent machines proposed by (Meystel, 1991), the general structure of the intelligent controller is described by a set of feedback loops. Each one of these loops is declared for a particular resolution level and works with a different time-scale. Resolution of a given level is defined by (Meystel, 1994) as "the size of undistinguishability zone for the representation of goal, plan and feedback law."

3.2 KBC as a generalized hybrid control methodology

3.2.1 Hybrid control

KBC can alternatively be considered with respect to hybrid control. In the early sixties, the discipline of hybrid control referred to controlled systems using both discrete and continuous parts. This discipline spanned a substantial area of research from basic switched linear systems to full-scale hybrid automata. Later, symbolic control methods came to include abstracting continuous dynamics to symbolic descriptions, instruction selection and coding in finite-bandwidth control applications, and applying formal language theory to the continuous systems domain. A number of results have emerged in this area with a conventional control-theoretic orientation, including optimal control, stability, system identification, observers, and well-posedness of solutions. At the same time, symbolic control provides faithful descriptions of the continuous level performance of the actual system, and as a result, provides a formal bridge between its continuous and the discrete characteristics (Egerstedt *et al.*, 2006).

3.2.2 Generalized hybrid control

Generalized hybrid control is meant to incorporate logic and control, whether discrete or continuous. For our KBC concern, we will consider KBC as an integration of pure control and logical inference as expressed by either propositional logic or first-order logic (FOL). As a result, KBC addresses questions at the highest level, *i.e.*, at the level of symbols, and as such stands half-way between computer science and logic, on the one hand and control theory, on the other hand. A whole research area is to be investigated whereby results from hybrid control are to be mapped onto generalized hybrid control. As for now, a new line of research in hybrid systems has been initiated that studies issues not quite standard to the controls community, including formal verification, abstractions, model expressiveness, computational tools, and specification languages. These issues were usually addressed in other areas, such as software engineering and formal languages (Hamdi-Cherif, 2010).

3.3 Overall architecture for intelligent control

Intelligent machines are those that perform anthropomorphic tasks, autonomously or interactively and/or proactively with a human operator in structured or unstructured, familiar or unfamiliar environments. The intelligent controller represents the driving force that allows intelligent machines achieve their goals autonomously. It embodies functions of inferences as well as conventional control based on numeric processing. When such environments treat more than one state of the process to be controlled, as in the case of RA control, then it is careful to separate between control and inference, both functionally and architecturally. To this end we propose, in Figure 1, an overall architecture for intelligent control which considers the following levels:

3.3.1 Formulation level

At the formulation level, we find a hierarchical task formulation / task negotiation process. In the worst case situation, this formulation elaborates a model of an imprecise and incomplete plant.

3.3.2 Controller-plant matching level

This level uses knowledge from the KB to decide which controller algorithm is suitable for a given plant when operating under some prescribed user-defined specifications and other additional constraints. This level is further expanded in Figure 2.

3.3.3 Reasoning level

At this level, a KB contains the necessary knowledge to solve the controller-plant matching problem on the basis of the formulation. For the obtainment of the final controller-plant matching, a hybrid numeric / symbolic system representation has to be used. Some trade-off tasks as part of control process have to be considered.

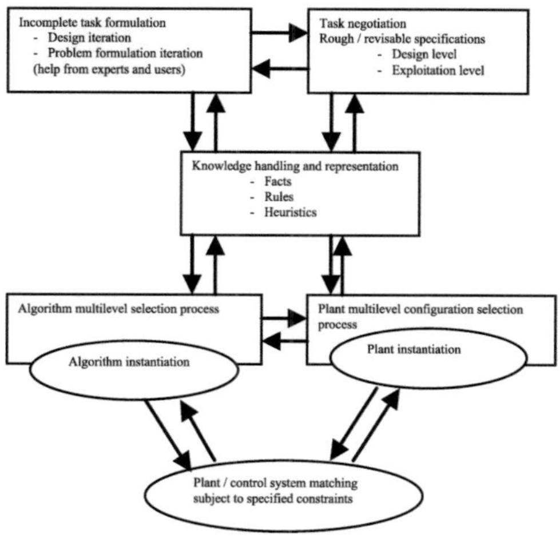

Fig. 1. Overall architecture for intelligent control

3.4 Inference issues

In evaluating any knowledge base system (KBS), and therefore any knowledge-based control (KBC) system, a wide range of criteria can be considered. We will define a generic framework for a description of the inferential part intervening in the KBC system. There are thousands of such systems ranging from free software (*e.g.* CLIPS, http://clipsrules.sourceforge.net/) to large industrial advanced packages such as G2™ from Gensym™ (http://www.gensym.com)

3.4.1 Knowledge base structure

Under this heading, we describe whether the system provides the representation by frames, messages, object-oriented languages, semantic nets, among others.

3.4.2 Type of logic involved

The usual types of logic available in KBS shells/systems are :

- *Propositional*: Boolean with no variables.
- *Predicate or first order logic (FOL):* Boolean with variables.
- *Temporal*: involves time in reasoning.
- *Fuzzy*: handles uncertainty, imprecision.

- *Non-monotone*: handles changing data.
- *Default*: handles situations like "most controllers are acceptable for these specifications".
- *Modal*: handles situations like "it's possible that", "it has been shown that this type of controller does not fit".

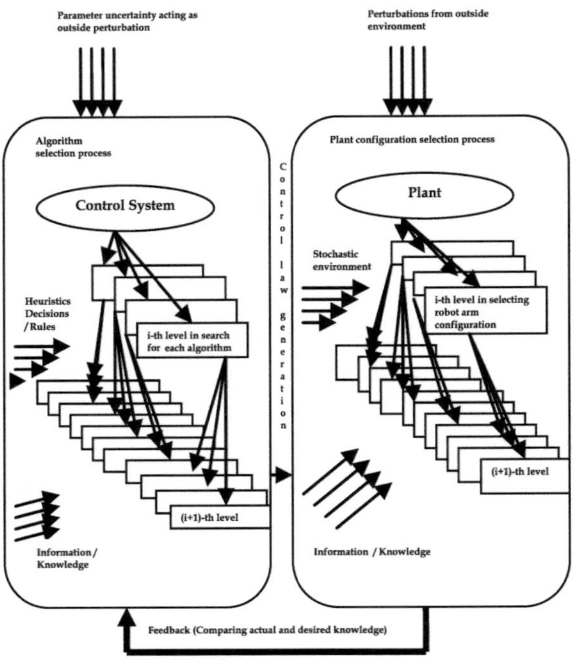

Fig. 2. Controller-plant matching problem

3.4.3 Reasoning strategy

- *Forward chaining*: hypothesis-driven.
- *Backward chaining*: goal-driven.
- *Hybrid chaining*: combining both forward and backward chaining.
- *Blackboards*: for keeping set of hypotheses of partial and final solutions.

3.4.4 Knowledge issues

- *Knowledge management* : browsers, editors, workspaces, workspaces security.
- *Knowledge validation* : "what if's" simulation capabilities.
- *Knowledge building tools* : human interface quality, KB construction quality, natural language environment.
- *Knowledge debugging* : levels of tracing, rules reporting, quality in entry and knowledge management.

3.4.5 Explanation, truth and uncertainty

- *Explanation of reasoning*: why's, natural language explanations, messages, variables values representation.
- *Truth maintenance* : forward update, backward update.

- *Uncertainty management* : certainty factors, fuzzy-oriented management.

3.4.6 Miscellaneous

- *Interface with outside world*: data acquisition, data bases, other specialized software interfacing.
- *Other performance*: stand-alone off-line, real-time performance, networking, other advanced special features.

4. RA Standard control problem

4.1 Brief history

On the control side, we concentrate on some classes of control methods such as adaptive control and passivity-based control. The development of RA control algorithms has gone through at least three historical phases. The first is the model reference adaptive control and self-tuning control followed by the passivity approach and then by the soft computing methods. We report here the first two phases while the soft computing methods have been described in Section 3.1.3 above..

4.1.1 Model Reference Adaptive Control (MRAC) and Self-Tuning Control (STC)

The first phase (1978-1985) concentrated its efforts on the approximation approach. The methods developed during this period are well-documented in the literature and some review papers have been written for that period (*e.g.* Hsia, 1986). Researches were concentrated on issues expanded below.

a. *Model reference adaptive control approach* (MRAC) guided by the minimization of the error between the actual system and some conveniently chosen model of it. At the methodological level, this represents a traditional example of supervised learning based on comparison between the actual and desired outputs while trying to minimize the error between desired and actual values.
b. *Self-tuning control* based on performance criteria minimization.

4.1.2 Parametrization approach

The methods developed during the second period that followed with some time overlaps with the previous period, concentrated on the parameterization approach. The methods developed within this period can be further separated in two broad classes, namely inverse dynamics and passivity-based control.

a. Inverse dynamics

The first set of methods treats the inverse dynamics-based control or computed torque method. It relies on the exact cancellation of all the nonlinearities in the system. In the ideal case, the closed-loop system is decoupled and linear. Stability in this case is based on the Lyapunov direct method. A dynamical system is said to be stable in the sense of Lyapunov if it has the characteristics that when it loses an un-restored energy over time, then it will stabilize at some final state, called the attractor. In Lord Kelvin's terms this means that conservative systems in the presence of dissipative forcing elements will decay to a local minimum of their potential energy. However, finding a function that gives the precise energy of a given physical system can be extremely difficult. On the other hand, for some systems (*e.g.* econometric and biological systems), the Lyapunov function has no physical meaning.

b. *Passivity-based control*

The second set of methods deals with passivity-based control. The aim is to find a control law that preserves the passivity of the rigid RA in closed-loop. Stability here is based on the Popov hyperstability method (Popov, 1973). One of the main motivations for using these control laws, as far as stability is concerned, is that they avoid looking for complex Lyapunov functions - a bottleneck of the Lyapunov-based design. These laws also lead, in the adaptive case, to error equations where the regressor is independent of the joint acceleration. The difficult issue of inertia matrix inversion is also avoided. At the opposite of inverse dynamics methods, passivity-based methods do not look for linearization but rather for the passivity of the closed-loop system. Stability is granted if the energy of the closed-loop system is dissipated. The resulting control laws are therefore different for the two previous classes.

4.2 Issues in adaptive and passivity RA control

From the vast literature on adaptive control, only a small portion is applicable to RA control. One of the first approaches to adaptive control, based on the assumption of decoupled joint dynamics, is presented in (Craig, 1988). In general, multi-input multi-output (MIMO) adaptive control provides the means of solving problems of coupled motion, though nonlinear robot dynamics with rapidly changing operating conditions complicate the adaptive control problem involved, even if there are also advantages when compared with the adaptive control of linear systems. Specialized literature has appeared in the field, *e.g.*, the interesting tutorial reported in (Ortega & Spong, 1989). As far as adaptive control is concerned, some methods assume that acceleration is available for measurement and that the inertia matrix inverse is bounded. Others avoid at least the boundedness constraint (*e.g.* Amestegui *et al.*, 1987) while passivity-based control avoids both limitations. We propose to classify the specialized contributions in the field as follows:

a. *Parameter estimation:* such as the linear estimation models suitable for identification of the payload of a partially known robot, going back to (Vukabratovic *et al.*, 1984).

b. *Direct adaptive control* of robot motion as studied by :
    1. (Craig et al., 1987) in conjunction with model reference adaptive control (MRAC). Here stability is studied using strictly positive real transfer functions (SPR-TF).
    2. (Slotine and Li, 1987) in conjunction with the so-called "MIT rule". Here the regulator is independent of the acceleration measurement and linear in the parameters.
    3. Johansson has still improved the work of (Craig et al., 1987) in terms of stability. This method avoids matrix inversion and SPR-TF requirements (Johansson, 1990).

c. *Decentralized control* for adaptive independent joint control as proposed by (Seraji, 1989).

d. *Control and stability analysis* such as passivity-based control developed by (Landau and Horowitz, 1989).

4.3 RA dynamics

A standard mathematical model is needed for any RA control problem. The RA dynamics are modeled as a set of n linked rigid bodies (Craig, 2005). The model is given by the following standard ordinary differential equation in matrix form.

$$\tau(t) = M(q)\ddot{q} + C(q,\dot{q})\dot{q} + G(q) + V(\dot{q}) \tag{1}$$

Time arguments are omitted for simplicity. The notations used have the following meaning:
q : joint angular position, *nx1* real vector.
$\dot{q}$: joint angular velocity, *nx1* real vector.
$\ddot{q}$: joint angular acceleration, *nx1* real vector.
$\tau(t)$: joint torque, *nx1* real vector.
$M(q)$: matrix of moment of inertia or inertia matrix, *nxn* real matrix.
$C(q,\dot{q})\dot{q}$: Coriolis, centrifugal and frictional forces. C is *nxn* real matrix.
$G(q)$: gravitational forces. G is an *nx1* real vector describing gravity.
$V(\dot{q})$: *nx1* real vector for viscous friction. It is neglected in our forthcoming treatment.

4.4 RA PID control

Proportional integral and derivative (PID) control is one of the simplest control schemes. It has been successfully used for the last six decades, or so, in many diversified applications of control. Despite its simplicity, PID is still active as an applied research field. In February 2006, a special issue of *IEEE Control Systems Magazine* has been devoted to the subject to account for its importance and actuality. Insofar as automatically-tuned PIDs (or autotuners) are concerned, commercial products became available around the early eighties. Since the Ziegler-Nichols rules of thumb developed in the 1940's, many attempts have been made in the "intelligent" choice of the three gains (*e.g.* Åström *et al.* 1992). The intelligent approach also helps in explanation of control actions usage. Indeed, in many real-life applications, explanation of control actions is desirable, *e.g.*, why derivative action is necessary. On the numerical level, the PID control *u(t)* is given by:

$$u(t) = K_p e(t) + K_v \dot{e}(t) + K_i \int_0^t e(n)dn \tag{2}$$

$$e(t) = q(t) - q_d(t) \tag{3}$$

$$\dot{e}(t) = \dot{q}(t) - \dot{q}_d(t) \tag{4}$$

Equation (1) describes the control *u(t)*. K_p, K_i, K_v are the gains for the proportional (P), integral (I) and derivative (D) actions, respectively.
Equation (3) defines the position error *e(t)*, *i.e.*, the difference between the actual system position *q(t)* and the desired position $q_d(t)$.
Equation (4) defines the velocity error and is simply the time-derivative of the error given in Equation (3) above. Equation (4) describes the difference between the actual system velocity and the desired velocity. The PID scheme block-diagram is given in Figure 3.

4.5 RA adaptive control

4.5.1 Purpose of adaptive control

The general adaptive controller design problem is as follows : given the desired trajectory $q_d(t)$, with some (perhaps all) manipulator parameters being unknown, derive a control law for the actuator torques and an estimation law for the unknown parameters such that the manipulator output $q(t)$ tracks the desired trajectories after an initial adaptation process. Adaptive control laws may be classified on the basis of their control objective and the signal

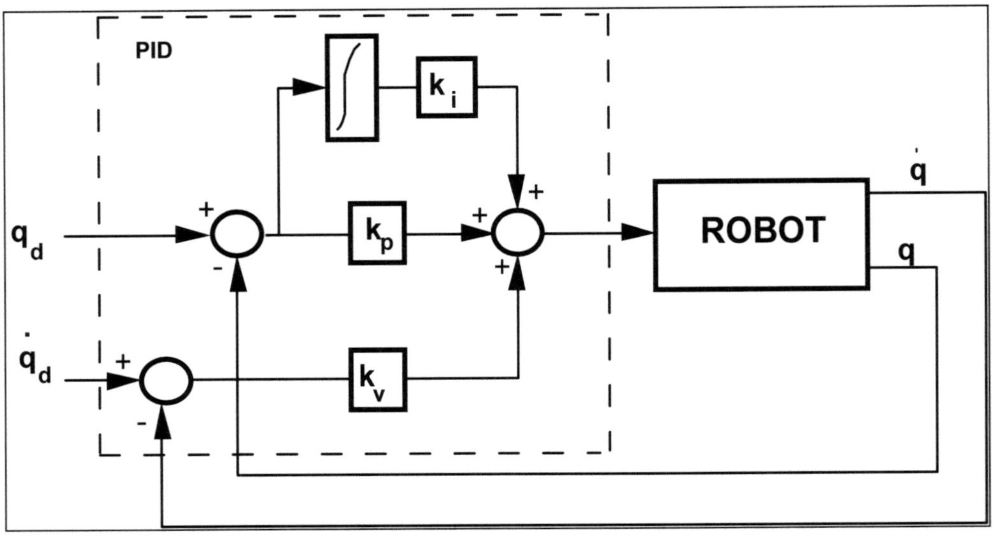

Fig. 3. RA PID Control

that drives the parameter update law. This latter can either be driven by the error signal between the estimated parameters and the true parameters (prediction or parametric error) or by the error signal between the desired and actual outputs (tracking error). Stability investigations are at the basis of acceptability of the proposed scheme.

4.5.2 Example of adaptive control scheme

As an example, the method due to (Amestegui *et al.*, 1987) compensates the modeling errors by a supplementary control $\delta\tau$. First, the computed torque approach is used whereby the linearizing control is obtained by a suitable choice of the torque. This amounts to simply replacing the acceleration $\ddot{q}$ by the control u in (1) above resulting in:

$$\tau(t) = M(q)u + C(q,\dot{q})\dot{q} + G(q) + V(\dot{q}) \tag{5}$$

Combining (1) and (5) yields:

$$M(q)(\ddot{q} - u) = 0 \tag{6}$$

Which amounts to n decoupled integrators $(\ddot{q} = u)$. In this case, the control u can be expressed in terms of the desired acceleration as a PD compensator.

Now compensate the modeling errors by a supplementary control $\delta\tau$ and neglect viscous friction.

$$\tau(t) = M_0(q)(u) + C_0(q,\dot{q})\dot{q} + G(q) + \delta\tau \tag{7}$$

Using the linear parametrization property, we obtain:

$$M_0(q)(u-\ddot{q})+\delta\tau=\psi(q,\dot{q},\ddot{q})\Delta\theta \tag{8}$$

The compensating control is then given by :

$$\delta\tau=\psi(q,\dot{q},\ddot{q})\Delta\hat{\theta} \tag{9}$$

and the estimated parametric error vector is solution of :

$$\Delta\dot{\hat{\theta}}=-\Gamma\psi^T(q,\dot{q},\ddot{q})\hat{M}_0(q)(u-\ddot{q}) \tag{10}$$

In the previous equations, the following notations are used:
$\psi(q,\dot{q},\ddot{q})$ represents the regressor matrix, of appropriate dimensions.
The parametric error vector:

$$\Delta\theta=\theta_0-\theta \tag{11}$$

where θ is the actual parameter vector and
θ_0 a constant and linear vector with respect to the nominal robot model.
$\Delta\hat{\theta}$ is the estimate of $\Delta\theta$ and

$$\Gamma=diag(\gamma_1,\gamma_2...,\gamma_n) \tag{12}$$

is a positive-definite diagonal matrix with $\gamma_i>0$, representing the adaptation gain for the gradient parametric estimation method. Note that this last scheme avoids the inversion of the inertia matrix. It reduces the calculations complexity. However the measurement of the acceleration is always required. The block-diagram is given in Figure 4.

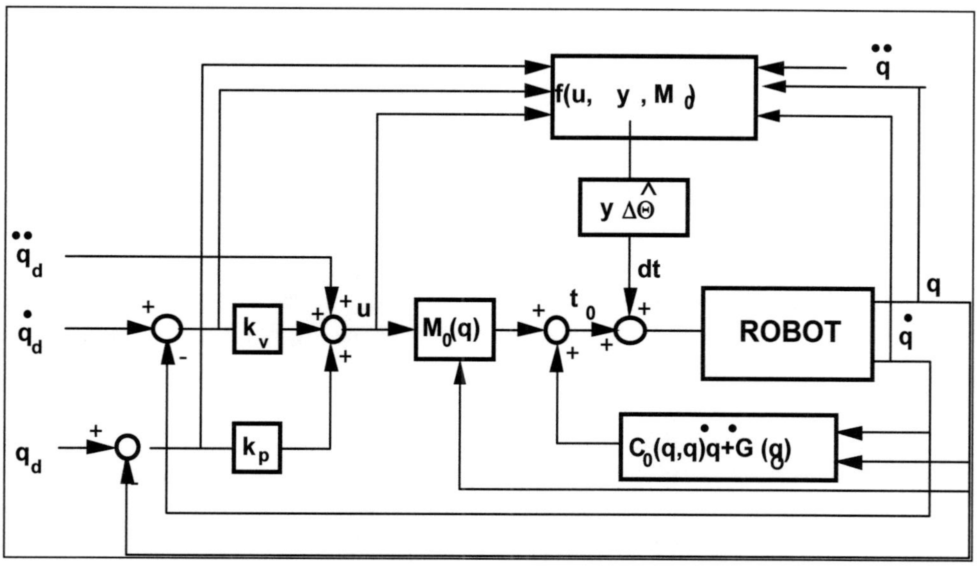

NB : In Figure 4, the following notations are used: $y=\psi$; $t=\tau$ $t_0=\tau_0$

Fig. 4. Amestegui's adaptive compensation scheme

4.6 RA robust control

Robust control approach considers adding a correcting term to the control signal. This compensates the parametric error. This supplementary signal gives better tracking and makes the system more robust with respect to parametric error. We can classify the robust methods as Lyapunov-based methods, variable structure methods and non-chattering high gains methods.

4.6.1 Lyapunov-based methods

This class of methods is based on the Lyapunov direct method and is based on (Spong and Vidyasagar, 2006). The main problem encountered by Lyapunov-based class of RA control algorithms is the so-called *chattering effect* which results from commutation of the supplementary signal. This behavior creates control discontinuities. Research efforts have been accomplished that cater for this undesirable chattering effect. The algorithm proposed by (Cai and Goldenberg 1988) is a tentative answer to the problem of chattering. The issue of chattering represents a predilection area for the applicability of KB methods, since chattering can be modeled using human expertise.

4.6.2 Variable structure methods

Variable structure methods, such as the one proposed by (Slotine, 1985) are based on high-speed switching feedback control where the control law switches to different values according to some rule. This class of methods drives the nonlinear plant's trajectory onto an adequately designed sliding surface in the phase space independently of modeling errors. In (Chen and Papavassilopoulos, 1991) four position control laws have been analyzed and compared for a single-arm RA dynamics with bounded disturbances, unknown parameter, and unmodeled actuator dynamics. Although very robust to system's disturbance and simplifying the complexity of control laws implementation, these methods suffer from undesirable control chattering at high frequencies.

4.6.3 Non-chattering high gains methods

The non-chattering high gains class of methods is based on the singular perturbation theory and considers two time scales. This class avoids the chattering effect (Samson, 1987). However, robustness in this case is guaranteed by the choice of a nonlinear gain which is calculated from the *a priori* knowledge of the parametric uncertainties and from the model chosen for control calculation. The resulting control can be considered as a regulator which automatically adapts the gains in accordance with the displacement errors (Seraji, 1989) and uses high gains only when these are needed, for instance when displacement error is large.

4.6.4 Example of robust control scheme

In this case, the parameters are not known but their range of variations is known. The basic idea of this method is to add a compensating term to the control which is obtained from an *a priori* estimated model. This compensation term takes into account the parameters bounds and tries to compensate the difference between the estimated and the real parameters of the robot. This makes possible an improved trajectory tracking and provides robustness with respect to the parametric errors. Several schemes of RA robust control have been studied and compared (Abdallah *et al.*, 1991). As an example, only one robust algorithm is described here, whose control law is given by :

$$\tau(t) = M_0(q)(u + \delta u) + C_0(q,\dot{q})\dot{q} + G_0(q) \tag{13}$$

where
* M_0, C_0 and G_0 are the *a priori* estimates of *M*, *C* and *G*, respectively.
* δu is the compensating control supplement.
* u is given by a PD compensator of the form:

$$u(t) = \ddot{q}_d(t) - K_p e(t) - K_v \dot{e}(t) \tag{14}$$

The additional control δu is chosen so as to ensure robustness of the control by compensating the parametric errors. Stability must be guaranteed. A reformulation of this control gives:

$$\dot{x} = Ax + B(\delta u + \eta(u,q,\dot{q})) \tag{15}$$

$$E_1 = Cx \tag{16}$$

where *A, B, C* and *x* are given by

$$A = \begin{bmatrix} 0 & I \\ -K_p & -K_v \end{bmatrix} \quad B = \begin{bmatrix} 0 \\ I \end{bmatrix} \quad C = \begin{bmatrix} \alpha & I \end{bmatrix} \quad x = \begin{bmatrix} e \\ \dot{e} \end{bmatrix} \tag{17}$$

with α is a diagonal constant positive-definite matrix of rank *n*, and

$$\eta(u,q,\dot{q}) = E(q)\delta u + E_1 u + M^{-1}(q)\Delta H(q,\dot{q}) \tag{18}$$

$$E(q) = M^{-1}(q)M_0(q) - I \tag{19}$$

$$\Delta H(q,\dot{q}) = [C_0(q,\dot{q}) - C(q,\dot{q})]\dot{q} + [G_0(q) - G] \tag{20}$$

Stability is granted only if the vector $\eta(u,q,\dot{q})$ is bounded. These bounds are estimated on the worst-case basis. Furthermore, under the assumption that there exists a function ρ such that:

$$\|\delta u\| < \rho(e,\dot{e},t) \tag{21}$$

$$\|\eta\| \le \rho(e,\dot{e},t) \tag{22}$$

the compensating control δu can be obtained from :

$$\delta u = \begin{cases} -\rho(e,\dot{e},t)\dfrac{E_1}{\|E_1\|} & if \quad \|E_1\| \neq 0 \\ 0 \quad if & \|E_1\| = 0 \end{cases} \tag{23}$$

This last control δu presents a chattering effect due to the discontinuities in (23). This phenomenon can cause unwanted sustained oscillations. Another control has been proposed which reduces these unwanted control jumps, (Cai and Goldenberg, 1988) as given in equation (24).

$$\delta u = \begin{cases} -\rho(e,\dot{e},t)\dfrac{E_1}{\|E\|} & if \quad \|E_1\| > \varepsilon \\ \dfrac{-\rho(e,\dot{e},t)}{\varepsilon}E_1 & if \quad \|E_1\| \leq \varepsilon \end{cases} \tag{24}$$

The robust control scheme is represented in Figure 5.

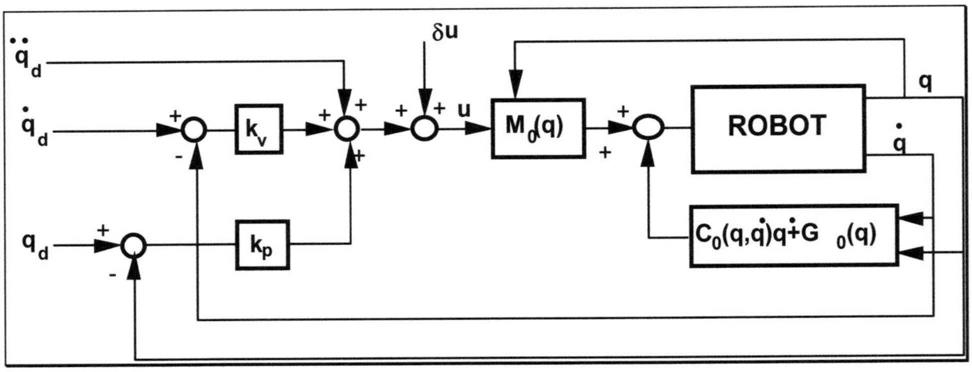

Fig. 5. Spong and Vidyasagar's robust control algorithm

5. Implementation

5.1 Basic architecture

The basic architecture is described in Figure 6. The general menus are described in Figure 7. The main program is started from the Matlab™ workspace window. Simulation triggers the Simulink™ environment and results can be obtained under the Matlab™ graphics window or in the Simulink™ environment (*e.g.* through scopes). Results can also be stored in *.MAT data files to be later handled by the knowledge base, through the interface.

The overall system is written in the Matlab™/ Simulink™ environment (http://www.mathworks.com). One of the main reasons for this choice is the possibility of interfacing it with the developed knowledge base using higher programming language, such as Microsoft Visual C++™ (MVC++™), under Windows™. The knowledge base is developed under a commercial expert system generator that supports interfacing with external MVC++™ executable programs. The other fundamental reason is the Matlab™ control systems library functions and specialized toolboxes, *e.g.* control systems toolbox and identification toolbox needed for adaptive control. Although, many languages / environments can be identified as suitable for the solution to our RA problem, we do not know, however whether any of these is interfaceable with the chosen expert system generator.

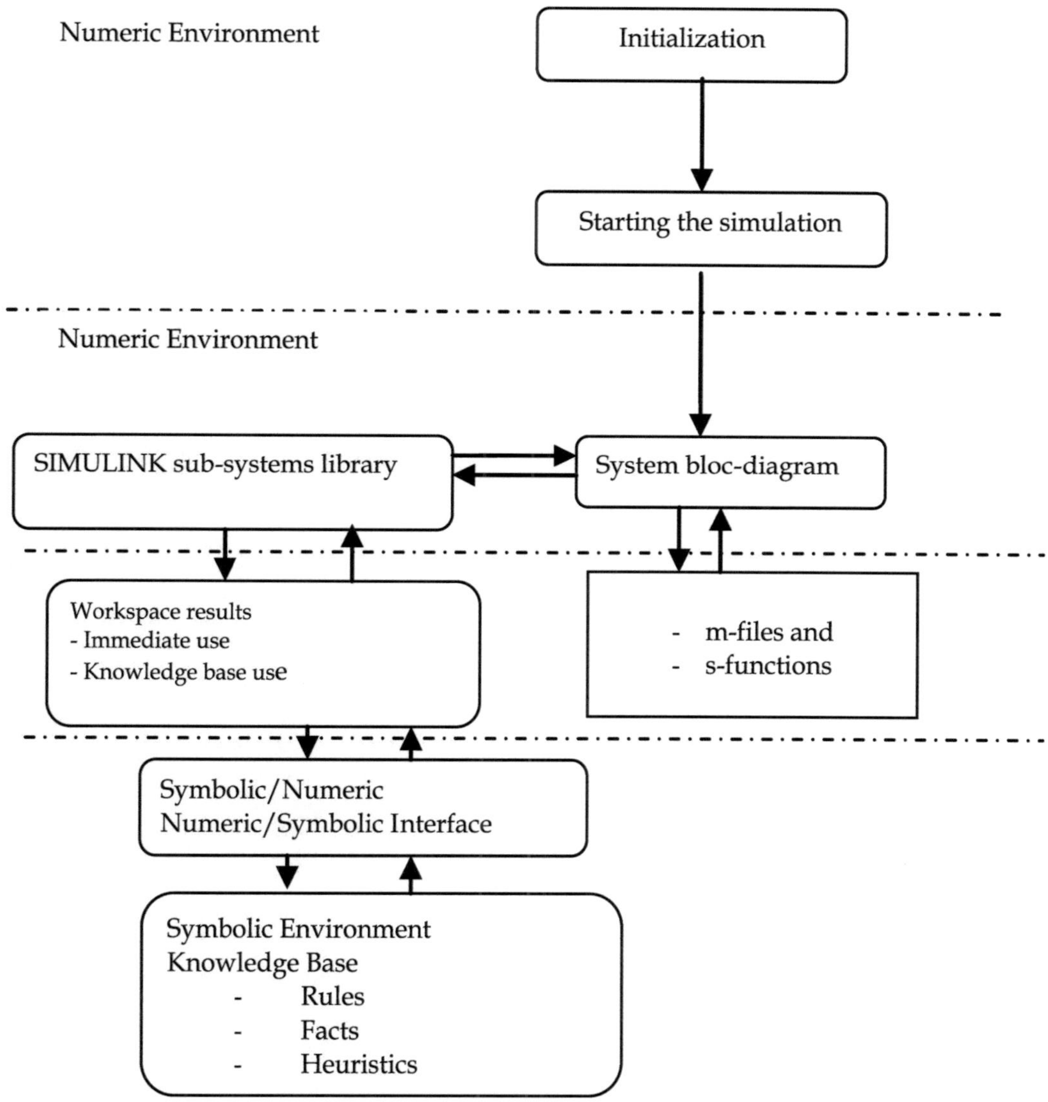

Fig. 6. Implemented Architecture

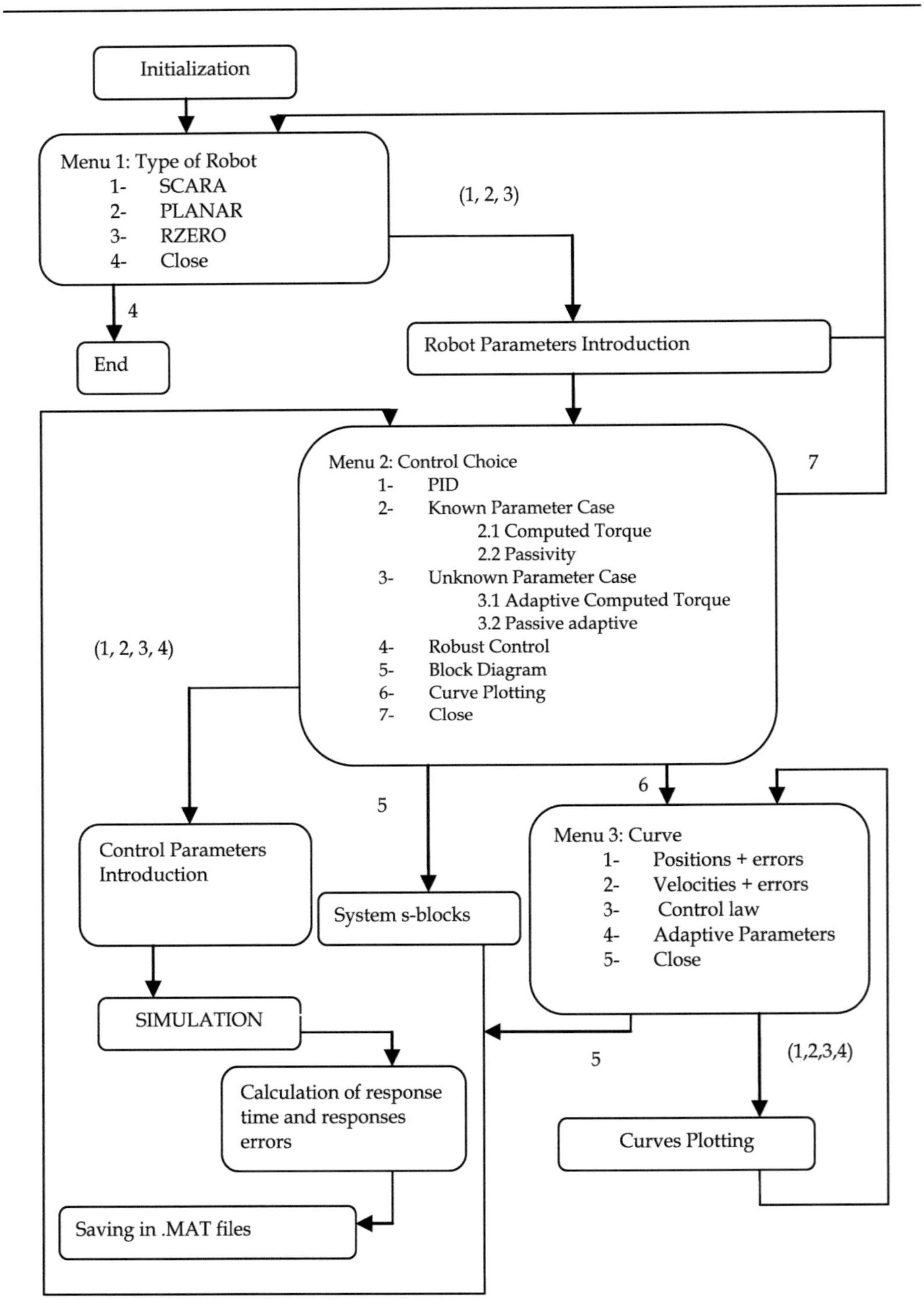

Fig. 7. Exploitation Environment

5.2 Plant configuration

Some of the available RA configuration have been used in the implementation, as examples. PLANAR, SCARA and RZERO are chosen because they are widely used and they represent different classes of configurations, as described in Figure 8,9,10 below. Of course the system is open to other configurations through the Matlab™ environment.

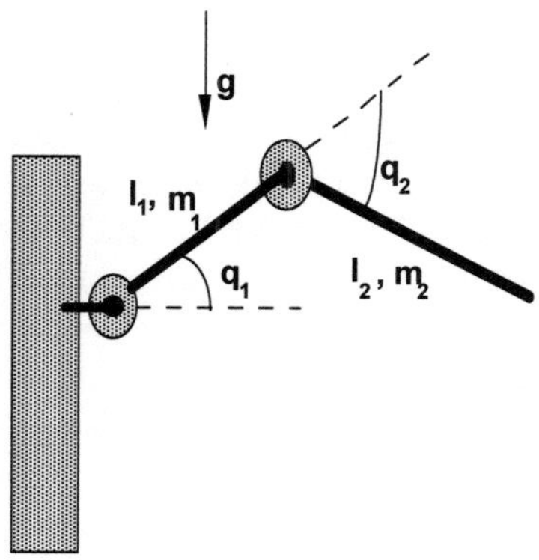

Fig. 8. PLANAR Robot

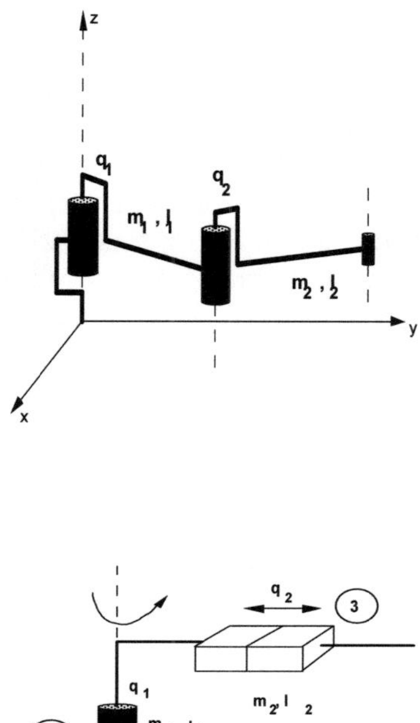

Fig. 9. SCARA Robot

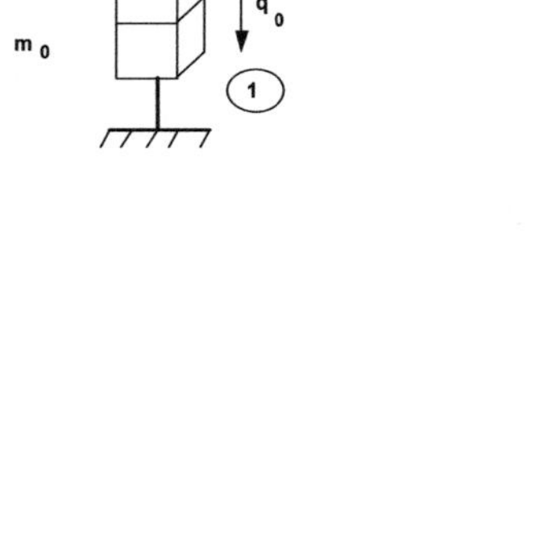

Fig. 10. RZERO Robot

5.3 Knowledge organization: from worlds to objects

Knowledge organization is handled by the inferential (exploration) environment. The knowledge is organized in different levels :

5.3.1 Global *vs.* local search

i. *At the global level* : this is done through the partitioning of the KB in coherent thematic sets of rules (each of these sets is called a *world*). These worlds can be hierarchically organized offering the possibility of describing global knowledge (ascending worlds, fathers worlds) and local knowledge (descending worlds, descendants worlds).
ii. *At the local level* : this is done through the expertise structuring using a network of classes and objects. A class is defined as an abstract object maker. Objects represent a declarative knowledge described by sets of particular data (called attributes) and the corresponding attributes values. Rules allow description of the expert knowledge using objects and / or classes. They are expressed in the conventional IF-THEN form. We only operate the KB as a stand-alone module to test its behaviour against that of human experts for further refinement.

5.3.2 The existing worlds

Worlds are coherent sets of rules and represent independent and encapsulated entities ensuring a high degree of knowledge modularity and maintenance. A world can be created according to the type of knowledge that is handled (*e.g.* set of rules dealing with PID controller). Hierarchical representation is available. This allows the organization of knowledge from the more general to the more specific (top-down fashion).

i. The meta-level nucleus (MLN)

The meta-level nucleus (MLN) represents the world that governs the navigation from one world to the other (or from one individual KB to the other). It is placed at the highest level of the hierarchy. All remaining worlds are sub-worlds of the MLN. All the rules of the MLN world (called a father world) are applicable in all the other inheriting sub-worlds (called descendants). Its description is given by the following structure:

```
World name : MLN
Father World : None
Descendants Worlds : %List of all the other remaining worlds%
```

ii. Pruning worlds

Based on the data and specifications provided by the user, the pruning worlds help in guiding the search towards the specialized individual knowledge base (IKB) as soon as possible. This pruning process is efficient in concentrating on the specific knowledge of interest at this or that particular step of the reasoning process. Once these worlds are selected, they become active while all other worlds remain inactive. The pruning worlds consist of the following :

```
W1 World : DynamicModelKnown
          % Describing the rules applicable for this case%
     W11 Sub-World : SpeedSlow
               W111 Sub-sub-World : ParametersKnown
               W112 Sub-sub-World : ParametersUnknown
               W113 Sub-sub-World : ParametersOthers
               W114 Sub-sub-World :ParametersNoImportance
     W12 Sub-World : SpeedHigh
```

```
        (Same sub-worlds as in W11 above)
        W13 Sub-World : SpeedUnknown
        (Same sub-worlds as in W11 above)
        W14 Sub-World : SpeedOthers
        (Same sub-worlds as in W11 above)
        W15 Sub-World : SpeedNoImportance
        (Same sub-worlds as in W11 above)


W2 World : DynamicModelUnknown
            % Describing the rules applicable for this case%

        W21 Sub-World : SpeedSlow
                        W211 Sub-sub-World : ParametersUnknown
                        W212 Sub-sub-World : ParametersOthers
                        W213 Sub-sub-World :ParametersNoImportance
        W22 Sub-World : SpeedHigh
        (Same sub-worlds as in W21 above)
        W23 Sub-World : SpeedUnknown
        (Same sub-worlds as in W21 above)
        W14 Sub-World : SpeedOthers
        (Same sub-worlds as in W21 above)
        W24 Sub-World : SpeedNoImportance
        (Same sub-worlds as in W21 above)

W3 World : DynamicModelOthers
            % Describing the rules applicable for this case%
        W31 Sub-World : SpeedSlow
                        W311 Sub-sub-world : ParametersKnown
                        W312 Sub-sub-World : ParametersUnknown
                        W313 Sub-sub-World : ParametersOthers
                        W314 Sub-sub-World :ParametersNoImportance

        W32 Sub-World : SpeedSlow
        (Same sub-worlds as in W31 above)
        W33 Sub-World : SpeedUnknown
        (Same sub-worlds as in W31 above)
        W14 Sub-World : SpeedOthers
        (Same sub-worlds as in W31 above)
        W34 Sub-World : SpeedNoImportance
        (Same sub-worlds as in W31 above)

W4 World : DynamicModelNoImportance
            % Describing the rules applicable for this case%
        W41 Sub-World : SpeedSlow
                        W411 Sub-sub-World : ParametersKnown
                        W412 Sub-sub-World : ParametersUnknown
                        W413 Sub-sub-World : ParametersOthers
                        W414 Sub-sub-World :ParametersNoImportance
        W42 Sub-World : SpeedSlow
        (Same sub-worlds as in W41 above)
        W43 Sub-World : SpeedUnknown
        (Same sub-worlds as in W41 above)
        W14 Sub-World : SpeedOthers
        (Same sub-worlds as in W41 above)
        W44 Sub-World : SpeedNoImportance
        (Same sub-worlds as in W41 above)
```

For each sub-world *Wij* (*i* = 1 to 4 ; *j* = 1 to 4), there correspond a sub-sub-world for RA parametric description. These pruning worlds give a preliminary guide to a world (corresponding to the chosen algorithm) where the initial search is to be started. If the results given by this algorithm are satisfactory then choose this algorithm as a solution. Otherwise, either fine-tune the obtained solution within the same world (or other eventual specialized sub-worlds) or go back to the meta-level nucleus (MLN) for further search.

iii. The worlds describing the RA algorithms
For each RA algorithm, we have developed a world. Each of these worlds can be considered as an independent KB (IKB). Some of the worlds have very few rules. Each IKB can obviously be incremented, provided the expertise is available. We have considered worlds and sub-worlds partially describing the following algorithms.

```
World PID
        Sub-worlds : Basic PID, Gravitational PID, Adaptive PID, Robust PID.
World Computed Torque (known parameters)
        Sub-worlds : PD control, Predictive control
World Compensators
        Sub-worlds : Spong's adaptive compensator, Amestegui's adaptive
        compensator
World Adaptive Control
        Sub-worlds : linearized adaptive, passive adaptive
World Robust Control
        Sub-worlds : Robust PID, large gains, variable structure control (VSC)
```

5.4 Example : Fuzzy rule involving fuzzy attributes in its conclusion.

If the user does not know the RA parameters but knows the dynamic model and that the RA is slow, then a tentative algorithm is the passive adaptive or the linear adaptive. In the conclusion, we can therefore translate this by a certainty factor (CF) of 50 meaning that either algorithm can be used with a degree of equal certainty. The CF can of course be changed according to the available knowledge and refined expertise. This rule can be expressed by :

```
WORLD : MLN % New world %
DESCENDANTS WORLDS % Here is a list of all worlds %

Rule TryPassvAdaptCF60  % This is the name of the rule %
CHAINING : forward
PRIORITY : 40  % can be changed from 0 to 100 %
CONTENT
        IF Guide.DynamicModelKnown_VelocitySlow = TRUE
        AND RA.Parameters = "don't know"
        AND Algorithm.AlgoActivation = "Activable"
        THEN TryAlgorithm.PassivAdaptivFuzzy = TRUE CF 50
        AND Guide.PassivAdaptivCF50 = TRUE
```

Other situations can be described in a similar manner.

6. Conclusion

We have described some foundational steps to solve the RA control using knowledge base systems approach. More specifically, this research work reports some features of KBC approach as applied to some RA control algorithms spanning PID through adaptive, and robust control. As such, this research represents an early contribution towards an objective evaluation of the effectiveness of KBC as applied to RA control. A unification of the diversified works dealing with RAs, while concentrating on KBC as an alternative control method, is therefore made possible. The adopted knowledge base systems approach is known for its flexibility and conveys a solution better than that provided by numerical means alone since it incorporates codified human expertise on top of the algorithms. The fundamental constraints of the proposed method is that it requires an elicitation of human expertise or extensive off-line trials to construct this expertise. This expertise codification has a direct impact on the size of the KB and on the rapidity of the user-defined problem solution. Like any KBS method, the proposed procedure also requires a diversified coverage of the working domain during the elicitation stage to obtain a richer KB. As a consequence, the results report only some aspects of the overall issue, since these describe only a fragment of the human expertise for a small class of control algorithms. Much work is still required on both sides, *i.e.,* robotics and KBS in order to further integrate these two entities within a single one while meeting the challenges of efficient real-life applications.

7. References

Abdallah, C.; Dawson, D.; Dorato, P. & Jamshidi, M. (1991). Survey of robust control for rigid robots, *IEEE Control Systems Magazine*, Vol. 11, No. 2 (February 1991) page 24–30, ISSN: 0272-1708.

Amestegui, M.; Ortega, R. & Ibarra, J.M. (1987). Adaptive linearizing and decoupling robot control : a comparative study of different parametrizations, *Proceedings of 5th Yale Workshop on Applications of Adaptive Systems Theory*, 1987, New Haven, CN, USA.

Åström, K.J. (1989). Towards intelligent control, *IEEE Control Systems Magazine*, Vol. 9, no.3, pp. 60-64, ISSN: 0272-1708.

Åström, K.J. & McAvoy, T.J. (1992). Intelligent control. *J. Process Control,* Vol. 2, no. 3, pp. 115-127, ISSN: 0959-1524.

Åström, K.J.; Hang, C.C.; Persson P. & Ho W.K. (1992). Towards intelligent PID control, *Automatica*, Vol. 28, No. 1 (January 1992) page 1-9, ISSN: 0005-1098.

Cai, L. & Goldenberg, A.A. (1988). Robust control of unconstrained maneuvers and collision for a robot manipulator with bounded parameter uncertainty. *Proceedings IEEE Conference on Robotics and Automation,* pp. 1010–1015, Vo. 2, 1988, Philadelphia, IEEE, NJ, USA.

Calangiu, G.A.; Stoica, M. & Sisak, F. A knowledge based system designed for making task execution more efficient for a robot arm. *IEEE 19th International Workshop on Robotics in Alpe-Adria-Danube Region (RAAD),* pp. 381-387, 24-26 June 2010, Budapest, Hungary.

Cellier, F.E.; Schooley, L.C.; Sundareshan, M.K. & Ziegler, B.P. (1992). Computer aided design of intelligent controllers : challenges of the nineties. In M.Jamshidi and C.J Herget (Eds.). *Recent Advances in Computer-Aided Control Systems Engineering.* Elsevier, Science Publishers, Amsterdam, Holland.

Chen, L.W. & Papavassilopoulos, G.P. (1991). Robust variable structure and adaptive control of single-arm dynamics. *Proceedings 30th Conference on Decision and Control,* pp. 367-372, Brighton, UK, 11–13 December 1991, IEEE, NJ, USA.

Corke, P.I. (1996). A robotics toolbox for MATLAB, *IEEE Robotics and Automation Magazine*, Vol. 3, No. (March 1996) page 24-32, ISSN: 1070-9932.

Craig, J.J. (2005). *Introduction to Robotics: Mechanics and Control* , 3rd Ed., Pearson Prentice Hall, ISBN: 0201-54361-3, Upper Saddle River, NJ, USA.

Craig, J.J. (1988). *Adaptive Control of Mechanical Manipulators*, Addison-Wesley, ISBN: -201-10490-3, Reading, MA, USA.

Craig, J.J., Hsu, P. & Sastry, S. (1987). Adaptive control of mechanical manipulators. *International Journal of Robotics Research*, Vol. 6, No. 2 (June 1987) page 16-28, ISSN: 0278-3649.

Egerstedt, M.; Frazzoli, E. & Pappas, G. (2006). Special section on symbolic methods for complex control systems, *IEEE Transactions On Automatic Control*. Vol. 51, No. 6 (June 2006) page 921-923, ISSN: 0018-9286.

Eldeib, H.K. & Tsai S. (1989). Applications of symbolic manipulation in control system analysis and design. *Proceedings of the IEEE Symposium on Intelligent Control* page 269-274, 1989, Albany, NY, USA.

Etxebarria, V. (1994). Animation of a simple planar robotic arm. *Proceedings of the European Simulation Multiconference (ESM'94)*, pp. 809-813, Barcelona, Spain, 1-3 June 1994.

Hamdi-Cherif, A. Towards robotic manipulator grammatical control, Invited Book Chapter In: Suraiya Jabin (Ed.) *"Robot Learning"*, SCIYO Pub., pp. 117-136, October 2010, ISBN: 978-953-307-104-6.

Hamdi-Cherif, A. & Kara-Mohammed (*alias* Hamdi-Cherif), C. (2009). Grammatical inference methodology for control systems. *WSEAS Transactions on Computers*, Vol. 8, No. 4 (April 2009) page 610-619, ISSN: 1109-2750.

Handelman, D.A.; Lane, S. H. & Gelfand, J. (1990). Integrating neural networks and knowledge-based systems for intelligent robotic control, *IEEE Control Systems Magazine*, Vol. 10, no.3, pp. 77-87, ISSN: 0272-1708.

Hanlei, W. (2010). On the recursive implementation of adaptive control for robot manipulators, *29th Chinese Control Conference (CCC10)*, pp. 2154 - 2161, 29-31 Jul. 2010, Beijing, China.

Hsia, T.C. (1986). Adaptive control of robot manipulators: a review. *IEEE International Conference on Robotics and Automation*, San Fransisco, CA, USA, 1986, IEEE, NJ, USA.

Johansson, R. (1990). Adaptive control of robot manipulator motion. *IEEE Transactions on Robotics and Automation*, Vol. 6 No. 4 (August 1990) pp. 483-490, ISSN: 1042-296X.

Klavins, E. (2007). Programmable self-assembly. *IEEE Control Systems Magazine*. Vol. 27, No. 4 (August 2007) page 43-56. ISSN: 0272-1708.

Kwan, C.; Dawson, D.M. & Lewis, F.L. (2001). Robust adaptive control of robots using neural network: global stability. *Asian Journal of Control*, Vol. 3, No. 2 (June 2001) page 111-121, ISSN: 1561-8625.

Landau, I.D. & Horowitz R. (1989). Applications of the passive systems approach to the stability analysis of the adaptive controllers for robot manipulators. *International Journal of Adaptive Control and Signal Processing*, Vol. 3, No. 1 (January 1989) pp. 23-38, ISSN: 0890-6327.

Lewis, F.L.; Dawson, D.M. & Abdallah, C.T. (2003). *Robot Manipulator Control: Theory and Practice, Control Engineering Series*, 2nd Ed., CRC Press, Taylor & Francis Group, ISBN: 978-0824740726, New York, USA.

Martins, J. F.; Dente, J.A.; Pires, A.J. & Vilela Mendes, R. (2001). Language identification of controlled systems: modeling, control, and anomaly detection. *IEEE Transactions On Systems Man and Cybernetics– Part C: Applications And Reviews* Vol. 31, No. 2 (April 2001) page 234-242, ISSN: 1094-6977.

Merchán-Cruz, E.A. & Morris, A.S. (2006). Fuzzy-GA-based trajectory planner for robot manipulators sharing a common workspace, *IEEE Transactions on Robotics*, Vol. 22, No. 4 (August 2006) page 613-624, ISSN: 1042-296X.

Meystel, A. (1991). Multiresolutional feedforward / feedback loops. *Proceedings of the IEEE International Symposium on Intelligent Control*, 13-15 August 1991, pp. 85-90, Arlington, Virginia, USA.

Meystel, A. (1994). Multiscale models and controllers. *Proceedings of the IEEE / IFAC Joint Symposium on Computer-Aided Control Systems Design*, Tucson, Arizona, 7-9 March 1994, pp. 13-26.

Ortega, R. & Spong, M.W. (1989). Adaptive motion control of rigid robots: A tutorial, *Automatica*, Vol. 25, No. 6 (November 1989) page 877–888, ISSN: 0005-1098.

Polyakov, V.; Ghanadan R. & Blackenship G.L. (1994). Symbolic numerical computation tools for nonlinear and adaptive control. *Proceedings of IEEE-IFAC Joint Symposium on CACSD*, pp. 117-122, Tucson, AZ, USA, 7-9 March 1994, IEEE, NJ, USA.

Popov, V.M. (1973). *Hyperstability of Control Systems.* Springer Verlag, ISBN: 0387063730, Berlin, Germany.

Quigley, M.; Gerkey B.; Conley, K.; Faust, J.; Foote, T., Leibs, J.; Berger, E.; Wheeler, R. & Ng, A. (2009). ROS: an open-source Robot Operating System. http://www.cs.stanford.edu/people/ang/papers/icraoss09-ROS.pdf

Rao, M. (1992) *Integrated System for Intelligent Control. LNCIS*, Vol. 167, Springer-Verlag, New York, USA.

Siciliano, B.; Sciavicco, L.; Villani, L. & Oriolo, G. (2009). *Robotics Modeling, Planning and Control.* Springer, ISBN 978-1-84628-641-4, e-ISBN 978-1-84628-642-1, Series published under ISSN 1439-2232, London, UK.

Samson, C. (1987). Robust control of a class of nonlinear systems and applications to robotics. *International Journal of Adaptive Control and Signal Processing*, Vol. 1, No. 1 (January 1987) page 49-68, ISSN: 0890-6327.

Saridis, G.N. (1987). Machine-intelligent robots : A hierarchical control approach. In T. Jorjanides and B. Torby (Eds.) *Expert Systems and Robotics.* NATO Series F, pp. 221-234.

Seraji, H. (1989). Decentralized adaptive control of manipulators: theory, simulation and experimentation. *IEEE Transactions on Robotics and Automation*, Vol. 5 No. 2 (April 1989) Page 183-201, ISSN: 1042-296X.

Slotine, J.J.E. (1985). The robust control of robot manipulators. *International Journal of Robotics Research.* Vol. 4, No. 2 (June 1985) page 465-492, ISSN: 0278-3649.

Slotine, J.J.E. & W. Li (1987). On the adaptive control of robot manipulators. *International Journal of Robotics Research*, Vol. 6, No. 3 (September 1987) page 49-59, ISSN: 0278-3649.

Spong, M.W.; Hutchinson, S. & Vidyasagar M. (2006). *Robot Modeling and Control*, Wiley, ISBN: 0471649902, New York, USA.

Vukabratovic, M.; Stoic D. & Kirchanski, N. (1984). Towards non-adaptive and adaptive control of manipulation robots. *IEEE Transactions on Automatic Control.* Vol. 29, No.9 (September 1984) page 841-844, ISSN: 0018-9286.

Yae, K.H.l; Lin, T.C. & Lin S.T. (1994). Constrained multibody library within EASY5. *Simulation,* Vol. 62, No. 5 (May 1994) page 329-336, ISSN (Online): 1741-3133, ISSN (Print): 0037-5497.

5

Distributed Nonlinear Filtering Under Packet Drops and Variable Delays for Robotic Visual Servoing

Gerasimos G. Rigatos
Industrial Systems Institute
Greece

1. Introduction

State estimation over communication networks is in use by many robotic applications in industry, in defense systems, as well as in several exploration and surveillance tasks. The incorporation of a communication network in the control loop has enabled to perform multi-sensor fusion and distributed information processing, thus improving significantly the autonomy and reliability of robotic systems (Medeiros et al., 2008), (Olfati-Saber, 2006), (Watanabe & Tzafestas, 1992). It has been shown that scalable distributed state estimation can be achieved for robotic models, when the measurements are linear functions of the state and the associated process and measurement noise models follow a Gaussian distribution (Mahler, 2007), (Nettleton et al., 2003). The results have been also extended to the case of nonlinear non-Gaussian dynamical systems (Rigatos, 2010a), (Makarenko & Durrant-Whyte, 2006).
An issue which is associated to the implementation of such networked control systems is how to compensate for random delays and packet losses so as to enhance the accuracy of estimation and consequently to improve the stability of the control loop. The idea of incorporating delayed measurements within a Kalman Filter framework is a possible solution for the compensation of network-induced delays and packet losses, and is also known as update with out-of-sequence measurements (Bar Shalom, 2002). The solution proposed in (Bar Shalom, 2002) is optimal under the assumption that the delayed measurement was processed within the last sampling interval (one-step-lag problem). There have been also some attempts to extend these results to nonlinear state estimation (Golapalakrishnan et al., 2011), (Jia et al., 2008). More recently there has been research effort in the redesign of distributed Kalman Filtering algorithms for linear systems so as to eliminate the effects of delays in measurement transmissions and packet drops, while also alleviating the one-step-lag assumption (Xia et al., 2009). This chapter presents an approach to distributed state estimation-based control of nonlinear systems, capable of incorporating delayed measurements in the estimation algorithm while being also robust to packet losses.
First, the chapter examines the problem of distributed nonlinear filtering over a communication/sensors network, and the use of the estimated state vector in a control loop. As a possible filtering approach, the Extended Information Filter is proposed (Rigatos, 2010a). In the Extended Information Filter the local filters do not exchange raw measurements but send to an aggregation filter their local information matrices (local inverse covariance matrices which can be also associated to Fisher Information Matrices) and their associated

local information state vectors (products of the local information matrices with the local state vectors) (Lee, 2008). The Extended Information Filter performs fusion of state estimates from local distributed Extended Kalman Filters which in turn are based on the assumption of linearization of the system dynamics by first order Taylor series expansion and truncation of the higher order linearization terms. Moreover, the Extended Kalman Filter requires the computation of Jacobians which in the case of high order nonlinear dynamical systems can be a cumbersome procedure. This approach introduces cumulative errors to the state estimation performed by the local Extended Kalman Filter recursion which is finally transferred to the master filter where the aggregate state estimate of the controlled system is computed. Consequently, these local estimation errors may result in the deterioration of the performance of the associated control loop or even risk its stability (Rigatos, 2009),(Rigatos et al., 2009).

To overcome the aforementioned weaknesses of the Extended Information Filter a derivative-free approach to Extended Information Filtering has been proposed (Rigatos & Siano, 2010), (Rigatos, 2010c). The system is first subject to a linearization transformation and next state estimation is performed by applying the standard Kalman Filter to the linearized model. At a second level, the standard Information Filter is used to fuse the state estimates obtained from local derivative-free Kalman filters running at the local information processing nodes. This approach has significant advantages because unlike the Extended Information Filter (i) is not based on local linearization of the system dynamics (ii) it does not assume truncation of higher order Taylor expansion terms thus preserving the accuracy and robustness of the performed estimation, (iii) it does not require the computation of Jacobian matrices.

At a second stage the chapter proposes a method for the compensation of random delays and packet drops which may appear during the transmission of measurements and state vector estimates, and which the can cause the deterioration of the performance of the distributed filtering-based control scheme (Xia et al., 2009), (Schenato, 2007), (Schenato, 2008). Two cases are distinguished: (i) there are time delays and packet drops in the transmission of information between the distributed local filters and the master filter, (ii) there are time delays and packet drops in the transmission of information from distributed sensors to each one of the local filters. In the first case, the structure and calculations of the master filter for estimating the aggregate state vector remain unchanged. In the second case, the effect of the random delays and packets drops has to be taken into account in the redesign of the local Kalman Filters, which implies a modified Riccati equation for the computation of the covariance matrix of the state vector estimation error, as well as the use of a correction (smoothing) term in the update of the state vector's estimate so as to compensate for delayed measurements arriving at the local Kalman Filters.

Finally, the chapter shows that the aggregate state vector produced by a derivative-free Extended Information Filter, suitably modified to compensate for communication delays and packet drops, can be used for sensorless control and robotic visual servoing. The problem of visual servoing over a network of synchronised cameras has been previously studied in (Schuurman & Capson, 2004). In this chapter, visual servoing over a cameras network is considered for the nonlinear dynamic model of a planar single-link robotic manipulator. It is assumed that the network on which the visual servoing loop relies, can be affected by disturbances, such as random delays or loss of frames during their transmission to the local processing vision nodes. The position of the robot's end effector in the cartesian space (and equivalently the angle of the robotic link) is measured through m cameras. In turn, these measurements are processed by m distributed derivative-free Kalman Filters thus providing

m different estimates of the robotic link's state vector. Next, the local state estimates are fused with the use of the standard Information Filter. After all, the aggregate estimation of the state vector is used in a control loop which enables the robotic link to perform trajectory tracking.
The structure of the chapter is as follows: In Section 2 the Extended Kalman Filter is introduced and its use for state estimation of nonlinear dynamical systems is explained. In Section 3 a derivative-free Kalman Filtering approach to state estimation of nonlinear systems is analyzed. In Section 4 the derivative-free Extended Information Filter is formulated as an approach to distributed state estimation for nonlinear systems, capable of overcoming the drawbacks of the standard Extended Information Filter. In Section 5 the problem of distributed filtering under random delays and packet drops is analyzed. The results are also applied to distributed state estimation with the use of the derivative-free Extended Information Filter. In Section 6 the previously described approach for derivative-free Extended Information Filtering under communication delays and packet drops is applied to the problem of state estimation-based control of nonlinear systems. As a case study the model of a planar robot is considered, while the estimation of its state vector is performed with the use of distributed filtering through the processing of measurements provided by vision sensors (cameras). In Section 7 simulation tests are presented, to confirm the efficiency of the proposed derivative-free Extended Information Filtering method. Finally, in Section 8 concluding remarks are given.

2. Extended Kalman Filtering for nonlinear dynamical systems

2.1 The continuous-time Kalman Filter for the linear state estimation model

First, the continuous-time dynamical system of Eq. (1) is assumed (Rigatos & Tzafestas, 2007), (Rigatos, 2010d):

$$\begin{cases} \dot{x}(t) = Ax(t) + Bu(t) + w(t), \ t \geq t_0 \\ z(t) = Cx(t) + v(t), \ t \geq t_0 \end{cases} \tag{1}$$

where $x \in R^{m \times 1}$ is the system's state vector, and $z \in R^{p \times 1}$ is the system's output. Matrices A,B and C can be time-varying and $w(t)$,$v(t)$ are uncorrelated white Gaussian noises. The covariance matrix of the process noise $w(t)$ is $Q(t)$, while the covariance matrix of the measurement noise is $R(t)$. Then, the Kalman Filter is a linear state observer which is given by

$$\begin{cases} \dot{\hat{x}} = A\hat{x} + Bu + K[z - C\hat{x}], \ \hat{x}(t_0) = 0 \\ K(t) = PC^T R^{-1} \\ \dot{P} = AP + PA^T + Q - PC^T R^{-1} CP \end{cases} \tag{2}$$

where $\hat{x}(t)$ is the optimal estimation of the state vector $x(t)$ and $P(t)$ is the covariance matrix of the state vector estimation error with $P(t_0) = P_0$. The Kalman Filter consists of the system's state equation plus a corrective term $K[z - C\hat{x}]$. The selection of gain K corresponds actually to the solution of an optimization problem. This is expressed as the minimization of a quadratic cost functional and is performed through the solution of a Riccati equation. In that case the observer's gain K is calculated by $K = PC^T R^{-1}$ considering an optimal control problem for the dual system (A^T, C^T), where the covariance matrix of the estimation error P is found by the solution of a continuous-time Riccati equation of the form

$$\dot{P} = AP + PA^T + Q - PC^TR^{-1}CP \tag{3}$$

where matrices Q and R stand for the process and measurement noise covariance matrices, respectively.

2.2 The discrete-time Kalman Filter for linear dynamical systems

In the discrete-time case a dynamical system is assumed to be expressed in the form of a discrete-time state model (Rigatos & Tzafestas, 2007), (Rigatos, 2010d):

$$\begin{aligned} x(k+1) &= A(k)x(k) + L(k)u(k) + w(k) \\ z(k) &= Cx(k) + v(k) \end{aligned} \tag{4}$$

where the state $x(k)$ is a m-vector, $w(k)$ is a m-element process noise vector and A is a $m \times m$ real matrix. Moreover the output measurement $z(k)$ is a p-vector, C is an $p \times m$-matrix of real numbers, and $v(k)$ is the measurement noise. It is assumed that the process noise $w(k)$ and the measurement noise $v(k)$ are uncorrelated.

Now the problem of interest is to estimate the state $x(k)$ based on the sequence of output measurements $z(1), z(2), \cdots, z(k)$. The initial value of the state vector $x(0)$, and the initial value of the error covariance matrix $P(0)$ is unknown and an estimation of it is considered, i.e. $\hat{x}(0)$= a guess of $E[x(0)]$ and $\hat{P}(0)$= a guess of $Cov[x(0)]$.

For the initialization of matrix P one can set $\hat{P}(0) = \lambda I$, with $\lambda > 0$. The state vector $x(k)$ has to be estimated taking into account $\hat{x}(0)$, $\hat{P}(0)$ and the output measurements $Z = [z(1), z(2), \cdots, z(k)]^T$, i.e. $\hat{x}(k) = \alpha_n(\hat{x}(0)), \hat{P}(0), Z(k))$. This is a linear minimum mean squares estimation problem (LMMSE) formulated as $\hat{x}(k+1) = a_{n+1}(\hat{x}(k), z(k+1))$. The process and output noise are white and their covariance matrices are given by: $E[w(i)w^T(j)] = Q\delta(i-j)$ and $E[v(i)v^T(j)] = R\delta(i-j)$.

Using the above, the discrete-time Kalman filter can be decomposed into two parts: i) time update (prediction stage), and ii) measurement update (correction stage). The first part employs an estimate of the state vector $x(k)$ made before the output measurement $z(k)$ is available (a priori estimate). The second part estimates $x(k)$ after $z(k)$ has become available (a posteriori estimate).

- When the set of measurements $Z^- = \{z(1), \cdots, z(k-1)\}$ is available. From Z^- an *a priori* estimation of $x(k)$ is obtained which is denoted by $\hat{x}^-(k)$= the estimate of $x(k)$ given Z^-.
- When $z(k)$ is available, the output measurements set becomes $Z = \{z(1), \cdots, z(k)\}$, where $\hat{x}(k)$= the estimate of $x(k)$ given Z.

The associated estimation errors are defined by $e^-(k) = x(k) - \hat{x}^-(k)$= the a priori error, and $e(k) = x(k) - \hat{x}(k)$= the a posteriori error. The estimation error covariance matrices associated with $\hat{x}(k)$ and $\hat{x}(k)$ are defined as $P^-(k) = Cov[e^-(k)] = E[e^-(k)e^-(k)^T]$ and $P(k) = Cov[e(k)] = E[e(k)e^T(k)]$ (Kamen & Su, 1999). From the definition of the trace of a matrix, the mean square error of the estimates can be written as $MSE(\hat{x}^-(k)) = E[e^-(k)e^-(k)^T] = tr(P^-(k))$ and $MSE(x(k)) = E[e(k)e^T(k) = tr(P(k))$.

Finally, the linear Kalman filter equations in cartesian coordinates are *measurement update*:

$$\begin{aligned} K(k) &= P^{-}(k)C^{T}[C{\cdot}P^{-}(k)C^{T}+R]^{-1} \\ \hat{x}(k) &= \hat{x}^{-}(k)+K(k)[z(k)-C\hat{x}^{-}(k)] \\ P(k) &= P^{-}(k)-K(k)CP^{-}(k) \end{aligned} \tag{5}$$

time update:

$$\begin{aligned} P^{-}(k+1) &= A(k)P(k)A^{T}(k)+Q(k) \\ \hat{x}^{-}(k+1) &= A(k)\hat{x}(k)+L(k)u(k) \end{aligned} \tag{6}$$

2.3 The extended Kalman Filter

State estimation can be also performed for nonlinear dynamical systems using the Extended Kalman Filter recursion (Ahrens & Khalil, 2005), (Boutayeb et al., 1997). The following nonlinear state model is considered (Rigatos, 2010a), (Rigatos & Tzafestas, 2007):

$$\begin{aligned} x(k+1) &= \phi(x(k))+L(k)u(k)+w(k) \\ z(k) &= \gamma(x(k))+v(k) \end{aligned} \tag{7}$$

where $x{\in}R^{m\times 1}$ is the system's state vector and $z{\in}R^{p\times 1}$ is the system's output, while $w(k)$ and $v(k)$ are uncorrelated, zero-mean, Gaussian zero-mean noise processes with covariance matrices $Q(k)$ and $R(k)$ respectively. The operators $\phi(x)$ and $\gamma(x)$ are vectors defined as $\phi(x)=[\phi_1(x),\phi_2(x),\cdots,\phi_m(x)]^T$, and $\gamma(x)=[\gamma_1(x),\gamma_2(x),\cdots,\gamma_p(x)]^T$, respectively. It is assumed that ϕ and γ are sufficiently smooth in x so that each one has a valid series Taylor expansion. Following a linearization procedure, ϕ is expanded into Taylor series about $\hat{x}$:

$$\phi(x(k)) = \phi(\hat{x}(k))+J_{\phi}(\hat{x}(k))[x(k)-\hat{x}(k)]+\cdots \tag{8}$$

where $J_{\phi}(x)$ is the Jacobian of ϕ calculated at $\hat{x}(k)$:

$$J_{\phi}(x) = \frac{\partial\phi}{\partial x}|_{x=\hat{x}(k)} = \begin{pmatrix} \frac{\partial\phi_1}{\partial x_1} & \frac{\partial\phi_1}{\partial x_2} & \cdots & \frac{\partial\phi_1}{\partial x_m} \\ \frac{\partial\phi_2}{\partial x_1} & \frac{\partial\phi_2}{\partial x_2} & \cdots & \frac{\partial\phi_2}{\partial x_m} \\ \vdots & \vdots & \vdots & \vdots \\ \frac{\partial\phi_m}{\partial x_1} & \frac{\partial\phi_m}{\partial x_2} & \cdots & \frac{\partial\phi_m}{\partial x_m} \end{pmatrix} \tag{9}$$

Likewise, γ is expanded about $\hat{x}^{-}(k)$

$$\gamma(x(k)) = \gamma(\hat{x}^{-}(k))+J_{\gamma}[x(k)-\hat{x}^{-}(k)]+\cdots \tag{10}$$

where $\hat{x}^{-}(k)$ is the estimation of the state vector $x(k)$ before measurement at the k-th instant to be received and $\hat{x}(k)$ is the updated estimation of the state vector after measurement at the k-th instant has been received. The Jacobian $J_{\gamma}(x)$ is

$$J_{\gamma}(x) = \frac{\partial\gamma}{\partial x}|_{x=\hat{x}^{-}(k)} = \begin{pmatrix} \frac{\partial\gamma_1}{\partial x_1} & \frac{\partial\gamma_1}{\partial x_2} & \cdots & \frac{\partial\gamma_1}{\partial x_m} \\ \frac{\partial\gamma_2}{\partial x_1} & \frac{\partial\gamma_2}{\partial x_2} & \cdots & \frac{\partial\gamma_2}{\partial x_m} \\ \vdots & \vdots & \vdots & \vdots \\ \frac{\partial\gamma_p}{\partial x_1} & \frac{\partial\gamma_p}{\partial x_2} & \cdots & \frac{\partial\gamma_p}{\partial x_m} \end{pmatrix} \tag{11}$$

The resulting expressions create first order approximations of ϕ and γ. Thus the linearized version of the system is obtained:

$$\begin{aligned} x(k+1) &= \phi(\hat{x}(k)) + J_{\phi}(\hat{x}(k))[x(k) - \hat{x}(k)] + w(k) \\ z(k) &= \gamma(\hat{x}^{-}(k)) + J_{\gamma}(\hat{x}^{-}(k))[x(k) - \hat{x}^{-}(k)] + v(k) \end{aligned} \tag{12}$$

Now, the EKF recursion is as follows: First the time update is considered: by $\hat{x}(k)$ the estimation of the state vector at instant k is denoted. Given initial conditions $\hat{x}^{-}(0)$ and $P^{-}(0)$ the recursion proceeds as:

- *Measurement update.* Acquire $z(k)$ and compute:

$$\begin{aligned} K(k) &= P^{-}(k)J_{\gamma}^{T}(\hat{x}^{-}(k))\cdot[J_{\gamma}(\hat{x}^{-}(k))P^{-}(k)J_{\gamma}^{T}(\hat{x}^{-}(k)) + R(k)]^{-1} \\ \hat{x}(k) &= \hat{x}^{-}(k) + K(k)[z(k) - \gamma(\hat{x}^{-}(k))] \\ P(k) &= P^{-}(k) - K(k)J_{\gamma}(\hat{x}^{-}(k))P^{-}(k) \end{aligned} \tag{13}$$

- *Time update.* Compute:

$$\begin{aligned} P^{-}(k+1) &= J_{\phi}(\hat{x}(k))P(k)J_{\phi}^{T}(\hat{x}(k)) + Q(k) \\ \hat{x}^{-}(k+1) &= \phi(\hat{x}(k)) + L(k)u(k) \end{aligned} \tag{14}$$

The schematic diagram of the EKF loop is given in Fig. 1.

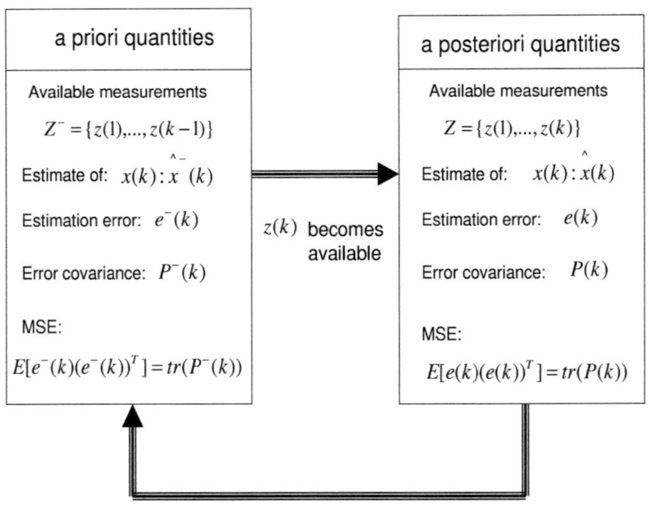

Fig. 1. Schematic diagram of the EKF loop

3. Derivative-free Kalman Filtering for a class of nonlinear systems

3.1 State estimator design through a nonlinear transformation

It will be shown that through a nonlinear transformation it is possible to design a state estimator for a class of nonlinear systems, which can substitute for the Extended Kalman Filter. The results will be generalized towards derivative-free Kalman Filtering for nonlinear systems. The following continuous-time nonlinear single-output system is considered (Marino, 1990),(Marino & Tomei, 1992)

$$\dot{x} = f(x) + q_0(x,u) + \sum_{i=1}^{p} \theta_i q_i(x,u), \text{ or}$$
$$\dot{x} = f(x) + q_0(x,u) + Q(x,u)\theta \; x \in R^n, \; u \in R^m, \; \theta \in R^p \tag{15}$$
$$z = h(x), \; z \in R$$

with $q_i : R^n \times R^m \to R^n$, $0 \le i \le p$, $f : R^n \to R^n$, $h : R^n \to R$, smooth functions, $h(x_0) = 0$, $q_0(x,0) = 0$ for every $x \in R^n$; x is the state vector, $u(x,t) : R^+ \to R^m$ is the control which is assumed to be known, θ is the parameter vector which is supposed to be constant and y is the scalar output.
The first main assumption on the class of systems considered is the linear dependence on the parameter vector θ. The second main assumption requires that systems of Eq.(15) are transformable by a parameter independent state-space change of coordinates in R^n

$$\zeta = T(x), \;\; T(x_0) = 0 \tag{16}$$

into the system

$$\dot{\zeta} = A_c\zeta + \psi_0(z,u) + \sum_{i=1}^{p} \theta_i \psi_i(z,u) \Rightarrow$$
$$\dot{\zeta} = A_c\zeta + \psi_0(z,u) + \Psi(z,u)\theta \tag{17}$$
$$z = C_c\zeta$$

with

$$A_c = \begin{pmatrix} 0 & 1 & 0 & \cdots & 0 \\ 0 & 0 & 1 & \cdots & 0 \\ \vdots & \vdots & \vdots & \vdots & \vdots \\ 0 & 0 & 0 & \cdots & 0 \end{pmatrix} \tag{18}$$

$$C_c = \begin{pmatrix} 1 & 0 & 0 & \cdots & 0 \end{pmatrix} \tag{19}$$

and $\psi_i : R \times R^m \to R^n$ smooth functions for $i = 0, \cdots, p$. The necessary and sufficient conditions for the initial nonlinear system to be transformable into the form of Eq.(17) have been given in (Marino, 1990),(Marino & Tomei, 1992), and are summarized in the following:

(i) rank$\{dh(x), d_{L_f}h(x), \cdots, d_{L_f^{n-1}}h(x)\} = n$, $\forall x \in R^n$ (which implies local observability). It is noted that $L_f h(x)$ stands for the Lie derivative $L_f h(x) = (\nabla h) f$ and the repeated Lie derivatives are recursively defined as $L_f^0 h = h$ for $i = 0$, $L_f^i h = L_f L_f^{i-1} h = \nabla L_f^{i-1} h f$ for $i = 1, 2, \cdots$.
(ii) $[ad_f^i g, ad_f^j g] = 0$, $0 \le i, \; j \le n-1$. It is noted that $ad_f^i g$ stands for a Lie Bracket which is defined recursively as $ad_f^i g = [f, ad_f^{i-1}]g$ with $ad_f^0 g = g$ and $ad_f g = [f, g] = \nabla g f - \nabla f g$.
(iii) $[q_i, ad_f^i g] = 0$, $0 \le i \le p$, $0 \le j \le n-2 \; \forall \; u \in R^m$.
(iv) the vector fields $ad_f^i g$, $0 \le i \le n-1$ are complete, in which g is the vector field satisfying

$$< \begin{pmatrix} dh \\ \vdots \\ d(L_f^{n-1}h) \end{pmatrix}, g >= \begin{pmatrix} 0 \\ \vdots \\ 1 \end{pmatrix} \tag{20}$$

Then for every parameter vector θ, the system

$$\begin{array}{l}\dot{\hat{\zeta}} = A_c\hat{\zeta} + \psi_0(z,u) + \sum_{i=1}^{p}\theta_i\psi_i(z,u) + K(z - C_c\hat{\zeta}) \\ \hat{x} = T^{-1}(\hat{\zeta})\end{array} \tag{21}$$

is an asymptotic observer for a suitable choice of K provided that the state $x(t)$ is bounded, with estimation error dynamics

$$\dot{e} = (A_c - KC_c)e = \begin{pmatrix} -k_1 & 1 & 0 & \cdots & 0 \\ -k_2 & 0 & 1 & \cdots & 0 \\ \vdots & \vdots & \vdots & \vdots & \vdots \\ -k_{n-1} & 0 & 0 & \cdots & 1 \\ -k_n & 0 & 0 & \cdots & 0 \end{pmatrix} e \tag{22}$$

The eigenvalues of $A_c - KC_c$ can be arbitrarily placed by choosing the vector K, since they coincide with the roots of the polynomial $s^n + k_1 s^{n-1} + \cdots + k_n$.

3.2 Derivative-free Kalman Filtering for nonlinear systems

Since Eq. (21) provides an asympotic observer for the initial nonlinear system of Eq. (15) one can consider a case in which the observation error gain matrix K can be provided by the Kalman Filter equations given initially in the continuous-time KF formulation, or in discrete-time form by Eq. (5) and Eq. (6). The following single-input single-output nonlinear dynamical system is considered

$$x^{(n)} = f(x,t) + g(x,t)u(x,t) \tag{23}$$

where $z = x$ is the system's output, and $f(x,t)$, $g(x,t)$ are nonlinear functions. It can be noticed that the system of Eq. (23) belongs to the general class of systems of Eq. (15). Assuming the transformation $\zeta_i = x^{(i-1)}$, $i = 1, \cdots, n$, and $x^{(n)} = f(x,t) + g(x,t)u(x,t) = v(\zeta,t)$, i.e. $\dot{\zeta}_n = v(\zeta,t)$, one obtains the linearized system of the form

$$\begin{array}{l}\dot{\zeta}_1 = \zeta_2 \\ \dot{\zeta}_2 = \zeta_3 \\ \cdots\ \cdots \\ \dot{\zeta}_{n-1} = \zeta_n \\ \dot{\zeta}_n = v(\zeta,t)\end{array} \tag{24}$$

which in turn can be written in state-space equations as

$$\begin{pmatrix} \dot{\zeta}_1 \\ \dot{\zeta}_2 \\ \cdots \\ \dot{\zeta}_{n-1} \\ \dot{\zeta}_n \end{pmatrix} = \begin{pmatrix} 0 & 1 & 0 & \cdots & 0 \\ 0 & 0 & 1 & \cdots & 0 \\ \vdots & \vdots & \vdots & \vdots & \vdots \\ 0 & 0 & 0 & \cdots & 1 \\ 0 & 0 & 0 & \cdots & 0 \end{pmatrix} \begin{pmatrix} \zeta_1 \\ \zeta_2 \\ \cdots \\ \zeta_{n-1} \\ \zeta_n \end{pmatrix} + \begin{pmatrix} 0 \\ 0 \\ \cdots \\ 0 \\ 1 \end{pmatrix} v(\zeta,t) \tag{25}$$

$$z = \begin{pmatrix} 1 & 0 & 0 & \cdots & 0 \end{pmatrix} \zeta \tag{26}$$

The system of Eq. (25) and Eq. (26) has been written in the form of Eq. (17), which means that Eq. (21) is the associated asymptotic observer. Therefore, the observation gain K appearing in Eq. (21) can be found using either linear observer design methods (in that case the elements of the observation error gain matrix K have fixed values), or the recursive calculation of the continuous-time Kalman Filter gain described in subsection 2.2. If the discrete-time Kalman Filter is to be used then one has to apply the recursive formulas of Eq. (5) and Eq. (6) on the discrete-time equivalent of Eq. (25) and Eq. (26).

4. Derivative-free Extended Information Filter

4.1 Calculation of local estimations in terms of EIF information contributions

Again the discrete-time nonlinear system of Eq. (7) is considered. The Extended Information Filter (EIF) performs fusion of local state vector estimates which are provided by local Extended Kalman Filters, using the *Information matrix* and the *Information state vector* (Lee, 2008), (Manyika & Durrant-Whyte, 1994). The Information Matrix is the inverse of the state vector covariance matrix, and can be also associated to the Fisher Information matrix (Rigatos & Zhang, 2009). The Information state vector is the product between the Information matrix and the local state vector estimate

$$\begin{aligned} Y(k) &= P^{-1}(k) = I(k) \\ \hat{y}(k) &= P^{-}(k)^{-1}\hat{x}(k) = Y(k)\hat{x}(k) \end{aligned} \tag{27}$$

The update equation for the Information Matrix and the Information state vector are given by

$$\begin{aligned} Y(k) &= P^{-}(k)^{-1} + J_{\gamma}^{T}(k)R^{-1}(k)J_{\gamma}(k) \\ &= Y^{-}(k) + I(k) \end{aligned} \tag{28}$$

$$\begin{aligned} \hat{y}(k) &= \hat{y}^{-}(k) + J_{\gamma}^{T}(k)R(k)^{-1}[z(k) - \gamma(x(k)) + J_{\gamma}(k)\hat{x}^{-}(k)] \\ &= \hat{y}^{-}(k) + i(k) \end{aligned} \tag{29}$$

where

$$\begin{aligned} I(k) &= J_{\gamma}^{T}(k)R(k)^{-1}J_{\gamma}(k) \text{ is the associated information matrix and} \\ i(k) &= J_{\gamma}^{T}(k)R(k)^{-1}[(z(k) - \gamma(x(k))) + J_{\gamma}\hat{x}^{-}(k)] \text{ is the information state contribution} \end{aligned} \tag{30}$$

The predicted information state vector and Information matrix are obtained from

$$\begin{aligned} \hat{y}^{-}(k) &= P^{-}(k)^{-1}\hat{x}^{-}(k) \\ Y^{-}(k) &= P^{-}(k)^{-1} = [J_{\phi}(k)P^{-}(k)J_{\phi}(k)^{T} + Q(k)]^{-1} \end{aligned} \tag{31}$$

The Extended Information Filter is next formulated for the case that multiple local sensor measurements and local estimates are used to increase the accuracy and reliability of the estimation. It is assumed that an observation vector $z^{i}(k)$ is available for N different sensor sites $i = 1, 2, \cdots, N$ and each sensor observes a common state according to the local observation model, expressed by

$$z^{i}(k) = \gamma(x(k)) + v^{i}(k),\ i = 1, 2, \cdots, N \tag{32}$$

where the local noise vector $v^{i}(k) \sim N(0, R^{i})$ is assumed to be white Gaussian and uncorrelated between sensors. The variance of a composite observation noise vector v_k is expressed in terms of the block diagonal matrix

$$R(k) = diag[R^{1}(k), \cdots, R^{N}(k)]^{T} \tag{33}$$

The information contribution can be expressed by a linear combination of each local information state contribution i^i and the associated information matrix I^i at the i-th sensor site

$$\begin{array}{c} i(k) = \sum_{i=1}^{N} {J_\gamma^i}^T(k) R^i(k)^{-1} [z^i(k) - \gamma^k(x(k)) + J_\gamma^i(k)\hat{x}^-(k)] \\ I(k) = \sum_{i=1}^{N} {J_\gamma^i}^T(k) R^i(k)^{-1} J_\gamma^i(k) \end{array} \quad (34)$$

Using Eq. (34) the update equations for fusing the local state estimates become

$$\begin{array}{c} \hat{y}(k) = \hat{y}^-(k) + \sum_{i=1}^{N} {J_\gamma^i}^T(k) R^i(k)^{-1} [z^i(k) - \gamma^k(x(k)) + J_\gamma^i(k)\hat{x}^-(k)] \\ Y(k) = Y^-(k) + \sum_{i=1}^{N} {J_\gamma^i}^T(k) R^i(k)^{-1} J_\gamma^i(k) \end{array} \quad (35)$$

It is noted that in the Extended Information Filter an aggregation (master) fusion filter produces a global estimate by using the local sensor information provided by each local filter.

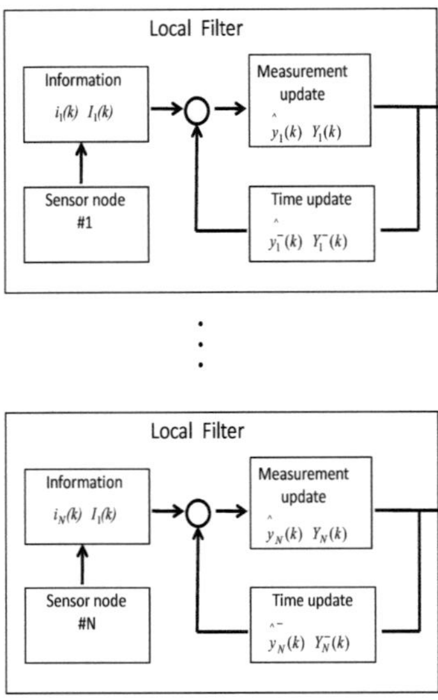

Fig. 2. Fusion of the distributed state estimates with the use of the Extended Information Filter

As in the case of the Extended Kalman Filter the local filters which constitute the Extended Information Filter can be written in terms of *time update* and a *measurement update* equation.

Measurement update: Acquire $z(k)$ and compute

$$\begin{array}{c} Y(k) = P^-(k)^{-1} + J_\gamma^T(k) R(k)^{-1} J_\gamma(k) \\ \text{or } Y(k) = Y^-(k) + I(k) \text{ where } I(k) = J_\gamma^T(k) R^{-1}(k) J_\gamma(k) \end{array} \quad (36)$$

$$\begin{array}{c} \hat{y}(k) = \hat{y}^-(k) + J_\gamma^T(k) R(k)^{-1} [z(k) - \gamma(\hat{x}(k)) + J_\gamma(k)\hat{x}^-(k)] \\ \text{or } \hat{y}(k) = \hat{y}^-(k) + i(k) \end{array} \quad (37)$$

Time update: Compute

$$Y^-(k+1) = P^-(k+1)^{-1} = [J_\phi(k) P(k) J_\phi(k)^T + Q(k)]^{-1} \quad (38)$$

$$y^-(k+1) = P^-(k+1)^{-1}\hat{x}^-(k+1) \quad (39)$$

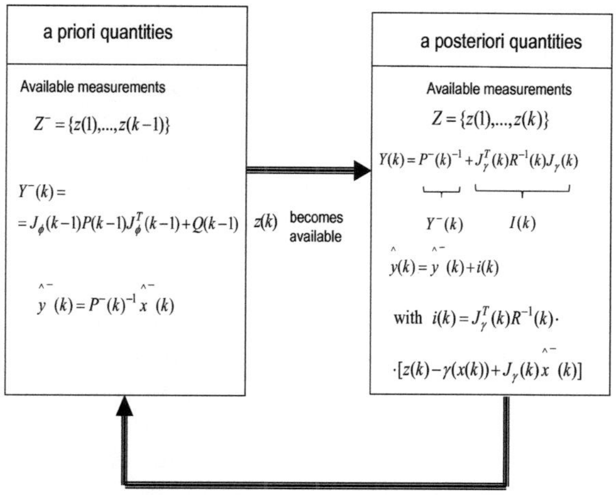

Fig. 3. Schematic diagram of the Extended Information Filter loop

4.2 Extended Information Filtering for state estimates fusion

In the Extended Information Filter each one of the local filters operates independently, processing its own local measurements. It is assumed that there is no sharing of measurements between the local filters and that the aggregation filter (Fig. 2) does not have direct access to the raw measurements giving input to each local filter. The outputs of the local filters are treated as measurements which are forwarded to the aggregation fusion filter (Lee, 2008). Then each local filter is expressed by its respective error covariance and estimate in terms of information contributions given in Eq.(48)

$$\begin{aligned} P_i^{-1}(k) &= P_i^-(k)^{-1} + J_\gamma^T(k)R(k)^{-1}J_\gamma(k) \\ \hat{x}_i(k) &= P_i(k)(P_i^-(k)^{-1}\hat{x}_i^-(k)) + J_\gamma^T(k)R(k)^{-1}[z^i(k) - \gamma^k(x(k)) + J_\gamma^i(k)\hat{x}_i^-(k)] \end{aligned} \quad (40)$$

It is noted that the local estimates are suboptimal and also conditionally independent given their own measurements. The global estimate and the associated error covariance for the aggregate fusion filter can be rewritten in terms of the computed estimates and covariances from the local filters using the relations

$$\begin{aligned} J_\gamma^T(k)R(k)^{-1}J_\gamma(k) &= P_i(k)^{-1} - P_i^-(k)^{-1} \\ J_\gamma^T(k)R(k)^{-1}[z^i(k) - \gamma^k(x(k)) + J_\gamma^i(k)\hat{x}^-(k)] &= P_i(k)^{-1}\hat{x}_i(k) - P_i(k)^{-1}\hat{x}_i(k-1) \end{aligned} \quad (41)$$

For the general case of N local filters $i = 1, \cdots, N$, the distributed filtering architecture is described by the following equations

$$\begin{aligned} P(k)^{-1} &= P^-(k)^{-1} + \sum_{i=1}^N [P_i(k)^{-1} - P_i^-(k)^{-1}] \\ \hat{x}(k) &= P(k)[P^-(k)^{-1}\hat{x}^-(k) + \sum_{i=1}^N (P_i(k)^{-1}\hat{x}_i(k) - P_i^-(k)^{-1}\hat{x}_i^-(k))] \end{aligned} \quad (42)$$

It is noted that again the global state update equation in the above distributed filter can be written in terms of the information state vector and of the information matrix

$$\begin{aligned}\hat{y}(k) &= \hat{y}^-(k) + \Sigma_{i=1}^N(\hat{y}_i(k) - \hat{y}_i^-(k)) \\ \hat{Y}(k) &= \hat{Y}^-(k) + \Sigma_{i=1}^N(\hat{Y}_i(k) - \hat{Y}_i^-(k))\end{aligned} \tag{43}$$

The local filters provide their own local estimates and repeat the cycle at step $k+1$. In turn the global filter can predict its global estimate and repeat the cycle at the next time step $k+1$ when the new state $\hat{x}(k+1)$ and the new global covariance matrix $P(k+1)$ are calculated. From Eq. (59) it can be seen that if a local filter (processing station) fails, then the local covariance matrices and the local state estimates provided by the rest of the filters will enable an accurate computation of the system's state vector.

4.3 Local estimations in terms information contributions for the derivative-free EIF

After applying the transformation described in Section 3, the nonlinear discrete-time model of the dynamical system given in Eq. (15) can be substituted by a linear model of the form given in Eq. (1). For this linearized model, the Information Filter (IF) performs fusion of the local state vector estimates which are provided by the local Kalman Filters, using again the *Information matrix* and the *Information state vector* (Rao & Durrant-Whyte, 1991). In place of the Jacobian matrix J_ϕ matrix A_d is used, (discretized equivalent of matrix A_c, which appears in Eq. (18)), while in place of the Jacobian matrix J_γ, matrix C_d is used (discretized equivalent of matrix C_c, which appears in Eq. (19)). As defined before, the Information Matrix is the inverse of the state vector covariance matrix, and can be also associated to the Fisher Information matrix (Rigatos & Zhang, 2009). The Information state vector is the product between the Information matrix and the local state vector estimate

$$\begin{aligned}Y(k) &= P^{-1}(k) = I(k) \\ \hat{y}(k) &= P^-(k)^{-1}\hat{x}(k) = Y(k)\hat{x}(k)\end{aligned} \tag{44}$$

The update equation for the Information Matrix and the Information state vector are given by

$$\begin{aligned}Y(k) &= P^-(k)^{-1} + {C_d}^T(k)R^{-1}(k)C_d(k) \\ &= Y^-(k) + I(k)\end{aligned} \tag{45}$$

$$\begin{aligned}\hat{y}(k) &= \hat{y}^-(k) + {C_d}^T(k)R(k)^{-1}[z(k) - \gamma(x(k)) + C_d\hat{x}^-(k)] \\ &= \hat{y}^-(k) + i(k)\end{aligned} \tag{46}$$

where

$$\begin{aligned}I(k) &= C_d^T(k)R(k)^{-1}C_d(k) \text{ is the associated information matrix and} \\ i(k) &= C_d^T(k)R(k)^{-1}[(z(k) - C_d(k)x(k)) + C_d\hat{x}^-(k)] \text{ is the information state contribution}\end{aligned} \tag{47}$$

The predicted information state vector and Information matrix are obtained from

$$\begin{aligned}\hat{y}^-(k) &= P^-(k)^{-1}\hat{x}^-(k) \\ Y^-(k) &= P^-(k)^{-1} = [A_d(k)P^-(k)A_d(k)^T + Q(k)]^{-1}\end{aligned} \tag{48}$$

The derivative-free Extended Information Filter is next formulated for the case that multiple local sensor measurements and local estimates are used to increase the accuracy and reliability of the estimation. It is assumed that an observation vector $z^i(k)$ is available for N different sensor sites $i = 1, 2, \cdots, N$ and each sensor observes a common state according to the local observation model, expressed by

$$z^i(k) = C_d(k)x(k) + v^i(k),\ i = 1, 2, \cdots, N \tag{49}$$

where the local noise vector $v^i(k) \sim N(0, R^i)$ is assumed to be white Gaussian and uncorrelated between sensors. The variance of a composite observation noise vector v_k is expressed in terms of the block diagonal matrix

$$R(k) = diag[R^1(k), \cdots, R^N(k)]^T \tag{50}$$

The information contribution can be expressed by a linear combination of each local information state contribution i^i and the associated information matrix I^i at the i-th sensor site

$$\begin{aligned} i(k) &= \Sigma_{i=1}^N {C_d^i}^T(k) R^i(k)^{-1}[z^i(k) - C_d^i(x(k)) + C_d^i(k)\hat{x}^-(k)] \\ I(k) &= \Sigma_{i=1}^N {C_d^i}^T(k) R^i(k)^{-1} C_d^i(k) \end{aligned} \tag{51}$$

Using Eq. (34) the update equations for fusing the local state estimates become

$$\begin{aligned} \hat{y}(k) &= \hat{y}^-(k) + \Sigma_{i=1}^N {J_\gamma^i}^T(k) R^i(k)^{-1}[z^i(k) - C_d(k)(x(k)) + C_d^i(k)\hat{x}^-(k)] \\ Y(k) &= Y^-(k) + \Sigma_{i=1}^N {C_d^i}^T(k) R^i(k)^{-1} C_d^i(k) \end{aligned} \tag{52}$$

It is noted that, as in the Extended Information Filter case, an aggregation (master) fusion filter produces a global estimate by using the local sensor information provided by each local filter. The local filters which constitute the Information Filter can be written in terms of *time update* and a *measurement update* equation.

Measurement update: Acquire $z(k)$ and compute

$$\begin{aligned} Y(k) &= P^-(k)^{-1} + C_d^T(k) R(k)^{-1} C_d(k) \\ \text{or } Y(k) &= Y^-(k) + I(k) \text{ where } I(k) = C_d^T(k) R^{-1}(k) C_d(k) \end{aligned} \tag{53}$$

$$\begin{aligned} \hat{y}(k) &= \hat{y}^-(k) + C_d^T(k) R(k)^{-1}[z(k) - C_d(\hat{x}(k)) + C_d\hat{x}^-(k)] \\ \text{or } \hat{y}(k) &= \hat{y}^-(k) + i(k) \end{aligned} \tag{54}$$

Time update: Compute

$$Y^-(k+1) = P^-(k+1)^{-1} = [A_d(k)P(k)A_d(k)^T + Q(k)]^{-1} \tag{55}$$

$$y^-(k+1) = P^-(k+1)^{-1}\hat{x}^-(k+1) \tag{56}$$

4.4 Derivative-free information filtering for state estimates fusion

The outputs of the local Kalman Filters described in subsection 4.3 are treated as measurements which are fed into the aggregation fusion filter (Rao & Durrant-Whyte, 1991). Then each local filter is expressed by its respective error covariance and estimate in terms of information contributions given in Eq.(48)

$$\begin{aligned} P_i^{-1}(k) &= P_i^-(k)^{-1} + C_d^T(k) R^{(}k)^{-1} C_d(k) \\ \hat{x}_i(k) &= P_i(k)(P_i^-(k)^{-1}\hat{x}_i^-(k)) + C_d^T(k) R(k)^{-1}[z^i(k) - C_d^i(k)x(k) + C_d^i(k)\hat{x}_i^-(k)] \end{aligned} \tag{57}$$

As explained in subsection 4.2, the local estimates are suboptimal and also conditionally independent given their own measurements. The global estimate and the associated error covariance for the aggregate fusion filter can be rewritten in terms of the computed estimates and covariances from the local filters using the relations

$$\begin{array}{c} C_d^T(k)R(k)^{-1}C_d(k) = P_i(k)^{-1} - P_i^-(k)^{-1} \\ C_d^T(k)R(k)^{-1}[z^i(k) - C_d^i(x(k)) + C_d^i(k)\hat{x}^-(k)] = P_i(k)^{-1}\hat{x}_i(k) - P_i(k)^{-1}\hat{x}_i(k-1) \end{array} \quad (58)$$

For the general case of N local filters $i = 1, \cdots, N$, the distributed filtering architecture is described by the following equations

$$\begin{array}{c} P(k)^{-1} = P^-(k)^{-1} + \sum_{i=1}^N [P_i(k)^{-1} - P_i^-(k)^{-1}] \\ \hat{x}(k) = P(k)[P^-(k)^{-1}\hat{x}^-(k) + \sum_{i=1}^N (P_i(k)^{-1}\hat{x}_i(k) - P_i^-(k)^{-1}\hat{x}_i^-(k))] \end{array} \quad (59)$$

It is noted that, once again, the global state update equation in the above distributed filter can be written in terms of the information state vector and of the information matrix

$$\begin{array}{c} \hat{y}(k) = \hat{y}^-(k) + \sum_{i=1}^N (\hat{y}_i(k) - \hat{y}_i^-(k)) \\ \hat{Y}(k) = \hat{Y}^-(k) + \sum_{i=1}^N (\hat{Y}_i(k) - \hat{Y}_i^-(k)) \end{array} \quad (60)$$

The local filters provide their own local estimates and repeat the cycle at step $k+1$. In turn the global filter can predict its global estimate and repeat the cycle at the next time step $k+1$ when the new state $\hat{x}(k+1)$ and the new global covariance matrix $P(k+1)$ are calculated. From Eq. (59) it can be seen again that if a local filter (processing station) fails, then the local covariance matrices and the local state estimates provided by the rest of the filters will enable an accurate computation of the system's state vector.

5. Distributed nonlinear filtering under random delays and packet drops

5.1 Networked Kalman Filtering for an autonomous system

The structure of networked Kalman Filtering is shown in Fig. 4. The problem of distributed filtering becomes more complicated if random delays and packet drops affect the transmission of information between the sensors and local processing units (filters), or between the local filters and the master filter where the fused state estimate is computed. First, results on the stability of the networked linear Kalman Filter will be presented (Xia et al., 2009). The general state-space form of a linear autonomous time-variant dynamical system is given by

$$x(k) = Ax(k-1) + w(k, k-1) \quad (61)$$

where $x(k) \in R^{m \times 1}$ is the system's state vector, $A \in R^{n \times n}$ is the system's state transition matrix, and $w(k, k-1)$ is the white process noise between time instants k and $k-1$.The sensor measurements are received starting at time instant $k \geq 1$ and are described by the measurement equation

$$z(k) = Cx(k) + v(k) \quad (62)$$

where $C \in R^{p \times m}$, $z(k) \in R^{p \times 1}$ and $v(k)$ is the white measurement noise. Measurements $z(k)$ are assumed to be transmitted over a communication channel.

To denote the arrival or loss of a measurement to the local Kalman Filter, through the communication network, one can use variable $\gamma_k \in \{0, 1\}$, where 1 stands for successful delivery of the packet, while 0 stands for loss of the packet.

Thus, in the case of packet losses, the discrete time Kalman Filter recursion that was described in Eq. (5) (measurement update) is modified as

$$K(k) = \gamma_k P^-(k)C^T[CP^-(k)C^T + R]^{-1} \tag{63}$$

where $\gamma_k \in \{0,1\}$. This modification implies that the value of the estimated state vector $\hat{x}(k)$ remains unchanged if the a packet drop occurs, i.e. when $\gamma_k = 0$.
It is assumed that the system $[A, C]$ is observable. Next, the following time sequences $\{\tau_k\}$ and $\{\beta_k\}$ are defined $\tau_1 = \inf\{k : k > 1, \gamma_k = 0\}$. Time τ_1 denotes the first time instant when the transmission over the communication channel is interrupted (loss of connection). On the other hand, time sequence β_k is defined as $\beta_1 = \inf\{k : k > \tau_1, \gamma_k = 1\}$. Time β_k denotes the k-th time instant in which the transmission over the communication channel is restored (reestablishment of connection). Therefore, for time sequences τ_k and β_k it holds $1 < \tau_1 < \beta_1 < \tau_2 < \beta_2 < \cdots < \tau_k < \beta_k < \cdots$.
Thus, 1 is the beginning of tranmission, τ_1 is the time instant at which the connection is lost for the first time, β_1 is the time instant at which the connection is re-established after first interruption, τ_2 is the time instant at which the connection is lost for second time, β_2 is the time instant at which the connection is re-established after second interruption,etc. The following variable is also defined $\beta_k^- = \beta_k - 1$, where β_k^- is the last time instant in a period of subsequent packet losses. Time β_k^- is useful for analyzing the behavior of the Kalman Filter in case of a sequence of packet losses (deterioration of the estimation error covariance matrix). It is noted that in the case of the filtering procedure over the communication network, the sequence of covariance matrices P_{β_k} is stable if $sup_{k>1} E||P_{\beta_k}|| < \infty$ (Xia et al., 2009). Equivalently, it can be stated that the networked system satisfies the condition of *peak covariance stability* (Xia et al., 2009).

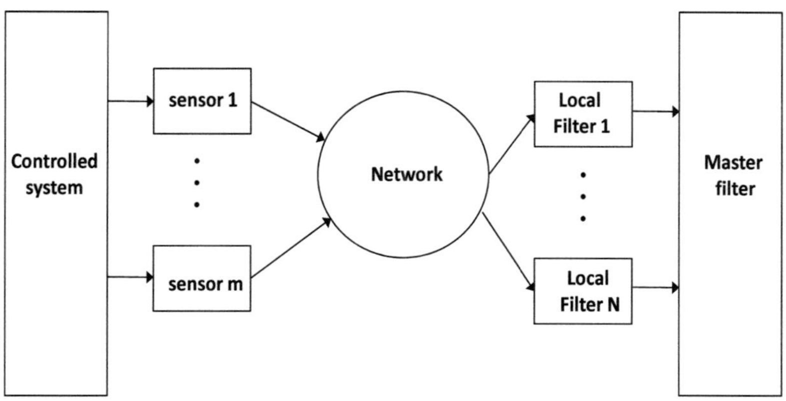

Fig. 4. Distributed filtering over sensors network with communication delays and packet drops

5.2 Processing of the delayed measurements for an autonomous system

Now, the processing of the delayed measurements for the networked linear Kalman Filter proceeds as follows: it is assumed that for all local filters the packet losses and time delays have the same statistical properties. It is also assumed that measurement $z_i(k-N)$ should have arrived at the i-th local filter at time instant $k-N$. Instead of this, the measurement arrives at time instant $k+1$. The delayed measurement $z_i(k-N)$ must be integrated in the estimation which has been performed by each local Kalman Filter (see Fig. 5 and Fig. 6).

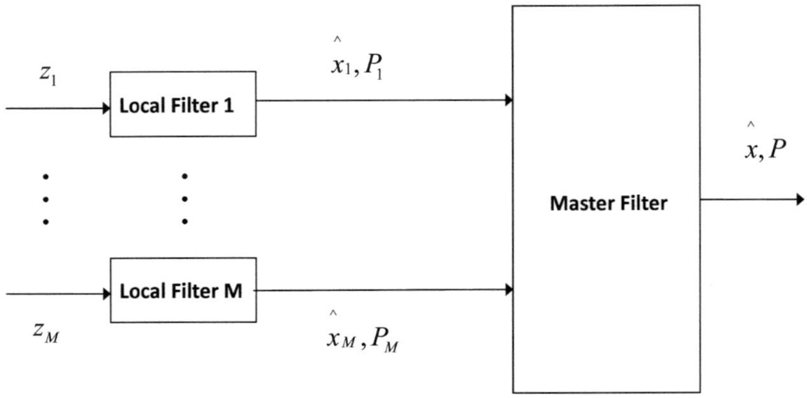

Fig. 5. Distributed filtering diagram implemented with the use of local filters and a master (aggregation) filter

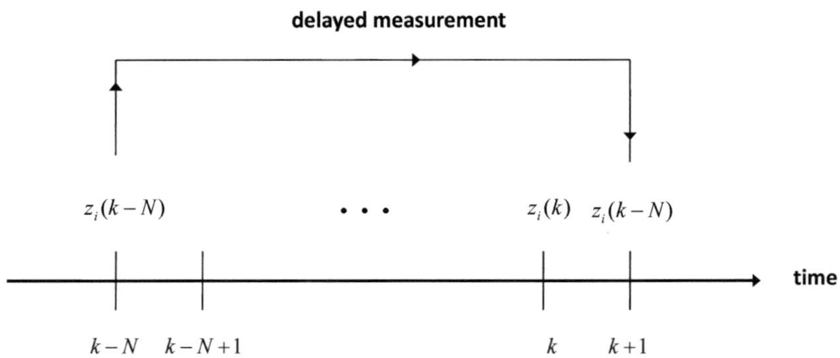

Fig. 6. Delayed measurement over the communication channel

This means that the estimation $\hat{x}_i(k|k)$ and the associated state estimation error covariance matrix $P_i(k|k)$ have to be modified. The transition matrices between different time instances of the discrete-time system of Eq. (61) are defined

$$A(k,k-j) = A(k,k-1)\cdots A(k-j+1,k-j),\ j \in Z^+ \tag{64}$$

Using the system's dynamic equation in transition matrix form, i.e

$$\begin{aligned} x(k) &= A(k,k-1)x(k-1) + w(k,k-1) \\ z_i(k) &= C_i(k)x(k) + v_i(k) \end{aligned} \tag{65}$$

one has

$$x(k) = A(k,k-N)x(k-N) + w(k,k-N) \tag{66}$$

where

$$w(k,k-N) = \Sigma_{j=1}^{N} A(k,k-j+1)w(k-j+1,k-j) \quad (67)$$

which means that knowing the state estimation $x(k-N)$ and the sequence of noises from time instant $k-N$ to time instant k one can calculate an estimation of the state vector at time instant k. Denoting $\Phi_1(k-N,k) = A(k,k-N)^{-1}$ and $w_a(k-N,k) = -A(k,k-N)^{-1}w(k,k-N)$ then, from Eq. (66) one obtains

$$x(k-N) = \Phi_1(k-N,k)x(k) + w_a(k-N,k) \quad (68)$$

To incorporate the delayed measurement $z_i(k-N)$ which arrives at the i-th local filter at time instant $k+1$, a state estimation is created first for instant $k-N$ using Eq. (68), i.e.

$$\hat{x}_i(k-N,k) = \Phi_1(k-N,k)\hat{x}_i(k|k) + \hat{w}_a(k-N,k|k) \quad (69)$$

where $\hat{x}_i(k|k)$ is the state estimation of the i-th local filter at time instant k and $\hat{w}_i(k-N,k|k)$ is the noise sequence for the i-th local filter, at time instant $k-N$. For the measurement (output) equation one has from Eq. (65)

$$z_i(k-N) = C_i(k-N)x(k-N) + v_i(k-N) \quad (70)$$

while substituting $x(k-N)$ from Eq. (68) one gets

$$z_i(k-N) = C_i(k-N)\Phi_1(k-N,k)x(k) + C_i(k-N)w_{a_i}(k-N,k) + v_i(k-N) \quad (71)$$

Next, using the current state estimate $\hat{x}(k|k)$ and Eq. (71) one can find the measurement estimate $\hat{z}_i(k-N|k)$ for the i-th local filter, $i = 1,\cdots,M$:

$$\hat{z}_i(k-N) = C_i(k-N)\Phi_{1_i}(k-N,k)\hat{x}(k|k) + C_i(k-N)\hat{w}_{\alpha_i}(k-N,k) \quad (72)$$

Defining, $\tilde{z}_i(k|j) = z_i(k) - \hat{z}_i(k|j)$ (innovation), $\tilde{x}_i(k|j) = x(k) - \hat{x}_i(k|j)$ (state estimation error), and $\tilde{w}_i(k-N,k|k) = w(k-N,k) - \hat{w}_i(k-N,k)$ (noise estimation error) one obtains

$$\tilde{z}_i(k-N|k) = C_i(k-N)\Phi_{1_i}(k-N,k)\tilde{x}_i(k|k) + C_i(k-N)\tilde{w}_{a_i}(k-N,k|k) + v_i(k-N) \quad (73)$$

The innovation $\tilde{z}_i(k-N,k)$ at time instant $k-N$ will be used to modify the estimation $\hat{x}_i(k|k)$ into

$$\hat{x}_i^*(k|k) = \hat{x}_i(k|k) + M_i\tilde{z}_i(k-N|k) \quad (74)$$

Thus, one can update (smooth) the state estimate at time instant k by adding to the current state estimate $\hat{x}_i(k|k)$ the corrective term

$$M_i\tilde{z}_i(k-N,k) \quad (75)$$

where M_i is a gain matrix to be defined in the sequel, and $\tilde{z}_i(k-N,k)$ is the innovation between the measurement $z_i(k-N)$ taken at time instant $k-N$ and the output estimate $\hat{z}_i(k-N)$ which has been calculated in Eq. (72).
The main difficulty in Eq. (74) is that one has to calculate first the noise estimation error $\tilde{w}_{a_i}(k-N,k|k)$, which means that one has to calculate an estimate of the process noise $\hat{w}_{a_i}(k-N,k)$.

The following theorem has been stated in (Xia et al., 2009), and is also applicable to the distributed filtering approach presented in this chapter:

Theorem 1: It is assumed that the observation error (innovation) at the i-th information processing unit (local filter), at time instant $k-n$ where $n \in [0,N]$, is given by

$$\tilde{z}(k-n) = z_i(k-n) - \hat{z}_i(k-n) \tag{76}$$

and that the covariance matrix of the white process noise $w_a(k-j+1,k-j)$ is

$$Q(k-j+1,k-j) = E\{w_a(k-j+1,k-j)w_a(k-j+1,k-j)^T\} \tag{77}$$

while the estimation error for the noise $w_{a_i}(k-N,k|k)$ is

$$\tilde{w}_{a_i}(k-N,k|k) = w_{a_i}(k-N,k) - \hat{w}_{a_i}(k-N,k|k) \tag{78}$$

Moreover, the covariance matrix for the error of the white estimated noise vector $\tilde{w}_{a_i}(k-N,k|k)$ is

$$Q_i^*(k-N,k) = E\{\tilde{w}_{a_i}(k-N,k|k)\tilde{w}_{a_i}(k-N,k|k)^T\} \tag{79}$$

Then, one can obtain the noise estimate $\hat{w}_{a_i}(k-N,k)$ from the relation

$$\begin{gathered}\hat{w}_{a_i}(k-N,k|k) = -\Phi_1(k-N,k)\Sigma_{n=0}^{N-1} \\ \tilde{C}_i(n)[C_i(k-n)P_i(k-n|k-n-1)C_i(k-n)^T + R_i(k-n)]^{-1}\tilde{z}_i(k-n)\end{gathered} \tag{80}$$

where

$$\begin{gathered}\tilde{C}_i(n) = \{A(k,k-n)Q(k-n,k-n-1) + \Sigma_{j=n+2}^{N}A(k,k-j+1)Q(k-j+1,k-j)\times \\ \times[\Pi_{m=n+1}^{j-1}A(k-m+1,k-m)[I-K_i(k-m)C_i(k-m)]]^T\}C_i(k-n)^T\end{gathered} \tag{81}$$

while the covariance matrix of the estimated white noise $w_{\alpha_i}(k-N,k)$ is calculated as

$$\begin{gathered}Q_i^*(k-N,k) = Q(k-N,k) - \Phi_1(k-N,k)\times \\ \times\Sigma_{n=0}^{N-1}\tilde{C}_i(n)[C_i(k-n)P_i(k-n|k-n-1)C_i(k-n)^T + R_i(k-n)]^{-1}\times \\ \times\tilde{C}_i(n)^T\Phi_1(k-N,k)^T\end{gathered} \tag{82}$$

where

$$\begin{gathered}Q(k-N,k) = \Phi_1(k-N,k)[\Sigma_{j=1}^{N}A(k,k-j+1)\times \\ \times Q(k-j+1,k-j)A(k,k-j+1)^T]\Phi_1(k-N,k)^T\end{gathered} \tag{83}$$

Next, a theorem is given about the calculation of covariance matrix M_i appearing in the modified state estimation of Eq. (74). The theorem comes from (Xia et al., 2009) and is also applicable to the distributed filtering approach which is presented in this chapter.

Theorem 2: It is assumed that the modified state estimation error at time instant k is

$$\tilde{x}_i^*(k|k) = x(k) - \hat{x}_i(k|k) \tag{84}$$

and that the covariance matrix of the modified state estimation error is

$$P_i^*(k|k) = E\{\tilde{x}_i^*(k|k)\tilde{x}_i^*(k|k)^T\} \tag{85}$$

and that the cross-covariance between $\tilde{x}_i(k|k)$ and $\tilde{w}_i(k-N,k|k)$ is

$$P_i^{\tilde{x}\tilde{w}}(k|k) = E\{\tilde{x}_i(k|k)\tilde{w}_i(k-N,k|k)^T\} \tag{86}$$

Then, the optimal filter for the processing of the delayed measurements is given by Eq. (74), i.e.

$$\hat{x}_i^*(k|k) = \hat{x}_i(k|k) + M_i[z_i(k-N) - \hat{z}_i(k-N|k)] \tag{87}$$

where

$$M_i = [P_i(k|k)\Phi_1(k-N,k)^T + P_i^{\tilde{x}\tilde{w}}]C_i(k-N)^T W_i^{-1} \tag{88}$$

In that case, the covariance matrix of the modified state estimation error becomes

$$\begin{aligned} P_i^*(k|k) = P_i(k|k) - [P_i^{\tilde{x}\tilde{w}} + P_i(k|k)\Phi_1(k-N,k)^T]\times \\ \times C_i(k-N)^T W_i^{-1} C_i(k-N)\times \\ \times[P_i^{\tilde{x}\tilde{w}} + P_i(k|k)\Phi_1(k-N,k)^T]^T \end{aligned} \tag{89}$$

where matrices W_i and $P_i^{\tilde{x}\tilde{w}}$ are defined as

$$\begin{aligned} W_i = C_i(k-N)\{\Phi_1(k-N,k)P_i(k|k)\Phi_1(k-N,k)^T + \\ +\Phi_1(k-N,k)P_i^{\tilde{x}\tilde{w}} + [A(k-N,k)P_i^{\tilde{x}\tilde{w}}]^T + Q_i^*(k-N,k)\} \\ \times C_i(k-N)' + R_i(k-N) \end{aligned} \tag{90}$$

$$\begin{aligned} P_i^{\tilde{x}\tilde{w}} = \Phi_1(k-N,k)\textstyle\sum_{n=0}^{N-1} P_i(k-N|k-N)D_i(n)^T\times \\ \times[C_i(k-n)P_i(k-n|k-n-1)C_i(k-n)^T + R_i(k-n)]^{-1}\times \\ \times\tilde{C}_i(n)^T\Phi_1(k-N,k)^T - A(k,k-N)Q_i^*(k-N,k) \end{aligned} \tag{91}$$

and matrix $D_i^T(n)$ is defined as

$$D_i(n) = \begin{cases} C_i(k-n)A(k-n,k-n-1), \text{ if } N=1 \\ C_i(k-n)A(k-n,k-n-1)\prod_{j=n}^{N-2}[I-K_i(k-j-1)C_i(k-j-1)]\times \\ \times A(k-j-1,k-j-2), \text{ if } N>1 \end{cases} \tag{92}$$

5.3 Processing of the delayed measurements for a linear non-autonomous system

5.3.1 The case of a time-variant linear system

In the case of a linear non-autonomous system, in place of Eq. (61) one has

$$x(k) = A(k,k-1)x(k-1) + B(k,k-1)u(k-1) + w(k,k-1) \tag{93}$$

Setting $w_1(k,k-1) = B(k,k-1)u(k-1) + w(k,k-1)$ one obtains

$$x(k) = A(k,k-1)x(k-1) + w_1(k,k-1) \tag{94}$$

and consequently it holds

$$\begin{aligned} x(k) = \textstyle\prod_{j=1}^{N} A(k-j+1,k-j)x(k-N) + \\ +\textstyle\sum_{m=1}^{N-1}\prod_{j=1}^{m} A(k-j+1,k-j)w_1(k-m,k-m-1) + w_1(k,k-1) \end{aligned} \tag{95}$$

where

$$w_1(k-m+1,k-m) = B(k-m+1,k-m)u(k-m) + w(k-m+1,k-m) \tag{96}$$

Thus, one can obtain a more compact form

$$x(k) = \Phi(k,k-N)x(k-N) + w_1(k,k-N) \tag{97}$$

with

$$\begin{array}{c} \Phi(k,k-N) = \prod_{j=1}^{N} A(k-j+1,k-j), \text{ and} \\ w_1(k,k-N) = \sum_{m=1}^{N-1}\prod_{j=1}^{m} A(k-j+1,k-j)w_1(k-m,k-m-1) + w_1(k,k-1) \end{array} \tag{98}$$

5.3.2 The case of a time-invariant linear system

For a linear time-invariant non-autonomous system

$$x(k) = Ax(k-1) + Bu(k-1) + w(k-1) \tag{99}$$

it holds

$$x(k) = A^N x(k-N) + \sum_{j=1}^{N} A^{N-j}Bu(k-N+j-1) + \sum_{j=1}^{N} A^{N-j}w(k-N+j-1) \tag{100}$$

Denoting $A^N = \Phi(k,k-N)$ one has

$$x(k) = \Phi(k,k-N)x(k-N) + \sum_{j=1}^{N} A^{N-j}Bu(k-N+j-1) + \sum_{j=1}^{N} A^{N-j}w(k-N+j-1) \tag{101}$$

Setting

$$w_1(k,k-N) = \sum_{j=1}^{N} A^{N-j}Bu(k-N+j-1) + \sum_{j=1}^{N} A^{N-j}w(k-N+j-1) \tag{102}$$

one has that Eq. (101) can be written in a more compact form as

$$x(k) = \Phi(k,k-N)x(k-N) + w_1(k,k-N) \tag{103}$$

Using that matrix $\Phi(k,k-N)$ is invertible, one has

$$x(k-N) = \Phi(k,k-N)^{-1}x(k) - \Phi(k,k-N)^{-1}w_1(k,k-N) \tag{104}$$

The following notation is used $\Phi_1(k-N,k) = \Phi(k,k-N)^{-1}$ while for the retrodiction of $w_1(k,k-N)$ it holds $w_a(k-N,k|k) = -\Phi(k,k-N)^{-1}w_1(k,k-N)$. Then, to smooth the state estimation at time instant $k-N$, using the measurement of output $z_i(k-N)$ received at time instant $k+1$ one has the state equation

$$\hat{x}(k-N,k) = \Phi_1(k-N,k)\hat{x}(k|k) + \hat{w}_a(k-N,k|k) \tag{105}$$

while the associated measurement equation becomes

$$z(k-N) = Cx(k-N) + v(k-N) \tag{106}$$

Substituting Eq. (104) into Eq. (106) provides

$$z(k-N) = C\Phi_1(k-N,k)x(k) + Cw_a(k-N,k) + v(k-N) \tag{107}$$

and the associated estimated-output at time instant $k-N$ is

$$\hat{z}(k-N) = C\Phi_1(k-N,k)\hat{x}(k|k) + C\hat{w}_a(k-N,k) \tag{108}$$

From Eq. (108) and Eq. (107) the innovation for the delayed measurement can be obtained

$$\tilde{z}(k-N) = z(k-N) - \hat{z}(k-N) \tag{109}$$

i.e. $\tilde{z}(k-N) = C\Phi_1(k-N)\tilde{x}(k|k) + C\tilde{w}_a(k-N)$, where $\tilde{x}(k|j) = x(k) - \hat{x}(k|j)$ is the state estimation error and $\hat{w}_a(k-N,k|k) = w_a(k-N,k) - \hat{w}_a(k-N,k)$ is the estimation error for w_α. With this innovation the estimation of the state vector $x(k|k)$ at time instant k is corrected. The correction (smoothing) relation is

$$\hat{x}^*(k|k) = x(k|k) + M\tilde{z}(k-N,k) \tag{110}$$

Therefore, again the basic problem for the implementation of the smoothing relation provided by Eq. (110) is the calculation of the term $w_a(k-N,k)$ i.e. $w_a(k-N) = \Phi(k,k-N)^{-1}w_1(k,k-N)$. This in turn requires the estimation of the term $w_1(k-N,k)$ which, according to Eq. (80), is provided by

$$\begin{aligned}\hat{w}_1(k-N,k) = -\Phi_1(k-N,k)\textstyle\sum_{n=0}^{N-1}\cdot \\ \cdot\tilde{C}(n)[C(k-n)P(k-n|k-n-1)C(k-n)]^T + R(k-n)]^{-1}\tilde{z}(k-n)\end{aligned} \tag{111}$$

where $\tilde{z}(k-n) = z(k-n) - \hat{z}(k-n)$ is the innovation for time-instant $k-n$, while, as given in Eq. (81)

$$\begin{aligned}\tilde{C}(n) = \{A(k,k-n)Q(k-n,k-n-1) + \textstyle\sum_{j=n+2}^{N}A(k,k-j+1)Q(k-j+1,k-j)\times \\ \times[\textstyle\prod_{m=n+1}^{j-1}A(k-m+1,k-m)[I-K_i(k-m)C(k-m)]^T]\}C(k-n)^T\end{aligned} \tag{112}$$

5.4 Derivative-free Extended Information Filtering under time-delays and packet drops

It has been shown that using a suitable transform (diffeomorphism), the nonlinear system of Eq. (15) can be transformed into the system of Eq. (17). Moreover, it has been shown that for the systems of Eq. (23) and Eq. (24) one can obtain a a state-space equation of the form

$$\begin{pmatrix}\dot{\zeta}_1\\ \dot{\zeta}_2\\ \cdots\\ \dot{\zeta}_{n-1}\\ \dot{\zeta}_n\end{pmatrix} = \begin{pmatrix}0 & 1 & 0 & \cdots & 0\\ 0 & 0 & 1 & \cdots & 0\\ \vdots & \vdots & \vdots & \vdots & \vdots\\ 0 & 0 & 0 & \cdots & 1\\ 0 & 0 & 0 & \cdots & 0\end{pmatrix}\begin{pmatrix}\zeta_1\\ \zeta_2\\ \cdots\\ \zeta_{n-1}\\ \zeta_n\end{pmatrix} + \begin{pmatrix}0\\ 0\\ \cdots\\ 0\\ 1\end{pmatrix}v(\zeta,t) \tag{113}$$

$$z = \begin{pmatrix}1 & 0 & 0 & \cdots & 0\end{pmatrix}\zeta \tag{114}$$

where $v(t) = f(x,t) + g(x,t)u(t)$, with $u(t)$ being the control input of the dynamical system. The description of the initial system of Eq. (17) in the form of Eq. (113) and Eq. (114) enables

the application of the previous analysis for the compensation of time-delays and packet-drops through smoothing in the computation of the linear Kalman Filter. The fact that the system of Eq. (113) and Eq. (114) is a time invariant one, facilitates the computation of the smoothing Kalman Filter given in Eq. (100) to Eq. (112). Thus, one has to use the time invariant matrices A_c and C_c defined in Eq. (18) and Eq. (19), while for matrix B_c it holds according to Eq. (25) that $B_c = [0,0,\cdots,0,1]^T$. The discrete-time equivalents of matrices A_c, B_c and C_c are noted as A_d, B_d and C_d, respectively. It is also noted that due to the specific form of matrix B_c, the term $Bu(k-1)$ appearing in Eq. (99) is a variable of small magnitude with mean value close to zero. Thus the term $w_1(k,k-1) = Bu(k-1) + w(k,k-1)$ differs little from $w(k,k-1)$.
It also becomes apparent that through the description of the initial system of Eq. (17) in the form of Eq. (113) and Eq. (114), the application of the derivative-free Extended Information Filter can be performed in a manner that enables the compensation of time-delays and packet drops. Writing the controlled system in the form of Eq. (113) and Eq. (114) permits to develop local linear Kalman Filters that smooth the effects of delayed sensor measurement or the loss of measurement packets. Moreover, the application of the standard Information Filter for fusing the estimates provided by the local Kalman Filters, permits to avoid the approximation errors met in the Extended Information Filter algorithm.

6. Distributed filtering under time-delays and packet drops for sensorless control

6.1 Visual servoing over a network of synchronized cameras

Visual servoing over a network of synchronized cameras is an example where the efficiency of the proposed distributed filtering approach under time delays and packet drops can be seen. Applications of vision-based robotic systems are rapidly expanding due to the increase in computer processing power and low prices of cameras, image grabbers, CPUs and computer memory. In order to satisfy strict accuracy constraints imposed by demanding manufacturing specifications, visual servoing systems must be fault tolerant. This means that despite failures in its components or the presence of disturbances, the system must continue to provide valid control outputs which will allow the robot to complete its assigned tasks (DeSouza & Kak, 2004),(Feng & Zeng, 2010),(Hwang & Shih, 2002),(Malis et al., 2000).
The example to be presented describes the control of a planar robot with the use of a position-based visual servo that comprises multiple fixed cameras. The chapter's approach relies on neither position nor velocity sensors, and directly sets the motor control current using only visual feedback. Direct visual servoing is implemented using a distributed filtering scheme which permits to fuse the estimates of the robot's state vector computed by local filters, each one associated to a camera in the cameras network (see Fig. 7). The cameras' network can be based on multiple RS-170 cameras connected to a computer with a frame grabber to form a vision node. Each vision node consists of the camera, the frame grabber and the filter which estimates motion characteristics of the monitored robot joint. The vision nodes are connected in a network to form a distributed vision system controlled by a master computer. The master computer is in turn connected to a planar 1-DOF robot joint and uses the vision feedback to perform direct visual servoing (see Fig. 7).
The master computer communicates video synchronization information over the network to each vision node. Typical sources of measurement noise include charge-coupled device (CCD) noise, analog-to-digital (A/D) noise and finite word-length effects. Under ideal conditions, the effective noise variance from these sources should remain relatively constant. Occlusions can be also considered as a noise source. Finally, communication delays and packet drops in the transmission of measurements from the vision sensors to the information processing

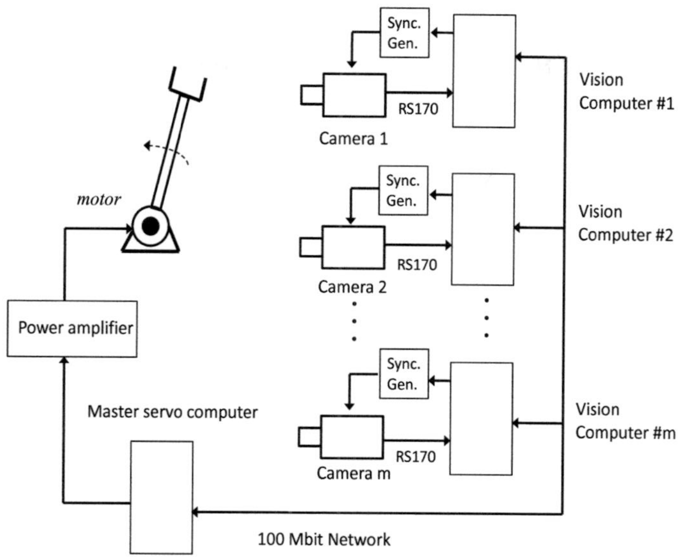

Fig. 7. Distributed cameras network and distributed information processing units for visual servoing

nodes induce additional disturbances which should be compensated by the virtual servoing control loop.

6.2 Distributed filtering-based fusion of the robot's state estimates

Fusion of the local state estimates which are provided by filters running on the vision nodes can improve the accuracy and robustness of the performed state estimation, thus also improving the performance of the robot's control loop (Sun et al., 2011),(Sun & Deng, 2005). Under the assumption of Gaussian noise, a possible approach for fusing the state estimates from the distributed local filters is the derivative-free Extended Information Filter (DEIF). As explained in Section 4, the derivative-free Extended Information Filter provides an aggregate state estimate by weighting the state vectors produced by local Kalman Filters with the inverse of the associated estimation error covariance matrices.

Visual servoing over the previously described cameras network is considered for the nonlinear dynamic model of a single-link robotic manipulator. The robot can be programmed to execute a manufacturing task, such as disassembly or welding (Tzafestas et al., 1997). The position of the robot's end effector in the cartesian space (and consequently the angle for the robotic link) is measured by the aforementioned m distributed cameras. The proposed multi-camera based robotic control loop can be also useful in other vision-based industrial robotic applications where the vision is occluded or heavily disturbed by noise sources, e.g. cutting. In such applications there is need to fuse measurements from multiple cameras so as to obtain redundancy in the visual information and permit the robot to complete safely and within the specified accuracy constraints its assigned tasks (Moon et al, 2006),(Yoshimoto et al., 2010).

The considered 1-DOF robotic model consists of a rigid link which is rotated by a DC motor, as shown in Fig. 8. The model of the DC motor is described by the set of equations: $L\dot{I} = -k_e\omega - RI + V$, $J\dot{\omega} = k_eI - k_d\omega - \Gamma_d$, with the following notations L : armature inductance, I : armature current, k_e : motor electrical constant, R : armature resistance, V : input voltage,

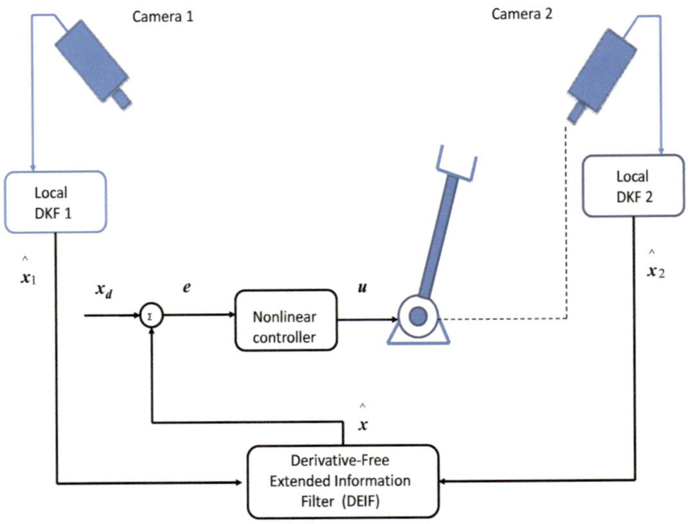

Fig. 8. Visual servoing based on fusion of state estimates provided by local derivative-free nonlinear Kalman Filters

taken as control input, J : motor inertia, ω : rotor rotation speed, k_d : mechanical dumping constant, Γ_d : disturbance or external load torque. It is assumed that $\Gamma_d = mgl{\cdot}sin(\theta)$, i.e. that the DC motor rotates a rigid robotic link of length l with a mass m attached to its end. Then, denoting the state vector as $[x_1, x_2, x_3]^T = [\theta, \dot{\theta}, \ddot{\theta}]^T$, a nonlinear model of the DC motor is obtained

$$\dot{x} = f(x,t) + g(x,t)u \tag{115}$$

where $f(x,t) = [f_1(x,t), f_2(x,t), f_3(x,t)]^T$ is a vector field function with elements: $f_1(x,t) = x_2$, $f_2(x,t) = x_3$, $f_3(x,t) = -\frac{k_e^2+k_dR}{JL}x_2 - \frac{RJ+K_dL}{JL}x_3 - \frac{Rmgl}{JL}sin(x_1) - \frac{mgl}{J}cos(x_1)x_2$. Similarly, for function $g(x,t)$ it holds that $g(x,t) = [g_1(x,t), g_2(x,t), g_3(x,t)]^T$, i.e. it is a vector field function with elements: $g_1(x,t) = 0$, $g_2(x,t) = 0$, $g_3(x,t) = \frac{k_e}{JL}$. Having chosen the joint's angle to be the system's output, the state space equation of the 1-DOF robot manipulator can be rewritten as

$$x^{(3)} = \bar{f}(x) + \bar{g}(x)u \tag{116}$$

where functions $\bar{f}(x)$ and $\bar{g}(x)$ are given by $\bar{f}(x) = -\frac{k_e^2+k_dR}{JL}x_2 - \frac{RJ+K_dL}{JL}x_3 - \frac{Rmgl}{JL}sin(x_1) - \frac{mgl}{J}cos(x_1)x_2$, and $\bar{g}(x) = \frac{k_e}{JL}$. This is a system in the form of Eq. (23), therefore a state estimator can be designed according to the previous results on derivative-free Kalman Filtering.
The controller has to make the system's output (angle θ of the motor) follow a given reference signal x_d. For measurable state vector x and uncertain functions $f(x,t)$ and $g(x,t)$ an appropriate control law for the 1-DOF robotic model is

$$u = \frac{1}{g(x,t)}[x_d^{(n)} - f(x,t) - K^T e + u_c] \tag{117}$$

with $e = x - x_d$, $e^T = [e, \dot{e}, \ddot{e}, \cdots, e^{(n-1)}]^T$, $K^T = [k_n, k_{n-1}, \cdots, k_1]$, such that the polynomial $e^{(n)} + k_1 e^{(n-1)} + k_2 e^{(n-2)} + \cdots + k_n e$ is Hurwitz. The previously defined control law results into $e^{(n)} = -K^T e + u_c + \tilde{d}$, where the supervisory control term u_c aims at the compensation of modeling errors as well as of the additive disturbance $\tilde{d}$ (Rigatos & Tzafestas, 2007). Suitable selection of the feedback gain K assures that the tracking error will converge to $\lim_{t \to \infty} e(t) = 0$. In case of state estimation-based (sensorless control), and denoting, $\hat{x}$ as the estimated state vector and $\hat{e} = \hat{x} - x_d$ as the estimated tracking error one has

$$u = \frac{1}{g(\hat{x}, t)} [x_d^{(n)} - f(\hat{x}, t) - K^T \hat{e} + u_c] \tag{118}$$

7. Simulation tests

The fusion of the distributed state estimates for the robotic model was performed with the use of the derivative-free Extended Information Filter. First, it was assumed that the transmission of measurements from the vision sensors (cameras) to the local information processing units, where the state estimators (filters) were running, was not affected by time delays or packet drops. At the local vision nodes, Kalman filters were used to produce estimations of the robot's state vector as well as the associated covariance matrices, after carrying out a linearization of the robot's nonlinear dynamic model through the transformation described in subsection 3.2 and processing the local xy position measurements. This standard Information Filter provided the overall estimate of the robot's state vector, through weighting of the local state vectors by the local covariance matrices. The obtained results are depicted in Fig. 9(a) and Fig. 9(b) in case of a sinusoidal and a see-saw reference trajectory (both reference trajectories are denoted with the red line).

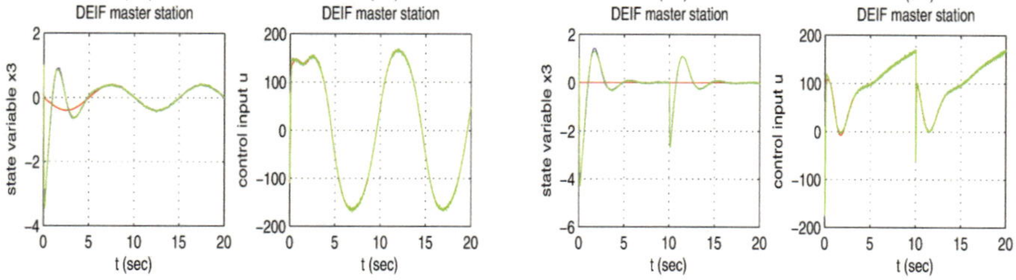

(a) (b)

Fig. 9. Control of the robotic manipulator with fusion of position measurements from distributed cameras through the use of the derivative-free Extended Information Filter (a) when tracking of a sinusoidal trajectory (b) when tracking of a see-saw trajectory

Next, time-delays were assumed in the transmission of image frames from the distributed cameras to the associated local vision nodes, where the local derivative-free Kalman Filters were running. For both vision nodes the delays in the transmission of measurements varied randomly between 6 and 25 sampling periods. Longer delays could be also handled by the proposed distributed filtering algorithm. The variation of measurement transmission delays with respect to time, is depicted in Fig. 10.

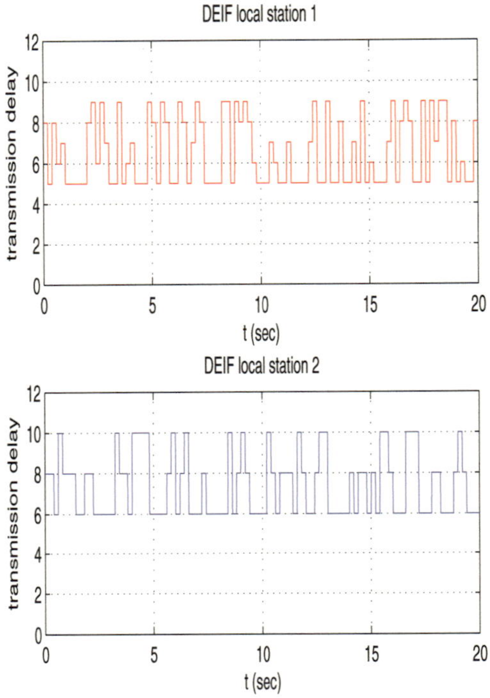

Fig. 10. Variation in time (in multiples of the sampling period) of the measurement delays appearing at the local information processing nodes 1 and 2.

The delayed measurements were processed by the Kalman Filter recursion according to the stages explained in subsection 5.3. The smoothing of the delayed measurements that was performed by the Kalman Filter was based on Eq. (74), i.e. $\hat{x}^*(k|k) = x(k|k) + M\tilde{y}(k-N,k)$. As explained in subsection 5.3, matrix M is a gain matrix calculated according to Eq. (88). The innovation is given by $\tilde{z}(k-N) = z(k-N) - \hat{z}(k-N)$. The tracking accuracy of the distributed filtering-based control loop is depicted in Fig. 11 to Fig. 13.
Additionally, some performance metrics were used to evaluate the distributed filtering-based control scheme. Table I, shows the variation of the traces of the covariance matrices at the local filters and at the master filter with respect to delay levels ($d_1, d_2 = k{\cdot}T_s$ i.e. multiples of the sampling period T_s), as well as with respect to the probability of delay occurrence in the transmission of the measurement packets ($p\in [0,1]$).
Moreover, the variation of the tracking error of the three state variables x_i, $i = 1, \cdots, 3$ with respect to delay levels as well as with respect to the probability of delay occurrence in the transmission of the measurement packets is given in Tables II to IV.

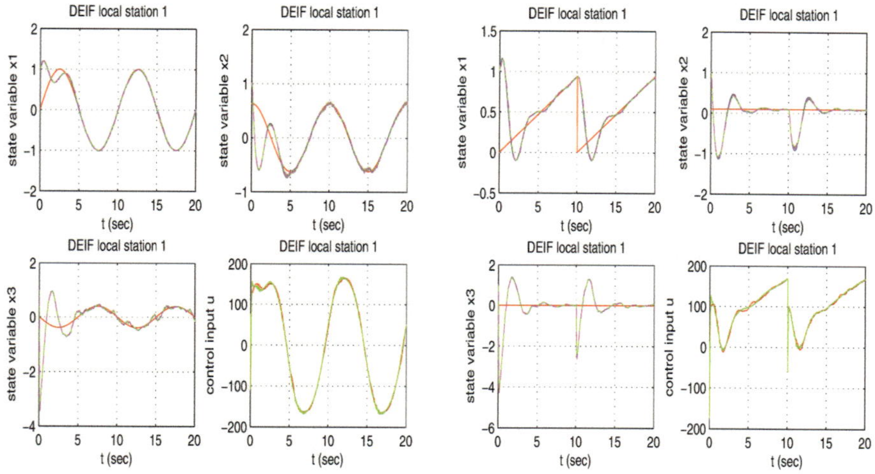

(a) **(b)**

Fig. 11. Estimation of the motion of the robotic manipulator under transmission delays at the first local measurement processing node, (a) when tracking a sinusoidal trajectory (b) when tracking a see-saw trajectory

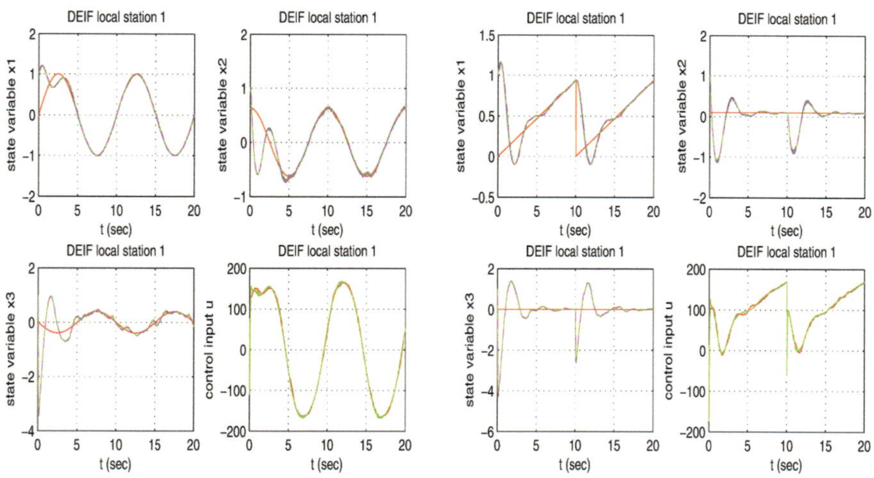

(a) **(b)**

Fig. 12. Estimation of the motion of the robotic manipulator under transmission delays at the second local measurement processing node, (a) when tracking a sinusoidal trajectory (b) when tracking a see-saw trajectory

It can be noticed that the smoothing performed by the distributed filtering algorithm, through the incorporation of out-of-sequence-measurements, enhances the robustness of the

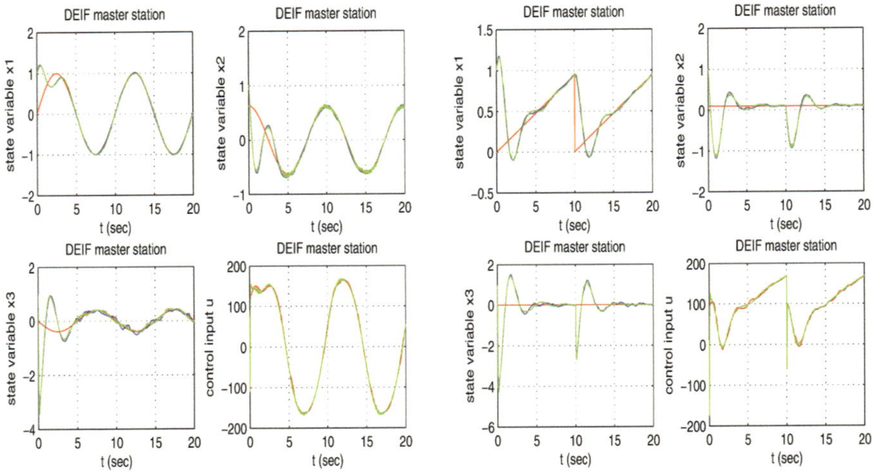

(a) **(b)**

Fig. 13. Control of the robotic manipulator under measurement transmission delays and using the derivative-free Extended Information Filter for state estimation, (a) tracking of a sinusoidal trajectory (b) tracking of a see-saw trajectory

Table I: Traces of the covariance error matrices for various delay levels

d_1	d_2	p	$Tr(P_1^*)$	$Tr(P_2^*)$	$Tr(P)$
0	0	0.0	$2.780 \cdot 10^{-2}$	$2.780 \cdot 10^{-2}$	$6.720 \cdot 10^{-3}$
6	8	0.8	$2.782 \cdot 10^{-2}$	$2.777 \cdot 10^{-2}$	$6.730 \cdot 10^{-3}$
9	10	0.8	$2.782 \cdot 10^{-2}$	$2.783 \cdot 10^{-2}$	$6.730 \cdot 10^{-3}$
12	15	0.8	$2.782 \cdot 10^{-2}$	$2.776 \cdot 10^{-2}$	$6.730 \cdot 10^{-3}$
18	20	0.6	$2.782 \cdot 10^{-2}$	$2.775 \cdot 10^{-2}$	$6.730 \cdot 10^{-3}$
25	30	0.6	$2.780 \cdot 10^{-2}$	$2.783 \cdot 10^{-2}$	$6.730 \cdot 10^{-3}$

Table II: RMSE tracking error at the 1st local filter for various delay levels

d_1	d_2	p	$x_1 - x_1^d$	$x_2 - x_2^d$	$x_3 - x_3^d$
0	0	0.0	$4.419 \cdot 10^{-3}$	$5.490 \cdot 10^{-3}$	$1.125 \cdot 10^{-2}$
6	8	0.8	$4.413 \cdot 10^{-3}$	$5.504 \cdot 10^{-3}$	$1.129 \cdot 10^{-2}$
9	10	0.8	$4.392 \cdot 10^{-3}$	$5.437 \cdot 10^{-3}$	$1.121 \cdot 10^{-2}$
12	15	0.8	$4.402 \cdot 10^{-3}$	$5.465 \cdot 10^{-3}$	$1.117 \cdot 10^{-2}$
18	20	0.6	$4.474 \cdot 10^{-3}$	$5.707 \cdot 10^{-3}$	$1.151 \cdot 10^{-2}$
25	30	0.6	$4.433 \cdot 10^{-3}$	$5.655 \cdot 10^{-3}$	$1.144 \cdot 10^{-2}$

estimation. Despite the raise of the delay levels in the transmission of measurements from the sensors (cameras) to the local information processing nodes (local derivative-free Kalman Filters) only slight variations of the tracking errors for state variables x_i, $i = 1, \cdots, 3$ were observed. Similarly, the changes of the traces of the estimation error covariance matrices, both at the local filters and at the master filter, were small.

Table III: RMSE tracking error at the 2nd local filter for various delay levels

d_1	d_2	p	$x_1 - x_1^d$	$x_2 - x_2^d$	$x_3 - x_3^d$
0	0	0.0	$4.390{\cdot}10^{-3}$	$5.453{\cdot}10^{-3}$	$1.116{\cdot}10^{-2}$
6	8	0.8	$4.380{\cdot}10^{-3}$	$5.468{\cdot}10^{-3}$	$1.114{\cdot}10^{-2}$
9	10	0.8	$4.441{\cdot}10^{-3}$	$5.495{\cdot}10^{-3}$	$1.118{\cdot}10^{-2}$
12	15	0.8	$4.451{\cdot}10^{-3}$	$5.521{\cdot}10^{-3}$	$1.125{\cdot}10^{-2}$
18	20	0.6	$4.508{\cdot}10^{-3}$	$5.744{\cdot}10^{-3}$	$1.161{\cdot}10^{-2}$
25	30	0.6	$4.432{\cdot}10^{-3}$	$5.755{\cdot}10^{-3}$	$1.150{\cdot}10^{-2}$

Table IV: RMSE tracking error at the master filter for various delay levels

d_1	d_2	p	$x_1 - x_1^d$	$x_2 - x_2^d$	$x_3 - x_3^d$
0	0	0.0	$4.416{\cdot}10^{-3}$	$5.452{\cdot}10^{-3}$	$1.101{\cdot}10^{-2}$
6	8	0.8	$4.504{\cdot}10^{-3}$	$5.505{\cdot}10^{-3}$	$1.130{\cdot}10^{-2}$
9	10	0.8	$4.473{\cdot}10^{-3}$	$5.493{\cdot}10^{-3}$	$1.106{\cdot}10^{-2}$
12	15	0.8	$4.408{\cdot}10^{-3}$	$5.423{\cdot}10^{-3}$	$1.094{\cdot}10^{-2}$
18	20	0.6	$4.533{\cdot}10^{-3}$	$5.785{\cdot}10^{-3}$	$1.139{\cdot}10^{-2}$
25	30	0.6	$4.529{\cdot}10^{-3}$	$5.755{\cdot}10^{-3}$	$1.149{\cdot}10^{-2}$

8. Conclusions

This chapter has proposed a solution to the problem of state estimation-based control under communication delays and packet drops. The considered approach was within the frame of distributed Kalman Filtering. First, the Extended Information Filter was presented as a basic approach to nonlinear distributed filtering. The Extended Information Filter (EIF) performs fusion of the the state estimates provided by the local monitoring stations, under the assumption of Gaussian noises. The Extended Information Filter is a generalization of the Information Filter in which the local filters do not exchange raw measurements but send to an aggregation filter their local information matrices (local inverse covariance matrices or differently known as Fisher Information Matrices) and their associated local information state vectors (products of the local information matrices with the local state vectors).

To improve the estimation accuracy and convergence properties of the Extended Information Filter, the derivative-free Extended Information Filter has been introduced. The derivative-free Extended Information Filter, has the following features (i) it is not based on local linearization of the controlled system dynamics, (ii) it does not assume truncation of higher order Taylor expansion terms, (iii) it does not require the computation of Jacobian matrices. In the proposed filtering method, the system is first subject to a linearization transformation and next state estimation is performed by applying local Kalman Filters to the linearized model. The class of systems to which the derivative-free Extended Information Filter can be applied has been also defined.

Next, distributed state-estimation under communication delays and packet drops was examined. First, results on networked linear Kalman Filtering were overviewed. These results were generalized in the case of the derivative-free Extended Information Filter, where the problem of communication delays and packet drops has again the following forms: (i) there are time delays and packet drops in the transmission of information between the distributed local filters and the master filter, (ii) there are time delays and packet drops in the transmission of information from distributed sensors to each one of the local filters. In the first case, the structure and calculations of the master filter for estimating the aggregate state vector remain unchanged. In the second case, the effect of the random delays and packets drops has to be

taken into account in the redesign of the local Kalman Filters, which implies (i) a modified Riccati equation for the computation of the covariance matrix of the state vector estimation error, (ii) the use of a correction term in the update of the state vector's estimate so as to compensate for delayed measurements arriving at the local Kalman Filters.
In the simulation experiments it was shown that the aggregate state vector produced by the derivative-free Extended Information Filter can be used for sensorless control and robotic visual servoing. Visual servoing over a cameras network was considered for the nonlinear dynamic model of a planar single-link robotic manipulator. The position of the robot's end effector in the cartesian space (and equivalently the angle of the robotic link) was measured through m cameras. In turn m distributed derivative-free Kalman Filters were used to estimate the state vector of the robotic link. Next, the local state estimates were fused with the use of the standard Information Filter. Finally, the aggregate estimation of the state vector was used in a control loop which enabled the robotic link to perform trajectory tracking. It was shown that the proposed redesign of the local derivative-free Kalman filters enabled to compensate for communication delays and packet drops, thus also improving the accuracy of the presented distributed filtering approach and the robustness of the associated control loop.

9. References

Ahrens, J.H. & Khalil, H.K. (2005). Asymptotic properties of Extended Kalman Filters for a class of nonlinear systems, *Proc. 44th IEEE Conference on Decision and Control*, Seville Spain, 2005.

Bar-Shalom, Y. (2002). Update with out-of-sequence measurements in tracking: exact solution, *IEEE Transactions on Aerospace and Electronic Systems*, Vol. 38, No. 3, pp. 769-778, 2002.

Boutayeb, M.; Rafaralahy, H.; & Darouach, M. (1997). Convergence analysis of the Extended Kalman Filter used as an observer for nonlinear deterministic discrete-time systems, *IEEE Transactions on Automatic Control*, Vol. 42, No.4, pp. 581-586, 1997.

DeSouza, G.N. & Kak, A.C. (2004). A subsumptive, hierarchical and distributed vision-based architecture for smart robotics, *IEEE Transactions on Systems, Man and Cybernetics - Part B*, Vol. 34, No. 5, pp. 1988-2002, 2004.

Feng,N. & Zeng, M. (2010). Optimal distributed Kalman filtering fusion for a linear dynamic system with cross-correlated noises, *International Journal of Systems Science, Taylor & Francis*, doi: 10.1080/00207721.2010.502601, 2010.

Golapalakhrishnan, A.; Kaisare, N. &Narasimhan, S. (2011). Incorporating delayed and infrequent measurements in Extended Kalman Filter-based nonlinear state estimation, *Journal of Process Control*, Elsevier, Vol. 21, pp. 119-129, 2011.

Hwang. C.L. & Shih, C.Y. (2002). A distributed active-vision network-space approach for the navigation of a car-like wheeled robot, *IEEE Transactions on Industrial Electronics*, Vol. 56, No. 3, pp. 846-855, 2002.

Jia, Z.; Balasuriyaa, & Challab, S. (2008). Sensor fusion-based visual target tracking for autonomous vehicles with the out-of-sequence measurements solution, *Robotics and Autonomous Systems*, Elsevier, vol. 56, pp. 157-176, 2008.

Kamen, E.W. & Su, J.K. (1999). *Introduction to Optimal Estimation*, Springer.

Lee, D.J. (2008). Nonlinear estimation and multiple sensor fusion using unscented information filtering, *IEEE Signal Processing Letters*, Vol. 15, pp. 861-864, 2008.

Mahler, R.P.S. (2007). *Statistical Multisource-Multitarget Information Fusion*, Artech House Inc, 2007.

Makarenko, A. & Durrany-Whyte, H. (2006). Decentralized Bayesian algorithms for active sensor networks, *Information Fusion*. Elsevier, Vol.7, pp. 418-433, 2006.

Malis, E.; Chaumette, F. & Boudet, S. (2000). Multi-cameras visual servoing, *IEEE Intrl. Conf. on Robotics and Automation*, ICRA 2000, pp. 3183-3188, 2000.

Manyika, J. & Durrant-Whyte, H. (1994). *Data fusion and sensor management: a decentralized information theoretic approach*, Prentice Hall, Englewood Cliffs, NJ, 1994.

Marino, R. (1990). Adaptive observers for single output nonlinear systems, *IEEE Transactions on Automatic Control*, Vol. 35, No.9, pp. 1054-1058, 1990.

Marino, R. & Tomei, P. (1992). Global asymptotic observers for nonlinear systems via filtered transformations, *IEEE Transactions on Automatic*, Vol. 37, No. 8, pp. 1239-1245, 1992.

Medeiros, H.; J. Park & A.C. Kak (2008). Distributed object tracking using a cluster-based Kalman Filter in wireless camera networks, *IEEE Journal of Selected Topics in Signal Processing*, Vol. 2, No.4, pp. 448-463, 2008.

Moon, H.S.; Kim, Y.B. & Beattie, R.J. (2006). Multi sensor data fusion for improving performance and reliability of fully automatic welding system, *International Journal of Advanced Manufacturing Technology*, Vol. 28, pp. 286-293, 2006.

Nettleton, E.; Durrant-Whyte, H. & Sukkarieh, S. (2003). A robust architecture for decentralized data fusion, *ICAR03, 11th International Conference on Advanced Robotics*, Coimbra, Portugal, 2003.

Olfati-Saber, R. (2006). Distributed Kalman Filtering and Sensor Fusion in Sensor Networks, *Lecture notes in control and information sciences*, Vol. 331, pp. 157-167, 2006.

Rao, B.S. & Durrant-Whyte, H.F. (1991). Fully decentralized algorithm for multisensor Kalman Filtering, *IEE Proceedings D*, Vol. 138, No.5, pp. 413-451, 1991.

Rigatos, G.G. (2010a). Distributed particle filtering over sensor networks for autonomous navigation of UAVs, *In: Robot Manipulators (A. Lazinica Ed.)*, InTech Publications.

Rigatos, G.G. (2010b). Distributed filtering over sensor networks for autonomous navigation of UAVs, *IEEE VTC 2010 Fall Conference*, Ottawa, Canada.

Rigatos, G.G. (2010c). A derivative-free Kalman Filtering approach for sensorless control of nonlinear systems, *IEEE ISIE 2010, IEEE International Symposium on Industrial Electronics*, Bari, Italy, July 2010.

Rigatos, G.G. (2010d). Extended Kalman and Particle Filtering for sensor fusion in motion control of mobile robots, *Mathematics and Computers in Simulation*, Elsevier, Vol. 81, pp. 590-607, 2010.

Rigatos, G. & Siano, P. (2010). A derivative-free Extended Information Filtering approach for sensorless control of nonlinear systems, *MASCOT 2010, IMACS Workshop on Scientific Computation, Italian Institute for Calculus Applications*, Gran Canaria Spain, Oct. 2010.

Rigatos, G.G. (2009). Particle Filtering for State Estimation in Nonlinear Industrial Systems, *IEEE Transactions on Instrumentation and Measurement*, Vol. 58, No. 11, pp. 3885-3900, 2009.

Rigatos, G. & Zhang, Q. (2009). Fuzzy model validation using the local statistical approach, *Fuzzy Sets and Systems, Elsevier*, Vol. 60, No.7, pp. 882-904, 2009.

Rigatos, G.G.; Siano, P. & Merola, E. (2009). Sensorless control of DC and induction motors using Kalman Filtering, *MASCOT 2009, IMACS Workshop on Scientific Computation, Italian Institute for Calculus Applications*, Roma Italy, Oct. 2009.

Rigatos, G.G. & Tzafestas, S.G. (2007). Extended Kalman Filtering for Fuzzy Modeling and Multi-Sensor Fusion, *Mathematical and Computer Modeling of Dynamical Systems, Taylor & Francis, Taylor and Francis*, Vol. 13, No. 3, 2007.

Rigatos, G.G. & Tzafestas, S.G. (2007). H_∞ tracking of uncertain SISO nonlinear systems: an observer-based adaptive fuzzy approach, *International Journal of Systems Sciences, Taylor & Francis*, Vol. 38, No.6, pp. 459-472, 2007.

Schenato, L. (2007). Optimal sensor fusion for distributed sensors subject to random delay and packet loss, *Proceedings of the 46th IEEE Conference on Decision and Control*, New Orleans, Luisiana, USA, Dec. 2007.

Schenato, L. (2008). Optimal estimation in networked control systems subject to random delay and pachet drop, *IEEE Transactions on Automatic Control*, Vol. 53, No 5., pp. 1311-1317, 2008.

Schuurman, D.C. & Capson, D.W. (2004). Robust direct visual servo using network-synchronized cameras, *IEEE Transactions on Robotics and Automation*, Vol 20, No. 2, pp. 319-334, 2004.

Sun, J.; Zhang, C. & Guo B. (2011). Distributed full-order optimal fusion filters and smoothers for discrete-time stochastic singular systems, *International Journal of Systems Science*, Taylor & Francis, Vol. 42, No. 3, pp. 507-516, 2011.

Sun, S.L. & Deng Z.L. (2005). Distributed optimal fusion steady-state Kalman filter for systems with coloured measurement noises, *International Journal of Systems Science*, Taylor & Francis, Vol. 36, No 3, pp. 113-118, 2005.

Tzafestas, S.G.; Rigatos, G.G. & Kyriannakis, E.J. (1997). Geometry and Thermal Regulation of GMA Welding via Conventional and Neural Adaptive Control, *Journal of Intelligent and Robotic Systems*, Springer, Vol. 19, pp. 153-186, 1997.

Watanabe, K. & Tzafestas, S.G. (1992). Filtering, Smoothing and Control in Discrete-Time Stochastic Distributed-Sensor Networks, *In: Stochastic Large-Scale Engineering Systems (S.G. Tzafestas and K. Watanabe Eds)*, pp. 229-252, Marcel Dekker, 1992.

Xia, Y.; Shang,J. ; Chen, J. & Liu, G.P. (2009). Networked Data Fusion with packet losses and variable delays, *IEEE Transactions on Systems, Man and Cybernetics - Part B: Cybernetics*, Vol. 39, No. 5, pp. 1107-1119, 2009.

Yoshimoto, Y.; Watanabe, K.; Iwatani, Y. & Hashimoto, K. (2010). Multi-camera visual servoing of a micro helicopter under occlusions, *in: Visual Servoing (Rong-Fong Fung Editor), InTech Publications*, 2010.

6

Cartesian Controllers for Tracking of Robot Manipulators under Parametric Uncertainties

R. García-Rodríguez and P. Zegers
College of Engineering and Applied Sciences, Universidad de los Andes
Chile

1. Introduction

The relevance of robot manipulators in different processes has created the need to design efficient controllers with low computational costs. Although several applications for this problem are defined in operational coordinates, a wide variety of controllers reported in the literature are defined in joint coordinates. Then, for a joint robot control the desired joint references are computed from desired Cartesian coordinates using inverse mappings and its derivatives up to second order. However, computing the inverse kinematics mappings is difficult due to the ill-posed nature of these mappings.
To circumvent the computation of inverse kinematics, a very old but not less important approach coined as Cartesian control can be used. Cartesian control deals with the problem of designing controllers in terms of desired Cartesian or operational coordinates. This allows saving a significant amount of time in real time applications due to the inherent simplification.

1.1 Cartesian control

Based on the seminal work of Miyazaki and Masutani [Miyazaki & Masutani (1990)] have been presented several approaches for regulating tasks, working with the assumption that the Jacobian is uncertain. Several approaches for setpoint control are presented [Yazarel & Cheah (2001)], [Chea et.al. (1999)], [Chea et.al. (2001)] [Huang et.al. (2002)], [Chea et.al. (2004)], assuming that the jacobian matrix can be parameterized linearly. Now, if we are interested that having the end effector of the robot manipulator follow a desired trajectory, Cartesian robot dynamics knowledge is required. However, Cartesian robot dynamics demands even more computational power than computing the inverse kinematics. Therefore, non-model based control strategies which guarantee convergence of the Cartesian tracking errors is desirable. In addition, Cartesian controllers should be robust and efficient with very low computational cost.
To differentiate this work from other approaches for tracking tasks [Chea et.al. (2006)], [Chea et.al. (2006)], [Moosavian & Papadopoulus (2007)], [Zhao et.al. (2007)] in this chapter it is assumed that the initial condition and desired trajectories belong to the Cartesian workspace Ω, which defines the hyperspace free of singular configurations, an standard assumption for joint robot control. However, this assumption is not evident for others Cartesian controllers [Huang et.al. (2002)], [Chea et.al. (2001)]. This assumption allows us to use a well posed inverse Jacobian for any initial condition. In addition, it is possible to prove that exponential

stability is guaranteed despite the fact the Jacobian is not exactly known and the Jacobian adaptive law is avoided.

Brief introduction to sliding mode control

The name *variable structure control* (sliding mode control) comes from the fact that the control signal is provided by one of two controllers. Which one? It depends on the sign of a scalar switching function **S** that in turn depends on the states of the system. If the outcome of this function is positive, one controller is used. If not, the other one. It is clear that the selection of the switching function is crucial for the control and that it allows to the designer to generate a rich family of behaviors.

If this switching function is designed such that the state velocity vectors in the vicinity of the switching surface (the geometric locus of the states that comply with $\mathbf{S} = 0$) points to the surface, then it is said that a *sliding surface* exists. Why this name? Because once the system intercepts such a surface it continues sliding within it until an equilibrium point is reached. Therefore, sliding mode control needs to comply with two conditions

- The control law has to provide with sufficient conditions to guarantee the existence and the reachability of the sliding surface.
- Once the state space behavior of the system is restricted to the sliding surface, the dynamics corresponds to the desired one, i.e. stability or tracking.

The properties of sliding mode control ensure that a properly controlled system will reach the sliding surface in a finite time $t_h < \infty$, beyond which the states of the system are ketp within the sliding surface and displaying the desired dynamics.

All the considerations given above rest on assuming ideal sliding modes. This implies having the capability of producing infinitely fast switchings, something of course impossible in the physical world. Therefore, the states of the system oscillate within a neighborhood of the sliding surface. This effect translates into a *chattering* signal [Utkin (1977)], [DeCarlo et.al. (1988)], [Hung et.al. (1993)] that looks like noise.

Contribution

In this chapter, free-chattering second order sliding mode control is presented in order to guarantee convergence of the tracking errors of the robot manipulator under parametric uncertainty. Specifically, a Cartesian second order sliding mode surface is proposed, which drives the sliding PID input. Therefore, the closed loop system renders a sliding mode for all time, whose solution converges to the sliding surface in finite time and a perfect tracking is guaranteed under assumption that the Jacobian is uncertain.

The main characteristics of the proposed scheme can be summarized as follows:

- The regressor is not required.
- Very fast tracking is guaranteed.
- The controller is smooth.
- An exact Jacobian is not required.
- A conservative tuning of feedback gains is required.

The chapter is organized as follows: Section II presents the dynamical model of a rigid n-link serial non-redundant robot manipulator and some useful properties. Section III presents a parameterization of the system in terms of the Cartesian coordinates. Furthermore, two

Cartesian controllers are presented assuming parametric uncertainty. In the first case, a traditional Cartesian controller based on the inverse Jacobian is presented. Now, assuming that the Jacobian is uncertain a Cartesian controller is proposed as a second case. In Section IV, numerical simulations using the proposed approaches are provided. Finally, some conclusions are presented in section V.

2. Dynamical equations of robot manipulator

The dynamical model of a non-redundant rigid serial n-link robot manipulator with all revolute joints is described as follows

$$\mathbf{H}(\mathbf{q})\ddot{\mathbf{q}} + \left(\frac{1}{2}\dot{\mathbf{H}}(\mathbf{q}) + \mathbf{S}(\mathbf{q}, \dot{\mathbf{q}})\right)\dot{\mathbf{q}} + \mathbf{g}(\mathbf{q}) = \mathbf{u} \tag{1}$$

where $\mathbf{q}, \dot{\mathbf{q}} \in \Re^n$ are the joint position and velocity vectors, $\mathbf{H}(\mathbf{q}) \in \Re^{nxn}$ denotes a symmetric positive definite inertial matrix, the second term in the left side represent the Coriolis and centripetal forces, $\mathbf{g}(\mathbf{q}) \in \Re^n$ models the gravitational forces, and $\mathbf{u} \in \Re^n$ stands for the torque input.
Some important properties of robot dynamics that will be used in this chapter are:

Property 1. *Matrix $\boldsymbol{H}(\boldsymbol{q})$ is symmetric and positive definite, and both $\boldsymbol{H}(\boldsymbol{q})$ and $\boldsymbol{H}^{-1}(\boldsymbol{q})$ are uniformly bounded as a function of $\boldsymbol{q} \in \Re^n$ [Arimoto (1996)].*

Property 2. *Matrix $S(q, \dot{q})$ is skew symmetric and hence satisface [Arimoto (1996)]:*

$$\dot{q}^T S(q, \dot{q})\dot{q} = 0 \qquad \forall q, \dot{q} \in \Re^n$$

Property 3. *The left-hand side of (1) can be parameterized linearly [Slotine & Li (1987)], that is, a linear combination in terms of suitable selected set of robot and load parameters, i.e.*

$$Y\Theta = \boldsymbol{H}(q)\ddot{q} + \left(\frac{1}{2}\dot{\boldsymbol{H}}(q) + S(q, \dot{q})\right)\dot{q} + g(q)$$

where $\boldsymbol{Y} = \boldsymbol{Y}(q, \dot{q}, \dot{q}, \ddot{q}) \in \Re^{nxp}$ is known as the regressor and $\Theta \in \Re^p$ is a vector constant parameters of the robot manipulator.

2.1 Open loop error equation

In order to obtain a useful representation of the dynamical equation of the robot manipulator for control proposes, equation (1) is represented in terms of the nominal reference $(\dot{\mathbf{q}}_r, \ddot{\mathbf{q}}_r) \in \Re^{2n}$ as follows, [Lewis (1994)]:

$$\mathbf{H}(\mathbf{q})\ddot{\mathbf{q}}_r + \left(\frac{1}{2}\dot{\mathbf{H}}(\mathbf{q}) + \mathbf{S}(\mathbf{q}, \dot{\mathbf{q}})\right)\dot{\mathbf{q}}_r + \mathbf{g}(\mathbf{q}) = \mathbf{Y}_r\Theta_r \tag{2}$$

where the regressor $\mathbf{Y}_r = \mathbf{Y}_r(\mathbf{q}, \dot{\mathbf{q}}, \dot{\mathbf{q}}_r, \ddot{\mathbf{q}}_r) \in \Re^{nxp}$ and $\Theta_r \in \Re^p$.
If we add and subtract equation (2) into (1) we obtain the open loop error equation

$$\mathbf{H}(\mathbf{q})\dot{\mathbf{S}}_r + \left(\frac{1}{2}\dot{\mathbf{H}}(\mathbf{q}) + \mathbf{S}(\mathbf{q}, \dot{\mathbf{q}})\right)\mathbf{S}_r = \mathbf{u} - \mathbf{Y}_r\Theta \tag{3}$$

where the joint error manifold $\mathbf{S}_r$ is defined as

$$\mathbf{S}_r = \dot{\mathbf{q}} - \dot{\mathbf{q}}_r \tag{4}$$

The robot dynamical equation (3) is very useful to design controllers for several control techniques which are based on errors with respect to the nominal reference [Brogliato et.al. (1991)], [Ge & Hang (1998)], [Liu et.al. (2006)].
Specially, we are interesting in to design controllers for tracking tasks without resorting on $\mathbf{H}(\mathbf{q}), \mathbf{S}(\mathbf{q}, \dot{\mathbf{q}}), \mathbf{g}(\mathbf{q})$. Also, to avoid the ill-posed inverse kinematics in the robot manipulator, a desired Cartesian coordinate system will be used rather than desired joint coordinates $(\mathbf{q}_d^T, \dot{\mathbf{q}}_d^T)^T \in \Re^{3n}$.
In the next section we design a convenient open loop error dynamics system based on Cartesian errors.

3. Cartesian controllers

3.1 Cartesian error manifolds

Let the forward kinematics be a mapping between joint space and task space (in this case Cartesian coordinates) given by [1]

$$\mathbf{X} = \mathbf{f}(\mathbf{q}) \tag{5}$$

where $\mathbf{X}$ is the end-effector position vector with respect to a fixed reference inertial frame, and $\mathbf{f}(\mathbf{q}) : \Re^n \rightarrow \Re^m$ is generally non-linear transformation. Taking the time derivative of the equation (5), it is possible to define a differential kinematics which establishes a mapping at level velocity between joint space and task space, that is

$$\dot{\mathbf{q}} = \mathbf{J}^{-1}(\mathbf{q})\dot{\mathbf{X}} \tag{6}$$

where $\mathbf{J}^{-1}(\mathbf{q})$ stands for the inverse Jacobian of $\mathbf{J}(\mathbf{q}) \in \Re^{n \times n}$.
Given that the joint error manifold $\mathbf{S}_r$ is defined at level velocities, equation (6) can be used to defined the nominal reference as

$$\dot{\mathbf{q}}_r = \mathbf{J}^{-1}(\mathbf{q})\dot{\mathbf{X}}_r \tag{7}$$

where $\dot{\mathbf{X}}_r$ represents the Cartesian nominal reference which will be designed by the user. Thus, a system parameterization in terms of Cartesian coordinates can be obtained by the equation (7). However an exact knowledge on the inverse Jacobian is required.
Substituting equations (6) and (7) in (4), the joint error manifold $\mathbf{S}_r$ becomes

$$\begin{aligned} \mathbf{S}_r &= \mathbf{J}^{-1}(\mathbf{q})(\dot{\mathbf{X}} - \dot{\mathbf{X}}_r) \\ &\triangleq \mathbf{J}^{-1}(\mathbf{q})\mathbf{S}_x \end{aligned} \tag{8}$$

where $\mathbf{S}_x$ is called as Cartesian error manifold. That is, the joint error manifold is driven by Cartesian errors through Cartesian error manifold.
Now two Cartesian controllers are presented, in order to solve the parametric uncertainty.
Case No.1
Given that the parameters of robot manipulator are changing constantly when it executes a task, or that they are sometimes unknown, then a robust adaptive Cartesian controller can be designed to compensate the uncertainty as follows [Slotine & Li (1987)]

$$\mathbf{u} = -\mathbf{K}_{d1}\mathbf{S}_{r1} + \mathbf{Y}_r\hat{\Theta} \tag{9}$$

$$\dot{\hat{\Theta}} = -\Gamma \mathbf{Y}_r^T \mathbf{S}_{r1} \tag{10}$$

[1] In this paper we consider that the robot manipulator is non-redundant, thus $m = n$.

where $\mathbf{K}_{d1} = \mathbf{K}_{d1}^T > 0 \in \Re^{n\times n}, \Gamma = \Gamma^T > 0 \in \Re^{p\times p}$.
Substituting equation (9) into (3), we obtain the following closed loop error equation

$$\mathbf{H}(\mathbf{q})\dot{\mathbf{S}}_{r1} + \left(\frac{1}{2}\dot{\mathbf{H}}(\mathbf{q}) + \mathbf{S}(\mathbf{q},\dot{\mathbf{q}})\right)\mathbf{S}_{r1} = -\mathbf{K}_{d1}\mathbf{S}_{r1} + \mathbf{Y}_r\Delta\Theta$$

where $\Delta\Theta = \hat{\Theta} - \Theta$. If the nominal reference is defined as $\dot{\mathbf{X}}_{r1} = \mathbf{x}_d - \alpha_1\Delta\mathbf{x}_1$ where α_1 is a positive-definite diagonal matrix, $\Delta\mathbf{x}_1 = \mathbf{x}_1 - \mathbf{x}_d$ and subscript d denotes desired trajectories, the following result can be obtained.

Assumption 1. *The desired Cartesian references $\mathbf{x}_d$ are assumed to be bounded and uniformly continuous, and its derivatives up to second order are bounded and uniformly continuous.*

Theorem 1. *[Asymptotic Stability] Assuming that the initial conditions and the desired trajectories are defined in a singularities-free space. The closed loop error dynamics used in equations (9), (10) guarantees that $\Delta\mathbf{x}_1$ and $\Delta\dot{\mathbf{x}}_1$ tends to zero asymptotically.*

Proof. Consider the Lyapunov function

$$V = \frac{1}{2}\mathbf{S}_{r1}^T\mathbf{H}(\mathbf{q})\mathbf{S}_{r1} + \frac{1}{2}\Delta\Theta^T\Gamma^{-1}\Delta\Theta$$

Differentiating V with respect to time, we get

$$\dot{V} = -\mathbf{S}_{r1}\mathbf{K}_{d1}\mathbf{S}_{r1} \leq 0$$

Since $\dot{V} \leq 0$, we can state that V is also bounded. Therefore, $\mathbf{S}_{r1}$ and $\Delta\Theta$ are bounded. This implies that $\hat{\Theta}$ and $\mathbf{J}^{-1}(\mathbf{q})\mathbf{S}_{x1}$ are bounded if $\mathbf{J}^{-1}(\mathbf{q})$ is well posed for all t. From the definition of $\mathbf{S}_{x1}$ we have that $\Delta\dot{\mathbf{x}}_1$, and $\Delta\mathbf{x}_1$ are also bounded. Since $\Delta\dot{\mathbf{x}}_1$, $\Delta\mathbf{x}_1$, $\Delta\Theta$, and $\mathbf{S}_{r1}$ are bounded, we have that $\dot{\mathbf{S}}_{r1}$ is bounded. This shows that $\ddot{V}$ is bounded. Hence, $\dot{V}$ is uniformly continuous. Using the Barbalat's lemma [Slotine & Li (1987)], we have that $\dot{V} \to 0$ at $t \to \infty$. This implies that $\Delta\mathbf{x}_1$ and $\Delta\dot{\mathbf{x}}_1$ tend to zero as t tends to infinity. Then, tracking errors $\Delta\mathbf{x}_1$ and $\Delta\dot{\mathbf{x}}_1$ are asymptotically stable [Lewis (1994)]. □

The properties of this controller can be numbered as:

a) On-line computing regressor and the exact knowledge of $\mathbf{J}^{-1}(\mathbf{q})$ are required.

b) Asymptotic stability is guaranteed assuming that $\mathbf{J}^{-1}(\mathbf{q})$ is well posed for all time. Therefore, the stability domain is very small because $\mathbf{q}(t)$ may exhibit a transient response such that $\mathbf{J}(\mathbf{q})$ losses rank.

In order to avoid the dependence on the inverse Jacobian, in the next case it is assumed that the Jacobian is uncertain. At the same time, the drawbacks presented in the Case No.1 are solved.

Case No.2 Considering that the Jacobian is uncertain, i.e. the Jacobian is not exactly known, the nominal reference proposed in equation (7) is now defined as

$$\hat{\dot{\mathbf{q}}}_r = \hat{\mathbf{J}}^{-1}(\mathbf{q})\dot{\mathbf{X}}_{r2} \quad (11)$$

where $\hat{\mathbf{J}}^{-1}(\mathbf{q})$ stands as an estimates of $\mathbf{J}^{-1}(\mathbf{q})$ such that $rank(\hat{\mathbf{J}}^{-1}(\mathbf{q})) = n$ for all t and for all $\mathbf{q} \in \Omega$ where $\Omega = \{\mathbf{q}|rank(\mathbf{J}(\mathbf{q})) = n\}$. Therefore, a new joint error manifold arises coined as uncertain Cartesian error manifold is defined as follows

$$\begin{aligned} \hat{\mathbf{S}}_{r2} &= \dot{\mathbf{q}} - \hat{\dot{\mathbf{q}}}_r \\ &= \mathbf{J}^{-1}(\mathbf{q})\dot{\mathbf{X}} - \hat{\mathbf{J}}^{-1}(\mathbf{q})\dot{\mathbf{X}}_{r2} \end{aligned} \tag{12}$$

In order to guarantee that the Cartesian trajectories remain on the manifold $\mathbf{S}_x$ although the Jacobian is uncertain, a second order sliding mode is proposed by means of tailoring $\dot{\mathbf{X}}_{r2}$. That is, a switching surface over the Cartesian manifold $\mathbf{S}_x$ should be invariant to changes in $\mathbf{J}^{-1}(\mathbf{q})$. Hence, high feedback gains can to ensure the boundedness of all closed loop signals and the exponential convergence is guaranteed despite Jacobian uncertainty.
Let the new nominal reference $\dot{\mathbf{X}}_{r2}$ be defined as

$$\begin{aligned} \dot{\mathbf{X}}_{r2} &= \dot{\mathbf{x}}_d - \alpha_2 \Delta \mathbf{x}_2 + \mathbf{S}_d - \gamma_p \sigma \\ \dot{\sigma} &= sgn(\mathbf{S}_e) \end{aligned} \tag{13}$$

where α_2 is a positive-definite diagonal matrix, $\Delta \mathbf{x}_2 = \mathbf{x}_2 - \mathbf{x}_d$, $\mathbf{x}_d$ is a desired Cartesian trajectory, γ_p is positive-definite diagonal matrix and function $sgn(*)$ stands for the signum function of $(*)$ and

$$\begin{aligned} \mathbf{S}_e &= \mathbf{S}_x - \mathbf{S}_d \\ \mathbf{S}_x &= \Delta \dot{\mathbf{x}}_2 + \alpha_2 \Delta \mathbf{x}_2 \\ \mathbf{S}_d &= \mathbf{S}_x(t_0) exp^{-\kappa(t-t_0)}, \quad \kappa > 0 \end{aligned}$$

Now, substituting equation (13) in (12) we have that

$$\hat{\mathbf{S}}_{r2} = \mathbf{J}^{-1}(\mathbf{q})\dot{\mathbf{X}} - \hat{\mathbf{J}}^{-1}(\mathbf{q})(\dot{\mathbf{x}}_d - \alpha_2 \Delta \mathbf{x}_2 + \mathbf{S}_d - \gamma_p \int_{t_0}^{t} sgn(\mathbf{S}_e(\tau))d\tau) \tag{14}$$

Uncertain Open Loop Equation
Using equation (11), the uncertain parameterization of $Y_r \Theta_r$ becomes

$$\mathbf{H}(\mathbf{q})\hat{\ddot{\mathbf{q}}}_r + \left(\frac{1}{2}\dot{\mathbf{H}}(\mathbf{q}) + \mathbf{S}(\mathbf{q}, \dot{\mathbf{q}})\right)\hat{\dot{\mathbf{q}}}_r + \mathbf{g}(\mathbf{q}) = \hat{Y}_r \Theta_r \tag{15}$$

If we add and subtract equation (15) to (1), the uncertain open loop error equation is defined as

$$\mathbf{H}(\mathbf{q})\dot{\hat{\mathbf{S}}}_{r2} + \left(\frac{1}{2}\dot{\mathbf{H}}(\mathbf{q}) + \mathbf{S}(\mathbf{q}, \dot{\mathbf{q}})\right)\hat{\mathbf{S}}_{r2} = \mathbf{u} - \hat{Y}_r \Theta_r \tag{16}$$

Theorem 2: [Local Stability] Assuming that the initial conditions and the desired trajectories are within a space free of singularities. Consider the uncertain open loop error equation (16) in closed loop with the controller given by

$$\mathbf{u} = -\mathbf{K}_{d2}\hat{\mathbf{S}}_{r2} \tag{17}$$

with $\mathbf{K}_{d2}$ an $n \times n$ diagonal symmetric positive-definite matrix. Then, for large enough gain $\mathbf{K}_{d2}$ and small enough error in initial conditions, local exponential tracking is assured provided that $\gamma_p \geq \|\dot{\mathbf{J}}(\mathbf{q})\hat{\mathbf{S}}_{r2} + \mathbf{J}(\mathbf{q})\dot{\hat{\mathbf{S}}}_{r2} + \dot{\mathbf{J}}(\mathbf{q})\Delta J\dot{\mathbf{X}}_{r2} + \mathbf{J}(\mathbf{q})\Delta\dot{J}\dot{\mathbf{X}}_{r2} + \mathbf{J}(\mathbf{q})\Delta J\ddot{\mathbf{X}}_{r2}\|$.

Proof. Substituting equation (17) into (16) we obtain the closed-loop dynamics given as

$$\mathbf{H}(\mathbf{q})\dot{\hat{\mathbf{S}}}_{r2} = -\left(\frac{1}{2}\dot{\mathbf{H}}(\mathbf{q}) + \mathbf{S}(\mathbf{q}, \dot{\mathbf{q}})\right)\hat{\mathbf{S}}_{r2} - \mathbf{K}_{d2}\hat{\mathbf{S}}_{r2} - \hat{Y}_r\Theta \tag{18}$$

The proof is organized in three parts as follows.
Part 1: Boundedness of Closed-loop Trajectories. Consider the following Lyapunov function

$$V = \frac{1}{2}\hat{\mathbf{S}}_{r2}^T\mathbf{H}(\mathbf{q})\hat{\mathbf{S}}_{r2} \tag{19}$$

whose total derivative of (19) along its solution (18) leads to

$$\dot{V} = -\hat{\mathbf{S}}_{r2}^T\mathbf{K}_{d2}\hat{\mathbf{S}}_{r2} - \hat{\mathbf{S}}_{r2}^T\hat{\mathbf{Y}}_r\Theta \tag{20}$$

Similarly to [Parra & Hirzinger (2000)], we have that $\hat{\mathbf{Y}}_r\Theta \leq \eta(t)$ with η a functional that bounds $\hat{\mathbf{Y}}_r$. Then, equation (20) becomes

$$\dot{V} \leq -\hat{\mathbf{S}}_{r2}^T\mathbf{K}_{d2}\hat{\mathbf{S}}_{r2} - \|\hat{\mathbf{S}}_{r2}\|\eta(t) \tag{21}$$

For initial errors that belong to a neighborhood ϵ_1 with radius $r > 0$ near the equilibrium $\hat{\mathbf{S}}_{r2} = 0$, we have that thanks to Lyapunov arguments, there is a large enough feedback gain $\mathbf{K}_{d2}$ such that $\hat{\mathbf{S}}_{r2}$ converges into a set-bounded ϵ_1. Thus, the boundedness of tracking errors can be concluded, namely

$$\hat{\mathbf{S}}_{r2} \to \epsilon_1 \qquad \text{as} \qquad t \to \infty \tag{22}$$

then

$$\hat{\mathbf{S}}_{r2} \in \mathcal{L}_\infty \Rightarrow \|\hat{\mathbf{S}}_{r2}\| < \epsilon_1 \tag{23}$$

where $\epsilon_1 > 0$ is a upper bounded.
Since desired trajectories are C^2 and feedback gains are bounded, we have that $(\dot{\mathbf{q}}_r, \ddot{\mathbf{q}}_r) \in \mathcal{L}_\infty$, which implies that $\dot{\mathbf{X}}_{r2} \in \mathcal{L}_\infty$ if $\hat{\mathbf{J}}^{-1}(\mathbf{q}) \in \mathcal{L}_\infty$. Then, the right hand side of (18) is bounded given that the Coriolis matrix and gravitational vector are also bounded. Since $\mathbf{H}(\mathbf{q})$ and $\mathbf{H}^{-1}(\mathbf{q})$ are uniformly bounded, it is seen from (18) that $\dot{\hat{\mathbf{S}}}_{r2} \in \mathcal{L}_\infty$. Hence there exists a bounded scalar $\epsilon_2 > 0$ such that

$$\|\dot{\hat{\mathbf{S}}}_{r2}\| < \epsilon_2 \tag{24}$$

So far, we conclude the boundedness of all closed-loop error signals.
Part 2. Sliding Mode. If we add and subtract $\mathbf{J}^{-1}(\mathbf{q})\dot{X}_r$ to (12), we obtain

$$\begin{aligned}\hat{\mathbf{S}}_{r2} &= \mathbf{J}^{-1}(\mathbf{q})\dot{\mathbf{X}} - \hat{\mathbf{J}}^{-1}(\mathbf{q})\dot{\mathbf{X}}_{r2} \pm \mathbf{J}^{-1}(\mathbf{q})\dot{\mathbf{X}}_{r2} \\ &= \mathbf{J}^{-1}(\mathbf{q})(\dot{\mathbf{X}} - \dot{\mathbf{X}}_{r2}) + (\mathbf{J}^{-1}(\mathbf{q}) - \hat{\mathbf{J}}^{-1}(\mathbf{q}))\dot{\mathbf{X}}_{r2} \\ &= \mathbf{J}^{-1}(\mathbf{q})\mathbf{S}_x - \Delta J\dot{\mathbf{X}}_{r2}\end{aligned} \tag{25}$$

which implies that $\Delta J = \mathbf{J}^{-1}(\mathbf{q}) - \hat{\mathbf{J}}^{-1}(\mathbf{q})$ is also bounded. Now, we will show that a sliding mode at $\mathbf{S}_e = 0$ arises for all time as follows.
If we premultiply (25) by $\mathbf{J}(\mathbf{q})$ and rearrange the terms, we obtain

$$\mathbf{S}_x = \mathbf{J}(\mathbf{q})\hat{\mathbf{S}}_{r2} + \mathbf{J}(\mathbf{q})\Delta J\dot{\mathbf{X}}_{r2} \tag{26}$$

Since $\mathbf{S}_x = \mathbf{S}_e + \gamma_p \int_{t_0}^{t} sgn(\mathbf{S}_e(\zeta))d\zeta$, we have that

$$\mathbf{S}_e = -\gamma_p \int_{t_0}^{t} sgn(\mathbf{S}_e(\zeta))d\zeta + \mathbf{J}(\mathbf{q})(\hat{\mathbf{S}}_{r2} + \Delta J \dot{\mathbf{X}}_{r2}) \tag{27}$$

Deriving (27), and then premultiplying by $\mathbf{S}_e^T$, we obtain

$$\begin{aligned}
\mathbf{S}_e^T \dot{\mathbf{S}}_e &= -\gamma_p |\mathbf{S}_e| + \mathbf{S}_e^T \frac{d}{dt}\left(\mathbf{J}(\mathbf{q})\hat{\mathbf{S}}_{r2} + \mathbf{J}(\mathbf{q})\Delta J \dot{\mathbf{X}}_{r2}\right) \\
&\leq -\gamma_p |\mathbf{S}_e| + \zeta |\mathbf{S}_e| \\
&\leq -(\gamma_p - \zeta)|\mathbf{S}_e| \\
&= -\mu |\mathbf{S}_e|
\end{aligned} \tag{28}$$

where $\mu = \gamma_p - \zeta$ and $\zeta = \dot{J}(q)\hat{\mathbf{S}}_{r2} + \mathbf{J}(\mathbf{q})\dot{\hat{\mathbf{S}}}_{r2} + \dot{J}(q)\Delta J \dot{\mathbf{X}}_{r2} + \mathbf{J}(\mathbf{q})\Delta \dot{J} \dot{\mathbf{X}}_{r2} + \mathbf{J}(\mathbf{q})\Delta J \ddot{\mathbf{X}}_{r2}$. Therefore, we obtain the sliding mode condition if

$$\gamma_p > \zeta \tag{29}$$

in such a way that $\mu > 0$ guarantees the existence of a sliding mode at $\mathbf{S}_e = 0$ at time $t_e \leq \frac{|\mathbf{S}_e(t_0)|}{\mu}$. However, notice that for any initial condition $\mathbf{S}_e(t_0) = 0$, and hence $t \equiv 0$ implies that a sliding mode in $\mathbf{S}_e = 0$ is enforced for all time without reaching phase.

Part 3: Exponential Convergence. Sliding mode at $\mathbf{S}_e = 0$ implies that $\mathbf{S}_x = \mathbf{S}_d$, thus

$$\Delta \dot{\mathbf{x}}_2 = -\alpha_2 \Delta \mathbf{x}_2 + \mathbf{S}_x(t_0)e^{-k_p t} \tag{30}$$

which decays exponentially fast toward $[\Delta \mathbf{x}_2, \Delta \dot{\mathbf{x}}_2] \to (0,0)$, that is

$$\mathbf{x}_2 \to \mathbf{x}_d \quad \text{and} \quad \dot{\mathbf{x}}_2 \to \dot{\mathbf{x}}_d \tag{31}$$

it is locally exponential. □

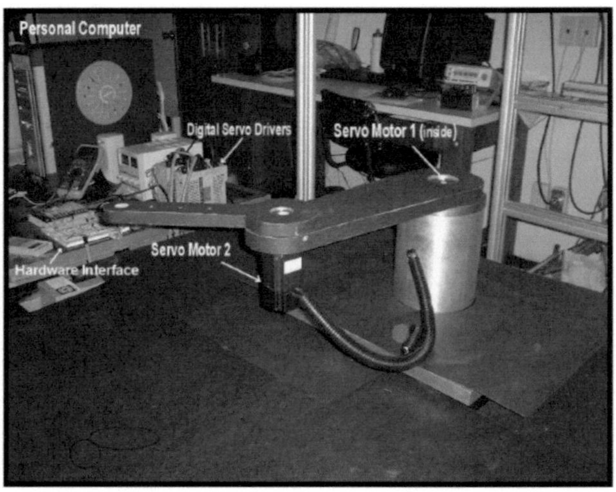

Fig. 1. Planar Manipulator of 2-DOF.

The properties of this controller can be numbered as

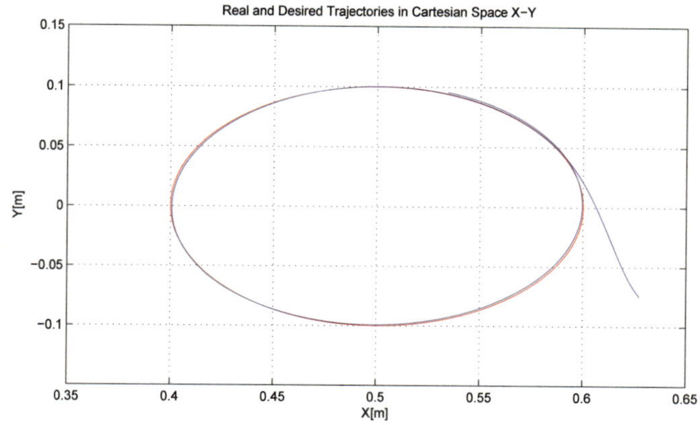

(a) Theorem 1: Plane Phase

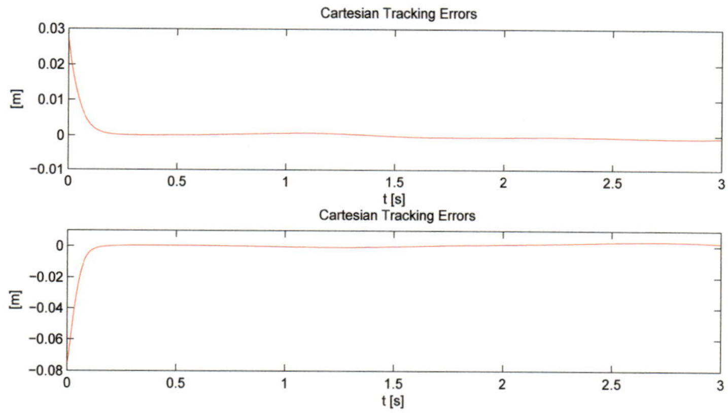

(b) Theorem 1: Cartesian Tracking Errors

Fig. 2. Cartesian Tracking of the Robot Manipulator using Theorem 1.

a) The sliding mode discontinuity associated to $\hat{\mathbf{S}}_{r2} = 0$ is relegated to the first order time derivative of $\dot{\hat{\mathbf{S}}}_{r2}$. Then, sliding mode condition in the closed loop system is induced by the $sgn(\mathbf{S}_e)$ and an exponential convergence of the tracking error is established. Therefore, the closed loop is robust due to the invariance achieved by the sliding mode, robustness against unmodeled dynamics, and parametric uncertainty. A difference of this approach from others [Lee & Choi (2004)], [Barambones & Etxebarria (2002)], [Jager (1996)], [Stepanenko et.al. (1998)], is that the closed loop dynamics does not exhibit chattering. Finally, notice that the discontinuous function $sgn(\mathbf{S}_e)$ is only used in the stability analysis.

c) The control synthesis does not depend on any knowledge of the robot dynamics: it is model free. In addition, a smooth control input is guaranteed.

d) Taking $\gamma_p = 0$ in equation (13), it is obtained the joint error manifold $\mathbf{S}_{r1}$ defined in the Case No.1, which is commonly used in several approaches. However under this sliding surface it is not possible to prove convergence in finite time as well as reaching the sliding condition. Then, a dynamic change of coordinates is proposed, where for a large enough feedback gain K_d in the control law, the passivity between η_1 and $\hat{\mathbf{S}}_{r2}$ is preserved with $\eta_1 = \dot{\hat{\mathbf{S}}}_{r2}$ [Parra & Hirzinger (2000)]. In addition, for large enough γ_p the dissipativity is established between $\mathbf{S}_e$ and η_2 with $\eta_2 = \dot{\mathbf{S}}_e$.

e) In order to differentiate from other approaches where the parametric uncertainty in the Jacobian matrix is expressed as a linear combination of a selected set of kinematic parameters [Chea et.al. (1999)], [Chea et.al. (2001)], [Huang et.al. (2002)], [Chea et.al. (2004)], [Chea et.al. (2006)], [Chea et.al. (2006)], in this chapter the Jacobian uncertainty is parameterized in terms of a regressor times as parameter vector. To get the parametric uncertainty, this vector is multiplied by a factor with respect to the nominal value.

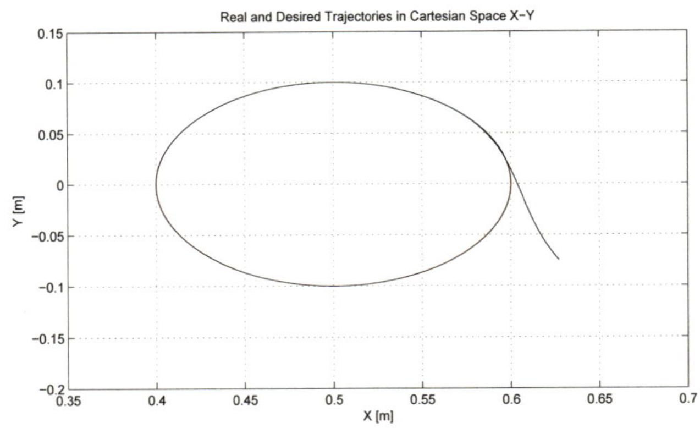

(a) Theorem 2: Plane Phase

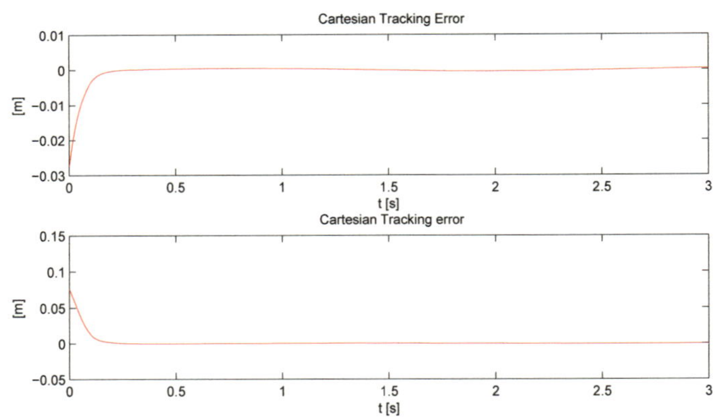

(b) Theorem 2: Cartesian Tracking Errors

Fig. 3. Cartesian Tracking of the Robot Manipulator using Theorem 2.

4. Simulation results

In this section we present simulation results carried out on 2 degree of freedom (DOF) planar robot arm, Fig. 1. The experiments were developed on Matlab 6.5 and each experiment has an average running of 3 [s]. Parameters of the robot manipulator used in these simulations are shown in Table 1.

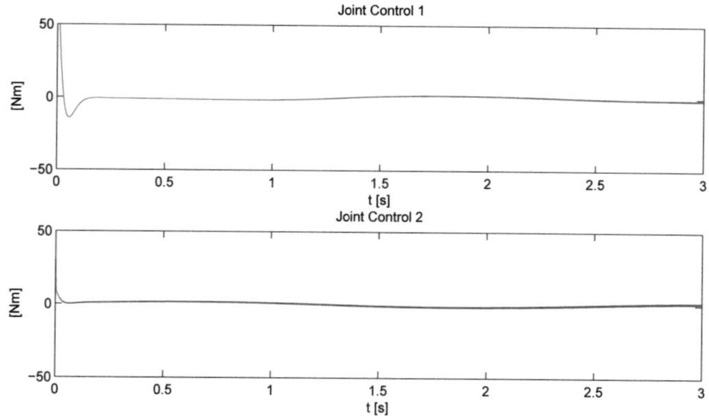

(a) Theorem 1: Control Inputs

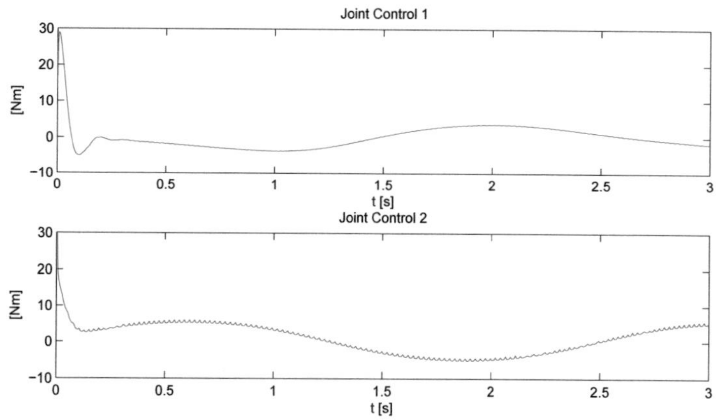

(b) Theorem 2: Control Inputs

Fig. 4. Control Inputs applied to Each Joint.

Parameters	m_1	m_2	l_1	l_2
Value	8 Kg	5 Kg	0.5 m	0.35 m
Parameters	l_{c1}	l_{c2}	I_1	I_2
Value	0.19 m	0.12 m	0.02 Kgm^2	0.16 Kgm^2

Table 1. Robot Manipulator Parameters.

The objective of these experiments is to given a desired trajectory, the end effector must follow it in a finite time. The desired task is defined as a circle of radius 0.1 [m] whose center located at X=(0.55,0) [m] in the Cartesian workspace. The initial condition is defined as $[q_1(0) = -0.5, q_2(0) = 0.9]^T$ [rad]. which is used for all experiments. In addition, we consider zero initial velocity and 95% of parametric uncertainty.
The performance of the robot manipulator using equations (9) and (10) defined in theorem 1 are presented in Fig. 2. In this case, the end-effector tracks the desired Cartesian trajectory once the Cartesian error manifold is reached, Fig. 2(a). In addition, as it is showed in Fig. 2(b),

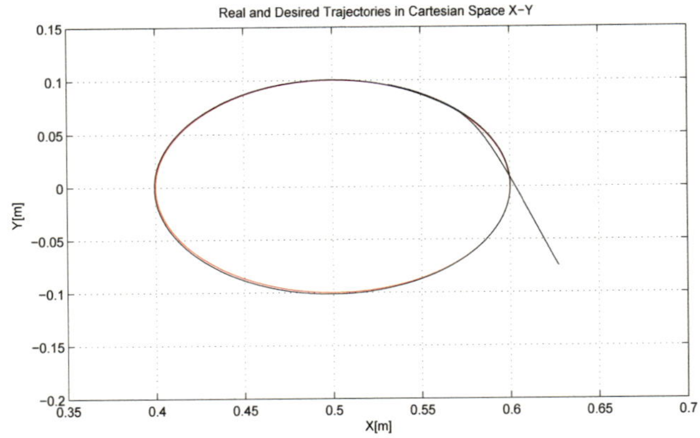

(a) TBG: Plane Phase

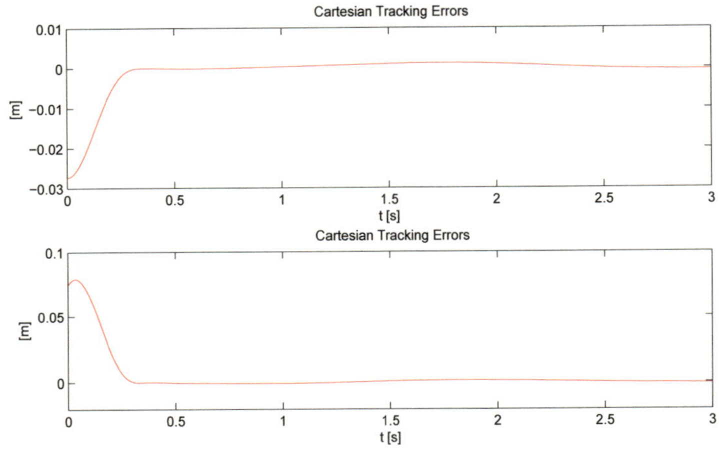

(b) TBG: Cartesian Tracking Errors

Fig. 5. Cartesian Tracking of the Robot Manipulator using TBG

the Cartesian tracking errors converge asymptotically to zero in few seconds. However, for practical applications it is necessary to know exactly the regressor and the inverse Jacobian. Now, assuming that the Jacobian is uncertain, there is no knowledge of the regressor, and there cannot be any overparametrization, then a Cartesian tracking of the robot manipulator using control law defined in equation (17) is presented in Fig 3(a). As it is expected, after a very short time, approximately 2 [s], the end effector of the robot manipulator follows the desired trajectory, Fig. 3(a) and Fig. 3(b). This is possible because in the proposed scheme all the time it is induced a sliding mode. Thus, it is more faster and robust.
On the other hand, in Fig. 4 are shown the applied input torques for each joint of the robot manipulator for the cases 1 and 2. It can be see that control inputs using the controller defined in equation (17) are more smooth and chattering free than controller defined in equation (9).
Given that in several applications, such as manipulation tasks or bipedal robots, it is not enough the convergence of the errors when t tends to infinity. Finite time convergence faster that exponential convergence has been proposed [Parra & Hirzinger (2000)]. To speed up the

response, a time base generator (TBG) that shapes a feedback gain α_2 is used. That is, it is necessary to modify the feedback gain α_2 defined in equation (13) by

$$\alpha_2(t) = \alpha_0 \frac{\dot{\xi}}{1 - \xi + \delta} \tag{32}$$

where $\alpha_0 = 1 + \epsilon$, for small positive scalar ϵ such that α_0 is close to 1 and $0 < \delta \ll 1$. The time base generator $\xi = \xi(t) \in C^2$ must be provided by the user so as to get ξ to go smoothly from 0 to 1 in finite time $t = t_b$, and $\dot{\xi} = \dot{\xi}(t)$ is a bell shaped derivative of ξ such that $\dot{\xi}(t_0) = \dot{\xi}(t_b) \equiv 0$ [Parra & Hirzinger (2000)]. Accordingly, given that the convergence speed of the tracking errors is increased by the TBG, a finite time convergence of the tracking errors is guaranteed.

In the Fig. 5 are shown simulation results using a finite time convergence at $t_b = 0.4$ [s]. As it is expected, the end effector follows exactly the desired trajectory at $t_b \geq 0.4$ [s], as shown in Fig. 5(a). At the same time, Cartesian tracking errors converge to zero in the desired time, Fig. 5(b).

The feedback gains used in these experiments are given in Table 2 where the subscript ji represents the joint of the robot manipulator with $i = 1, 2$.

K_{dj1}	K_{dj2}	α_{j1}	α_{j2}	γ_{pj1}	γ_{pj2}	k_p	Γ	t_b	Case
60	60	25	25	-	-	-	0.01	-	1
50	20	30	30	0.01	0.01	20	-	-	2
60	60	2.2	2.2	0.01	0.01	20	-	0.4s	TBG

Table 2. Feedback Gains

5. Conclusion

In this chapter, two Cartesian controllers under parametric uncertainties are presented. In particular, an alternative solution to the Cartesian tracking control of the robot manipulator assuming parametric uncertainties is presented. To do this, second order sliding surface is used in order to avoid the high frequency commutation. In addition, closed loop renders a sliding mode for all time to ensure convergence without any knowledge of robot dynamics and Jacobian uncertainty. Simulation results allow to visualize the predicted stability properties on a simple but representative task.

6. References

Arimoto, S. (1996). *Control Theory of Non-linear Mechanical Systems*, Oxford University Press.

Barambones, O. & Etxebarria, V. (2002). *Robust Neural Network for Robotic Manipulators*, Automtica, Vol. 38, pp. 235-242.

Brogliato, B., Landau, I-D. & Lozano-Leal, R. (1991).*Adaptive Motion Control of Robot Manipulators: A unified Approach based on Passivity*, International Journal of Robust and Nonlinear Control, Vol. 1, No. 3, pp. 187-202.

Cheah, C.C., Kawamura, S. & Arimoto, S. (1999). *Feedback Control for Robotics Manipulator with Uncertain Jacobian Matrix*, Journal of Robotics Systems, Vol. 16, No. 2, pp. 120-134.

Cheah, C.C., Kawamura, S., Arimoto, S. & Lee, K. (2001). *A Tuning for Task-Space Feedback Control of Robot with Uncertain Jacobian Matrix*, IEEE Trans. on Automat. Control, Vol. 46, No. 8, pp. 1313-1318.

Cheah, C.C., Hirano, M., Kawamura, S. & Arimoto, S. (2004). *Approximate Jacobian Control With Task-Space Damping for Robot Manipulators*, IEEE Trans. on Autom. Control, Vol. 19, No. 5, pp. 752-757.

Cheah, C.C., Liu, C. & Slotine, J. J. E. (2006). *Adaptive Jacobian Tracking Control of Robots with Uncertainties in Kinematics, Dynamics and Actuator Models*, IEEE Trans. on Autom. Control, Vol. 51, No. 6, pp. 1024-1029.

Cheah, C.C. & Slotine, J. J. E. (2006). *Adaptive Tracking Control for Robots with unknown Kinematics and Dynamics Properties*, International Journal of Robotics Research, Vol. 25, No. 3, pp. 283-296.

DeCarlo, R. A., Zak, S. H. & Matthews, G. P. (1988). *Variable Structure Control of Nonlinear Multivariable Systems: A Tutorial*, Proceedings of the IEEE, Vol. 76, No. 3, pp. 212-232.

Ge, S.S. & Hang, C.C. (1998). *Structural Network Modeling and Control of Rigid Body Robots*, IEEE Trans. on Robotics and Automation, Vol. 14, No. 5, pp. 823-827.

Huang, C.Q., Wang, X.G. & Wang, Z.G. (2002). *A Class of Transpose Jacobian-based NPID Regulators for Robot Manipulators with Uncertain Kinematics*, Journal of Robotic Systems, Vol. 19, No. 11, pp. 527-539.

Hung, J. Y., Gao, W. & Hung, J. C. (1993).*Variable Structure Control: A Survey*, IEEE Transactions on Industrial Electronics, Vol. 40, No. 1, pp. 2-22.

Jager, B., (1996). *Adaptive Robot Control with Second Order Sliding Component*, 13th Triennial World Congress, San Francisco, USA, pp. 271-276.

Lee, M.-J. & Choi, Y.-K. (2004). *An Adaptive Neurocontroller Using RBFN for Robot Manipulators*, IEEE Trans. on Industrial Electronics, Vol. 51 No. 3, pp. 711-717.

Lewis, F.L. & Abdallaah, C.T. (1994). *Control of Robot Manipulators*, New York: Macmillan.

Liu, C., Cheah, C.C. & Slotine, J.J.E. (2006). *Adaptive Jacobian Tracking Control of Rigid-link Electrically Driven Robots Based on Visual Task-Space Information*, Automatica, Vol. 42, No. 9, pp. 1491-1501.

Miyazaki, F. & Masutani, Y. (1990). *Robustness of Sensory Feedback Control Based on Imperfect Jacobian*, Robotic Research: The Fifth International Symposium, pp. 201-208.

Moosavian, S. A. & Papadopoulus, E. (2007). *Modified Transpose Jacobian Control of Robotics Systems*, Automatica, Vol. 43, pp. 1226-1233.

Parra-Vega, V. & Hirzinger, G. (2000). *Finite-time Tracking for Robot Manipulators with Singularity-Free Continuous Control: A Passivity-based Approach*, Proc. IEEE Conference on Decision and Control , 5, pp. 5085-5090.

Slotine, J.J.E. & Li, W. (1987). *On the Adaptive Control of Robot Manipulator*, Int. Journal of Robotics Research, Vol. 6, No. 3, pp. 49-59.

Stepanenko, Y., Cao, Y. & Su, A.C. (1998). *Variable Structure Control of Robotic Manipulator with PID Sliding Surfaces*, Int. Journal of Robust and Nonlinear Control, Vol. 8, pp. 79-90.

Utkin, V.J. (1977). *Variable Structure Systems with Sliding Modes*, IEEE Trans. on Autom. Control, Vol. 22, No. 1, pp. 212-222.

Yazarel, H.& Cheah, C.C. (2001). *Task-Space Adaptive Control of Robots Manipulators with Uncertain Gravity Regressor Matrix and kinematics*, IEEE Trans. on Autom. Control, Vol. 47, No. 9, pp. 1580-1585.

Zhao, Y., Cheah, C.C. & Slotine, J.J.E. (2007). *Adaptive Vision and Force Tracking Control of Constrained Robots with Structural Uncertainties*, Proc. IEEE International Conference on Robotics and Automation, pp.2349-2354.

7

Robotic Grasping of Unknown Objects

Mario Richtsfeld and Markus Vincze
Institute of Automation and Control Vienna University of Technology, Vienna Austria

1. Introduction

This work describes the development of a novel vision-based grasping system for unknown objects based on laser range and stereo data. The work presented here is based on *2.5D* point clouds, where every object is scanned from the same view point of the laser range and camera position. We tested our grasping point detection algorithm separately on laser range and single stereo images with the goal to show that both procedures have their own advantages and that combining the point clouds reaches better results than the single modalities. The presented algorithm automatically filters, smoothes and segments a *2.5D* point cloud, calculates grasping points, and finds the hand pose to grasp the desired object.

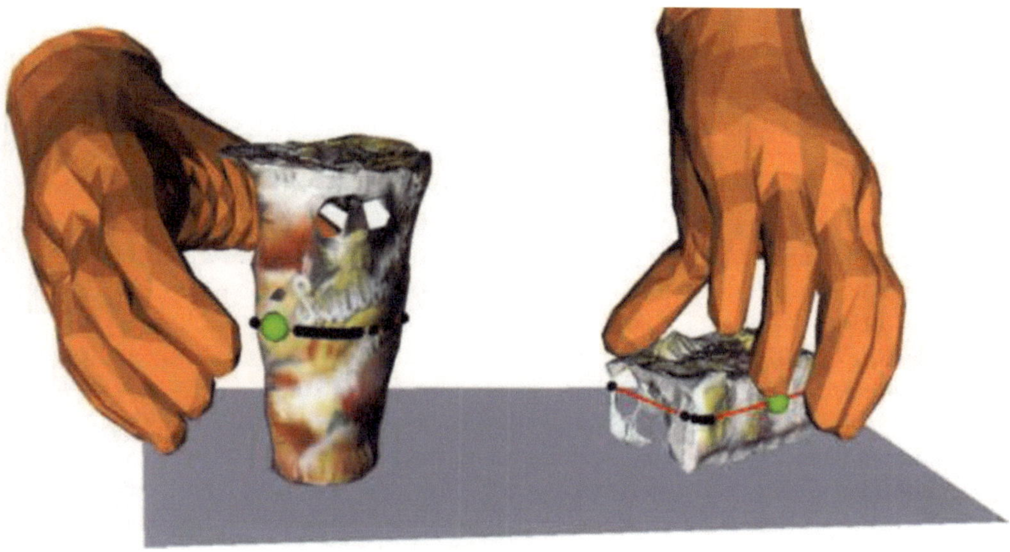

Fig. 1. Final detection of the grasping points and hand poses. The green points display the computed grasping points with hand poses.

The outline of the paper is as follows: The next Section introduces our robotic system and its components. Section 3 describes the object segmentation and details the analysis of the objects to calculate practical grasping points. Section 4 details the calculation of optimal hand poses to grasp and manipulate the desired object without any collision. Section 5 shows the achieved results and Section 6 finally concludes this work.

1.2 Problem statement and contribution

The goal of the work is to show a new and robust way to calculate grasping points in the recorded point cloud from single views of a scene. This poses the challenge that only the front side of objects is seen and, hence, the second grasp point on the backside of the object needs to be assumed based on symmetry assumptions. Furthermore we need to cope with the typical sensor data noise, outliers, shadows and missing data points, which can be caused by specular or reflective surfaces. Finally, a goal is to link the grasp points to a collision free hand pose using a full *3D* model of the gripper used to grasp the object. The main idea is depicted in Fig. 1[1].

The main problem is that *2.5D* point clouds do not represent complete *3D* object information. Furthermore stereo data includes measurement noise and outliers depending on the texture of the scanned objects. Laser range data includes also noise and outliers where the typical problem is missing sensor data because of absorption. The laser exhibits high accuracy while the stereo data includes more object information due to the better field of view. The contribution is to show in detail the individual problems of using both sensor modalities and we then show that better results can be obtained by merging the data provided by the two sensors.

1.3 Related work

In the last few decades, the problem of grasping novel objects in a fully automatic way has gained increasing importance in machine vision and robotics. There exist several approaches on grasping quasi planar objects (Sanz et al., 1999; Richtsfeld & Zillich, 2008). (Recatalá et al., 2008) developed a framework for the development of robotic applications based on a grasp-driven multi-resolution visual analysis of the objects and the final execution of the calculated grasps. (Li et al., 2007) presented a 2D data-driven approach based on a hand model of the gripper to realize grasps. The algorithm finds the best hand poses by matching the query object by comparing object features to hand pose features. The output of this system is a set of candidate grasps that will then be sorted and pruned based on effectiveness for the intended task. The algorithm uses a database of captured human grasps to find the best grasp by matching hand shape to object shape. Our algorithm does not include a shape matching method, because this is a very time intensive step. The *3D* model of the hand is only used to find a collision free grasp.

(Ekvall & Kragic, 2007) analyzed the problem of automatic grasp generation and planning for robotic hands where shape primitives are used in synergy to provide a basis for a grasp evaluation process when the exact pose of the object is not available. The presented algorithm calculates the approach vector based on the sensory input and in addition tactile information that finally results in a stable grasp. The only two integrated tactile sensors of the used robotic gripper in this work are too limited for additional information to calculate grasping points. These sensors are only used if a potential stick-slip effect occurs.

(Miller et al., 2004) developed an interactive grasp simulator "GraspIt!" for different hands and hand configurations and objects. The method evaluates the grasps formed by these hands. This grasp planning system "GraspIt!" is used by (Xue et al., 2008). They use the grasp planning system for an initial grasp by combining hand pre-shapes and automatically generated approach directions. The approach is based on a fixed relative position and

[1] All images are best viewed in colour.

orientation between the robotic hand and the object, all the contact points between the fingers and the object are efficiently found. A search process tries to improve the grasp quality by moving the fingers to its neighboured joint positions and uses the corresponding contact points to the joint position to evaluate the grasp quality and the local maximum grasp quality is located. (Borst et al., 2003) show that it is not necessary in every case to generate optimal grasp positions, however they reduce the number of candidate grasps by randomly generating hand configuration dependent on the object surface. Their approach works well if the goal is to find a fairly good grasp as fast as possible and suitable. (Goldfeder et al., 2007) presented a grasp planner which considers the full range of parameters of a real hand and an arbitrary object including physical and material properties as well as environmental obstacles and forces.
(Saxena et al., 2008) developed a learning algorithm that predicts the grasp position of an object directly as a function of its image. Their algorithm focuses on the task of identifying grasping points that are trained with labelled synthetic images of a different number of objects. In our work we do not use a supervised learning approach. We find grasping points according to predefined rules.
(Bone et al., 2008) presented a combination of online silhouette and structured-light *3D* object modelling with online grasp planning and execution with parallel-jaw grippers. Their algorithm analyzes the solid model, generates a robust force closure grasp and outputs the required gripper pose for grasping the object. We additionally analyze the calculated grasping points with a *3D* model of the hand and our algorithm obtains the required gripper pose to grasp the object. Another *3D* model based work is presented by (El-Khoury et al., 2007). They consider the complete *3D* model of one object, which will be segmented into single parts. After the segmentation step each single part is fitted with a simple geometric model. A learning step is finally needed in order to find the object component that humans choose to grasp. Our segmentation step identifies different objects in the same table scene. (Huebner et al., 2008) have applied a method to envelop given *3D* data points into primitive box shapes by a fit-and-split algorithm with an efficient minimum volume bounding box. These box shapes give efficient clues for planning grasps on arbitrary objects.
(Stansfield, 1991) presented a system for grasping *3D* objects with unknown geometry using a Salisbury robotic hand, where every object was placed on a motorized and rotated table under a laser scanner to generate a set of *3D* points. These were combined to form a *3D* model. In our case we do not operate on a motorized and rotated table, which is unrealistic for real world use, the goal is to grasp objects when seen only from one side.
Summarizing to the best knowledge of the authors in contrast to the state of the art reviewed above our algorithm works with *2.5D* point clouds from a single-view. We do not operate on a motorized and rotated table, which is unrealistic for real world use. The presented algorithm calculates for arbitrary objects grasping points given stereo and / or laser data from one view. The poses of the objects are calculated with a 3D model of the gripper and the algorithm checks and avoids potential collision with all surrounding objects.

2. Experimental setup

We use a fixed position and orientation between the AMTEC[2] robot arm with seven degrees of freedom and the scanning unit. Our approach is based on scanning the objects on the

---

[2] http://www.amtec-robotics.com

table by a rotating laser range scanner and a fixed stereo system and the execution of the subsequent path planning and grasping motion. The robot arm is equipped with a hand prosthesis from the company Otto Bock[3], which we are using as gripper, see Fig. 2. The hand prosthesis has integrated tactile force sensors, which detect a potential sliding of an object and enable the readjustment of the pressure of the fingers. This hand prosthesis has three active fingers the thumb, the index finger and the middle finger; the last two fingers are for cosmetic reasons. Mechanically it is a calliper gripper, which can only realize a tip grasp and for the computation of the optimal grasp only 2 grasping points are necessary. The middle between the fingertip of the thumb and the index finger is defined as tool centre point (TCP). We use a commercial path planning tool from AMROSE[4] to bring the robot to the grasp location.

The laser range scanner records a table scene with a pan/tilt-unit and the stereo camera grabs two images at -4° and +4°. (Scharstein & Szeliski, 2002) published a detailed description of the used dense stereo algorithm. To realize a dense stereo calibration to the laser range coordinate system as exactly as possible the laser range scanner was used to scan the same chessboard that is used for the camera calibration. At the obtained point cloud a marker was set as reference point to indicate the camera coordinate system. We get good results by the calibration most of the time. In some cases at low texture of the scanned objects and due to the simplified calibration method the point clouds from the laser scanner and the dense stereo did not correctly overlap, see Fig. 3. To correct this error of the calibration we used the iterative closest point (ICP) method (Besl & McKay, 1992) where the reference is the laser point cloud, see Fig. 4. The result is a transformation between laser and stereo data that can now be superimposed for further processing.

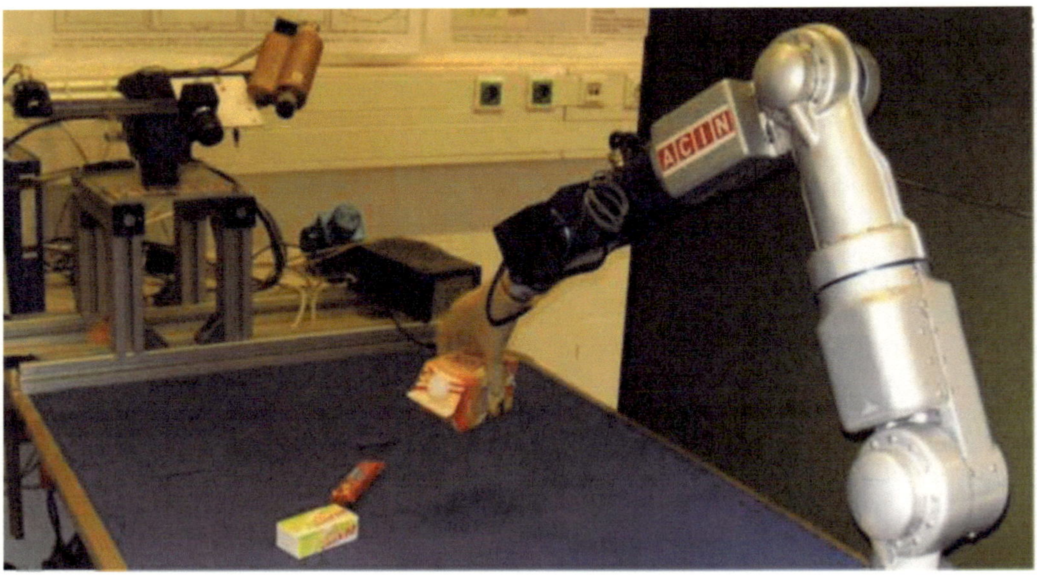

Fig. 2. Overview of the system components and their interrelations.

[3] http://www.ottobock.de

[4] http://www.amrose.dk

Fig. 3. Partially overlapping point clouds from the laser range scanner (white points) and dense stereo (coloured points). A clear shift between the two point clouds shows up.

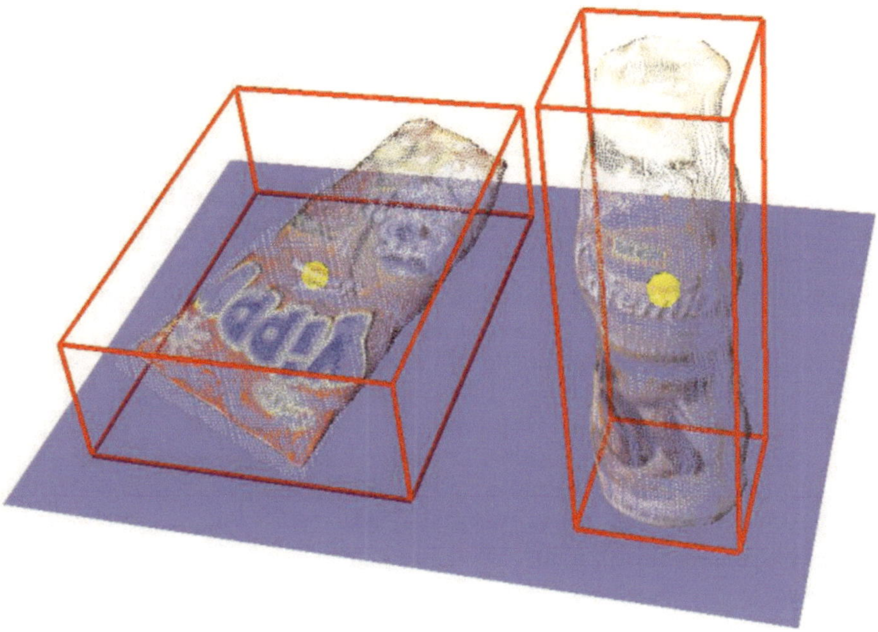

Fig. 4. Correction of the calibration error applying the iterative closest point (ICP) algorithm. The red lines represent the bounding boxes of the objects and the yellow points show the approximation to the centre of the objects.

3. Grasp point detection

The algorithm to find grasp points on the objects consists of four main steps as depicted in Fig. 5:

- Raw Data Pre-processing: The raw data points are pre-processed with a geometrical filter and a smoothing filter to reduce noise and outliers.
- Range Image Segmentation: This step identifies different objects based on a 3D DeLaunay triangulation, see Section 4.
- Grasp Point Detection: Calculation of practical grasping points based on the centre of the objects, see Section 4.
- Calculation of the Optimal Hand Pose: Considering all objects and the table surface as obstacles, find an optimal gripper pose, which maximizes distances to obstacles, see Section 5.

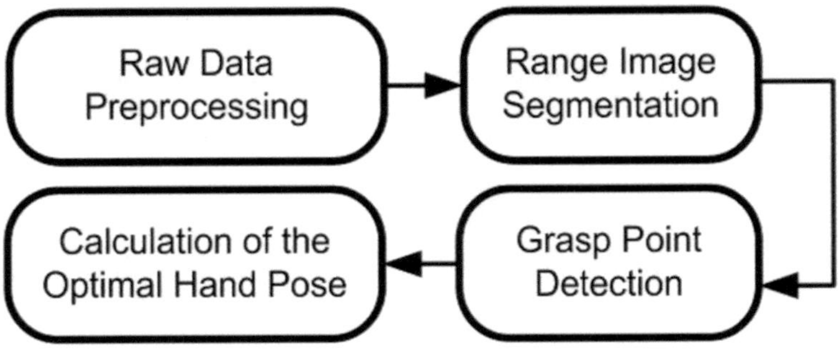

Fig. 5. Overview of our grasp point and gripper pose detection algorithm.

4. Segmentation and grasp point detection

There is no additional segmentation step for the table surface needed, because the red light laser of the laser range scanner is not able to detect the surface of the blue table and the images of the stereo camera were segmented and filtered directly. However, plane segmentation is a well known technique for ground floor or table surface detection and could be used alternatively, e.g., (Stiene et al., 2006).

The segmentation of the unknown objects will be achieved with a *3D* mesh generation, based on the triangles, calculated by a DeLaunay triangulation [10]. After mesh generation we look at connected triangles and separate objects.

In most grasping literature it is assumed that good locations for grasp contacts are actually at points of high concavity. That's absolutely correct for human grasping, but for grasping with a robotic gripper with limited DOF and only two tactile sensors a stick slip effect occurs and makes these grasp points rather unreliable.

Consequently to realize a possible, stable grasp the calculated grasping points should be near the centre of mass of the objects. Thus, the algorithm calculates the centre c of the objects based on the bounding box, Fig. 4, because with a *2.5D* point cloud no accurate centre of mass can be calculated. Then the algorithm finds the top surfaces of the objects with a RANSAC based plane fit (Fischler & Bolles, 1981). We intersect the point clouds with horizontal planes through the centre of the objects. If the object does not exhibit a top plane, the normal vector of the table plane will be used. From these n cutting plane points p_i we calculate the (planar) convex hull V, using Equ. 1 and illustrated in Fig. 6.

$$V = ConvexHull\left(\bigcup_{i=0}^{n-1} p_i\right) \tag{1}$$

With the distances between two neighbouring hull points to the centre of the object c we calculate the altitude d of the triangle, see Equ. 2. $\vec{v}$ is the direction vector to the neighbouring hull point and $\vec{w}$ is the direction vector to c. Then the algorithm finds the shortest normal distance d_{min} of the convex hull lines, illustrated in Fig. 6 as red lines, to the centre of the object c, where the first grasping point is located.

$$d = \frac{\|\vec{v} \times \vec{w}\|}{\|\vec{v}\|} \tag{2}$$

In *2.5D* point clouds it is only possible to view the objects from one side, however we assume a symmetry of the objects. Hence, the second grasping point is determined by a reflection of the first grasping point using the centre of the object. We check a potential lateral and above grasp of the object on the detected grasping points with a simplified *3D* model of the hand. If no accurate grasping points could be calculated with the convex hull of the cutting plane points p_i the centre of the object is displaced in *1mm* steps towards the top surface of the object (red point) with the normal vector of the top surface until a positive grasp could be detected. Another method is to calculate the depth of indentation of the gripper model and to calculate the new grasping points based on this information.
Fig. 6 gives two examples and shows that the laser range images often have missing data, which can be caused by specular or reflective surfaces. Stereo clearly correct this disadvantage, see Fig. 7.

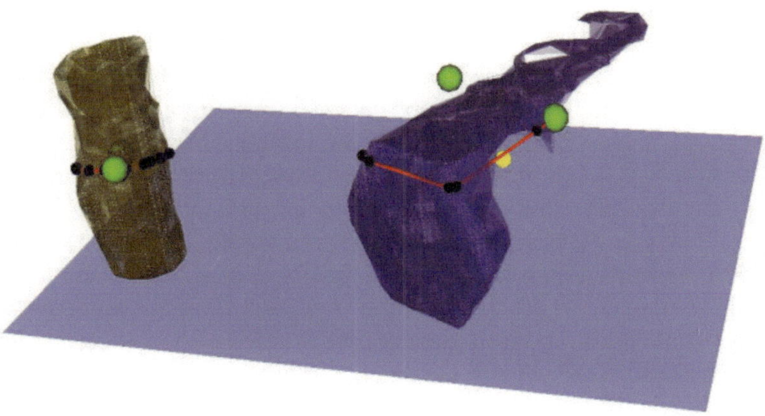

Fig. 6. Calculated grasping points (green) based on laser range data. The yellow points show the centre of the objects. If, through the check of the 3D gripper no accurate grasping points could be calculated with the convex hull (black points connected with red lines) the centre of the objects is displaced towards the top surface of the objects (red points).

Fig. 7 illustrates that with stereo data alone there are definitely better results possible then with laser range data alone given that object appearance has texture. This is also reflected in Tab. 2. Fig. 8 shows that there is a smaller difference between the stereo data alone (see Fig. 7) and the overlapped laser range and stereo data, which Tab. 2 confirms.

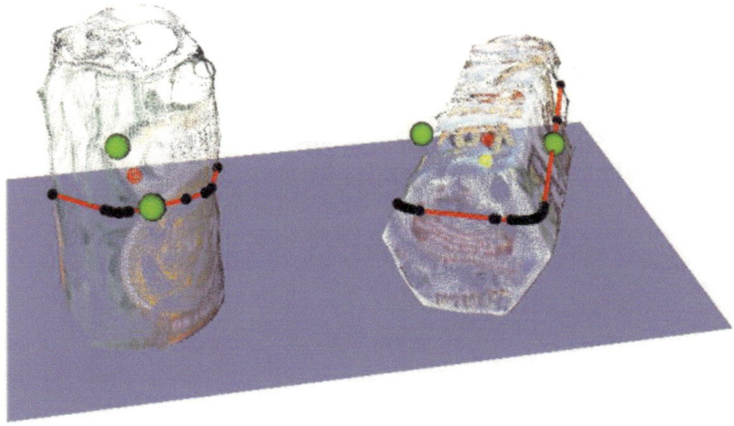

Fig. 7. Calculated grasping points (green) based on stereo data. The yellow points show the centre of mass of the objects. If, through the check of the 3D gripper no accurate grasping points could be calculated with the convex hull (black points connected with red lines) the centre of the objects is displaced towards the top surface of the objects (red points).

5. Grasp pose

To successfully grasp an object it is not always sufficient to find locally the best grasping points, the algorithm should also decide at which angle it is possible to grasp the selected object. For this step we rotate the *3D* model of the hand prosthesis around the rotation axis, which is defined by the grasping points. The rotation axis of the hand is defined by the fingertip of the thumb and the index finger of the hand, as illustrated in Fig. 9. The algorithm checks for a collision of the hand with the table, the object that shall be grasped and all obstacles around it. This will be repeated in *5°* steps to a full rotation by *180°*. The algorithm notes with each step whether a collision occurs. Then the largest rotation range where no collision occurs is found. We find the optimal gripper position and orientation by an averaging of the maximum and minimum largest rotation range. From this the algorithm calculates the optimal gripper pose to grasp the desired object.

The grasping pose depends on the orientation of the object itself, surrounding objects and the calculated grasping points. We set the grasping pose as a target pose to the path planner, illustrated in Fig. 9 and Fig. 1. The path planner tries to reach the target object on his part. Fig. 10 shows the advantage to calculate the gripper pose. The left Figure shows a collision free path to grasp the object. The right Figure illustrates a collision of the gripper with the table.

6. Experiments and results

To evaluate our method, we choose ten different objects, which are shown in Fig. 11. The blue lines represent the optimal positions for grasping points. Optimal grasping points are

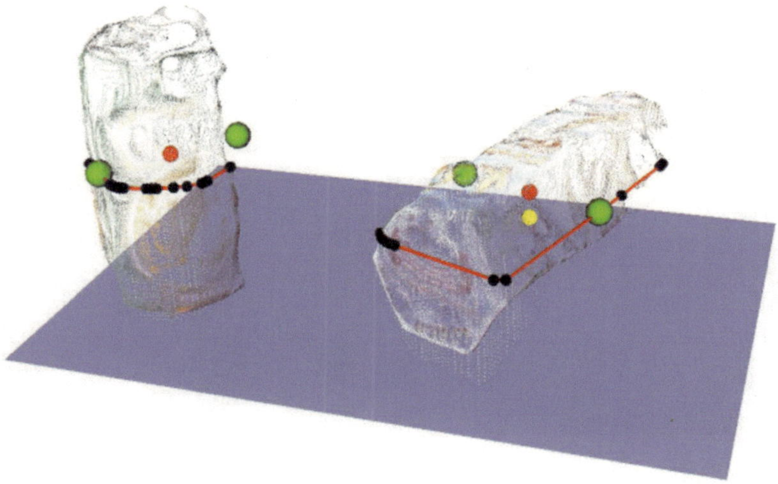

Fig. 8. Calculated grasping points (green) based on the combined laser range and stereo data.

required to be placed on parallel surfaces near the centre of the objects. To challenge the developed algorithm we included one object (Manner, object no. 6), which is too big for the used gripper. The algorithm should calculate realistic grasping points for object no. 6 in the pre-defined range, however it should recognize that the object is too large and the maximum opening angle of the hand is too small.

Fig. 9. The rotation axis of the hand is defined by the fingertip of the thumb and the index finger of the gripper. This rotation axis must be aligned with the axis defined by the grasping points. The calculated grasping pose of the gripper is by object no. 8 (Cappy) -32.5° and object no. 9 (Smoothie) -55°.

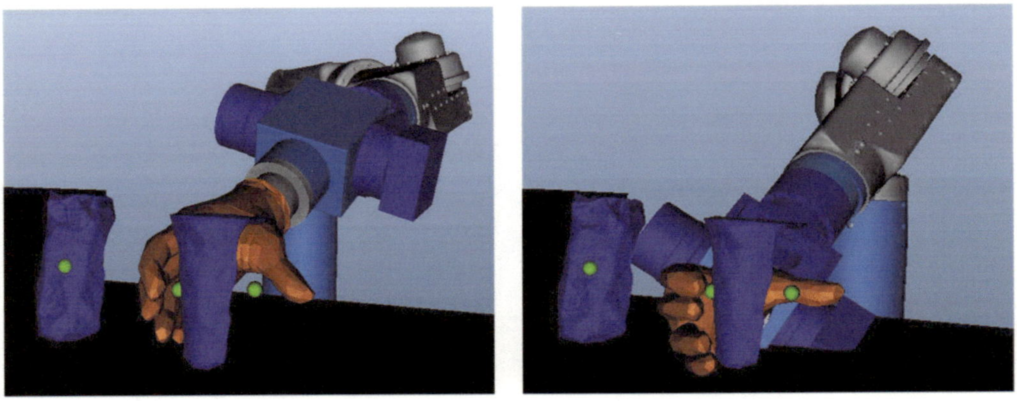

Fig. 10. The left Figure shows the calculated grasping points with an angle adjustment, where as the right Figure shows a collision with the table and a higher collision risk with the left object no. 8 (Cappy) as the left Figure with an angle adjustment of -55°.

In our work, we demonstrate that our grasping point detection algorithm and the validation with a *3D* model of the used gripper for unknown objects shows very good results, see Tab. 2. All tests were performed on a PC with *3.2GHz* Pentium dual-core processor and the average run time is about 463.78sec and the calculation of the optimal gripper pose needs about *380.63sec*, see Tab. 1 for the illustrated point cloud, see Fig. 9. The algorithm is implemented in C++ using the Visualization ToolKit (VTK)[5].

Calculation Steps	**Time [sec]**
Filter (Stereo Data)	*14sec*
Smooth (Stereo Data)	*4sec*
Mesh Generation	*58.81sec*
Segmentation	*2sec*
Grasp Point Detection	*4.34sec*
Grasp Angle	*380.63sec*
Overall	*463.78sec*

Table 1. Duration of calculation steps.

Tab. 2 illustrates the evaluation results of the detected grasping points by comparing them to the optimal grasping points as defined in Fig. 11. For the evaluation every object was scanned four times in combination with another object in each case. This analysis shows that a successful grasp based on stereo data with *82.5%* is considerably larger than with laser range data with *62.5%*. The combination of both data sets with *90%* definitely wins.
We tested every object with four different combined point clouds, as illustrated in Tab. 3. In no case the robot was able to grasp the test object no. 6 (Manner), because the size of the object is too big for the used gripper. This fact could be determined before with the computation of the grasping points, however the calculated grasping points are in the

[5] Open source software, http://public.kitware.com/vtk.

defined range of object no. 6. Thus the negative test object, as described in Section 4 was successfully tested.

No.	Objects	Laser [%]	Stereo [%]	Both [%]
1	Dextro	*100%*	*100%*	*100%*
2	Yippi	*0%*	*0%*	*25%*
3	Snickers	*100%*	*100%*	*100%*
4	Cafemio	*50%*	*100%*	*100%*
5	Exotic	*100%*	*100%*	*100%*
6	Manner	*75%*	*100%*	*100%*
7	Maroni	*75%*	*50%*	*75%*
8	Cappy	*25%*	*75%*	*100%*
9	Smoothie	*100%*	*100%*	*100%*
10	Koala	*0%*	*100%*	*100%*
Overall		*62.5%*	*82.5%*	*90%*

Table 2. Grasping rate of different objects on pre-defined grasping points.

Tab. 2 shows that the detected grasping points of object no. 2 (Yippi) are not ideal to grasp it. The *75%* in Tab. 3 were possible due to the rubber coating of the hand and the compliance of the object. For a grasp to be counted as successful, the robot had to grasp the object, lift it up and hold it without dropping it. On average, the robot picked up the unknown objects *85%* of the time, including the defined test object (Manner, object no. 6), which is too big for the used gripper. If object no. 6 is not regarded success rate is *95%*.

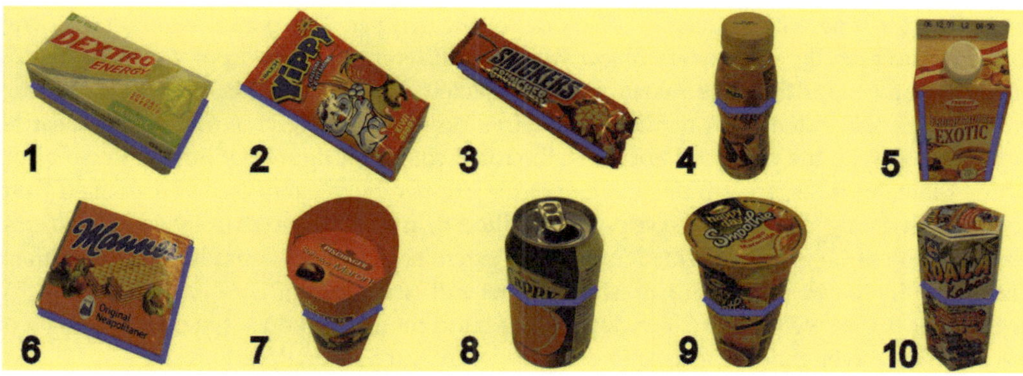

Fig. 11. Ten test objects. The blue lines represent the optimal positions for grasping points near the centre of the objects, depending on the used gripper. From left top: 1. Dextro, 2. Yippy, 3. Snickers, 4. Cafemio, 5. Exotic, 6. Manner, 7. Maroni, 8. Cappy, 9. Smoothie, 10. Koala.

For objects such as Dextro, Snickers, Cafemio, etc., the algorithm performed perfectly with a *100%* grasp success rate in our experiments. However, grasping objects such as Yippi or Maroni is more complicated, because of the strongly curved surfaces, and so its a greater challenge to successfully detect possible grasping points, so that even a small error in the grasping point identification, resulting in a failed grasp attempt.

No.	Objects	Grasp-Rate [%]
1	Dextro	*100%*
2	Yippi	*75%*
3	Snickers	*100%*
4	Cafemio	*100%*
5	Exotic	*100%*
6	Manner	*0%*
7	Maroni	*75%*
8	Cappy	*100%*
9	Smoothie	*100%*
10	Koala	*100%*
Overall		*85%*

Table 3. Successfully grasps with the robot based on point clouds from combined laser range and stereo data.

7. Conclusion and future work

In this work we present a framework to successfully calculate grasping points of unknown objects in *2.5D* point clouds from combined laser range and stereo data. The presented method shows high reliability. We calculate the grasping points based on the convex hull points, which are obtained from a plane parallel to the top surface plane in the height of the visible centre of the objects. This grasping point detection approach can be applied to a reasonable set of objects and for the use of stereo data textured objects should be used. The idea to use a *3D* model of the gripper to calculate the optimal gripper pose can be applied to every gripper type with a suitable *3D* model of the gripper. The presented algorithm was tested to successfully grasp every object with four different combined point clouds. In *85%* of all cases, the algorithm was able to grasp completely unknown objects.

Future work will extend this method to obtain more grasp points in a more generic sense. For example, with the proposed approach the robot could not figure out how to grasp a cup whose diameter is larger than the opening of the gripper. Such a cup could be grasped from above by grasping the rim of the cup. This method is limited to successfully convex objects. For this type of objects the algorithm must be extended, but with more heuristic functions the possibility to calculate wrong grasping points will be enhanced.

In the near future we plan to use a deformable hand model to reduce the opening angle of the hand, so we can model the closing of a gripper in the collision detection step.

8. References

Besl, P.J., McKay, H.D. (1992). A method for registration of 3-D shapes, IEEE Transactions on Pattern Analysis and Machine Intelligence, Vol. 14, No. 2, pp. 239-256.

Borst, C., Fischer, M., Hirzinger, G. (2003). Grasping the dice by dicing the grasp. IEEE/RSJ International Conference on Robotics and Systems, pp. 3692-3697.

Bone, G.M., Lambert, A., Edwards, M. (2008). Automated modelling and robotic grasping of unknown three-dimensional objects. IEEE International Conference on Robotics and Automation, pp. 292-298.

Castiello, U. (2005). The neuroscience of grasping. Nature Reviews Neuroscience, Vol. 6, No. 9, pp. 726-736.

Ekvall, S., Kragic, D. (2007). Learning and Evaluation of the Approach Vector for Automatic Grasp Generation and Planning. IEEE International Conference on Robotics and Automation, pp. 4715-4720.

El-Khoury, S., Sahbani A., Perdereau, V. (2007). Learning the Natural Grasping Component of an Unknown Object. IEEE/RSJ International Conference on Intelligent Robots and Systems, pp. 2957-2962.

Fischler, M.A., Bolles, R.C. (1981). Random Sample Consensus: A Paradigm for Model Fitting with Applications to Image Analysis and Automated Cartography, Communications of the ACM, Vol. 24, No. 6, pp. 381-395.

Goldfeder, C., Allen, P., Lackner, C., and Pelossof, R. (2007). Grasp Planning via Decomposition Trees. IEEE International Conference on Robotics and Automation, pp. 4679-4684.

Huebner, K., Ruthotto, S., and Kragic, D. (2008). Minimum Volume Bounding Box Decomposition for Shape Approximation in Robot Grasping. IEEE International Conference on Robotics and Automation, pp. 1628-1633.

Li, Y., Fu, J.L., Pollard, N.S. (2007). Data-Driven Grasp Synthesis Using Shape Matching and Task-Based Pruning. IEEE Transactions on Visualization and Computer Graphics, Vol. 13, No. 4, pp. 732-747.

Miller, A.T., Knoop, S. (2003). Automatic grasp planning using shape primitives. IEEE International Conference on Robotics and Automation, pp. 1824-1829.

Recatalá, G., Chinellato, E., Del Pobil, Á.P., Mezouar, Y., Martinet, P. (2008). Biologically-inspired *3D* grasp synthesis based on visual exploration. Autonomous Robots, Vol. 25, No. 1-2, pp. 59-70.

Richtsfeld, M., Zillich, M. (2008). Grasping Unknown Objects Based on *2.5D* Range Data. IEEE Conference on Automation Science and Engineering, pp. 691-696.

Sanz, P.J., Iñesta, J.M., Del Pobil, Á.P. (1999). Planar Grasping Characterization Based on Curvature-Symmetry Fusion. Applied Intelligence, Vol. 10, No. 1, pp. 25-36.

Saxena, A., Driemeyer, J., Ng, A.Y. (2008). Robotic Grasping of Novel Objects using Vision. International Journal of Robotics Research, Vol. 27, No. 2, pp. 157-173.

Scharstein, D., Szeliski, R. (2002). A Taxonomy and Evaluation of Dense Two-Frame Stereo Correspondence Algorithms, International Journal of Computer Vision, Vol. 47, No. 1-3, pp. 7-42.

Stansfield, S.A. (1991). Robotic grasping of unknown objects: A knowledge-based approach. International Journal of Robotics Research, Vol. 10, No. 4, pp. 314-326.

Stiene, S., Lingemann, K., Nüchter, A., Hertzberg, J. (2006). Contour-based Object Detection in Range Images, Third International Symposium on *3D* Data Processing, Visualization, and Transmission, pp. 168-175.

Xue, Z., Zoellner, J.M., Dillmann, R. (2008). Automatic Optimal Grasp Planning Based On Found Contact Points. IEEE/ASME International Conference on Advanced Intelligent Mechatronics, pp. 1053-1058.

8

Object-Handling Tasks Based on Active Tactile and Slippage Sensations

Masahiro Ohka[1], Hanafiah Bin Yussof[2] and Sukarnur Che Abdullah[1,2]
[1]*Nagoya University*
[2]*Universiti Teknologi MARA*
Japan
Malaysia

1. Introduction

Many tactile sensors have been developed to enhance robotic manufacturing tasks, such as assembly, disassembly, inspection and materials handling as described in several survey papers (Harmon, 1982; Nicholls & Lee 1989; Ohka, 2009a). In the last decade, progress has been made in tactile sensors by focusing on limited uses. Many examples of practical tactile sensors have gradually appeared. Using a Micro Electro Mechanical System, *MEMS-based tactile sensors* have been developed to incorporate pressure-sensing elements and piezoelectric ceramic actuators into a silicon tip for detecting not only pressure distribution but also the hardness of a target object (Hasegawa et al., 2004). Using PolyVinylidene DiFluoride, a *PVDF film-based tactile sensor* has been developed to measure the hardness of tumors based on comparison between the obtained sensor output and the input oscillation (Tanaka et al., 2003). A *wireless tactile sensor* using two-dimensional signal transmission has been developed to be stretched over a large sensing area (Chigusa et al., 2007). An *advanced conductive rubber-type tactile sensor* has been developed to be mounted on robotic fingers (Shimojo et al., 2004). Furthermore, *image based tactile sensors* have been developed using a charge-coupled device (CCD) and complementary metal oxide semiconductor (CMOS) cameras and image data processing, which are mature techniques (Ohka, 1995, 2004, 2005a, 2005b, Kamiyama et al., 2005).
In particular, the three-axis tactile sensor that is categorized as an *image based tactile sensor* has attracted the greatest anticipation for improving manipulation because a robot must detect the distribution not only of normal force but also of slippage force applied to its finger surfaces (Ohka, 1995, 2004, 2005a, 2005b, 2008). In addition to our three-axis tactile sensors, there are several designs of multi-axis force cells based on such physical phenomena as magnetic effects (Hackwood et al., 1986), variations in electrical capacity (Novak, 1989; Hakozaki & Shinoda 2002), PVDF film (Yamada & Cutkosky, 1994), and a photointerrupter (Borovac et al., 1996).
Our three-axis tactile sensor is based on the principle of an optical waveguide-type tactile sensor (Mott et al., 1984; Tanie et al., 1986; Nicholls et al., 1990; Kaneko et al., 1992; Maekawa et al., 1992), which is composed of an acrylic hemispherical dome, a light source, an array of rubber sensing elements, and a CCD camera (Ohka, 1995, 2004a, 2005a, 2005b, 2008). The sensing element of the silicone rubber comprises one columnar feeler and eight conical

feelers. The contact areas of the conical feelers, which maintain contact with the acrylic dome, detect the three-axis force applied to the tip of the sensing element. Normal and shearing forces are then calculated from integration and centroid displacement of the grayscale value derived from the conical feeler's contacts.

The tactile sensor is evaluated with a series of experiments using an x-z stage, a rotational stage, and a force gauge. Although we discovered that the relationship between the integrated grayscale value and normal force depends on the sensor's latitude on the hemispherical surface, it is easy to modify the sensitivity based on the latitude to make the centroid displacement of the grayscale value proportional to the shearing force.

To demonstrate the effectiveness of the three-axis tactile sensor, we designed a hand system composed of articulated robotic fingers sensorized with the three-axis tactile sensor (Ohka, 2009b, 2009c). Not only tri-axial force distribution directly obtained from the tactile sensor but also the time derivative of the shearing force distribution are used for the hand control algorithm: the time derivative of tangential force is defined as slippage; if slippage arises, grasping force is enhanced to prevent fatal slippage between the finger and an object. In the verification test, the robotic hand twists on a bottle cap completely.

In the following chapters, after the optical three-axis tactile sensor is explained, the robotic hand sensorized with the tactile sensors is described. The above cap-twisting task is discussed to show the effectiveness of tri-axial tactile data for robotic control.

2. Optical three-axis tactile sensor

2.1 Sensing principle

2.1.1 Structure of optical tactile sensors

Figure 1 shows a schematic view of the present tactile processing system to explain the sensing principle. The present tactile sensor is composed of a CCD camera, an acrylic dome, a light source, and a computer. The light emitted from the light source is directed into the optical waveguide dome. Contact phenomena are observed as image data, acquired by the CCD camera, and transmitted to the computer to calculate the three-axis force distribution.

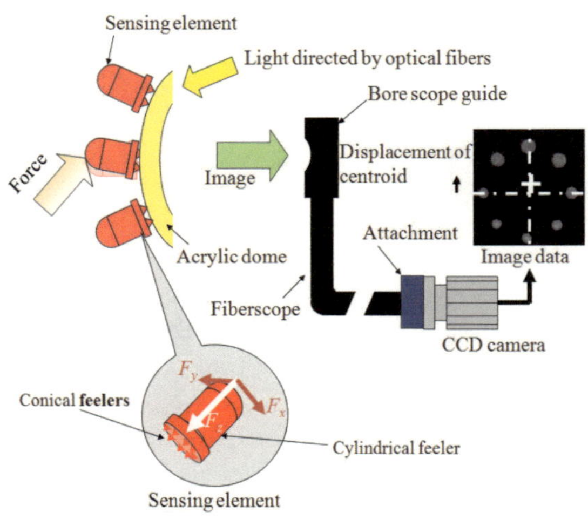

Fig. 1. Principle of the three-axis tactile sensor system

In this chapter, we adopt a sensing element comprised of a columnar feeler and eight conical feelers, as shown in Fig. 2, because the element showed wide measuring range and good linearity in a previous paper (Ohka, 2004b). Since a single sensing element of the present tactile sensor should carry a heavier load compared to a flat-type tactile sensor, the height of the columnar feeler of the flat-type tactile sensor is reduced from 5 to 3 mm. The sensing elements are made of silicone rubber (KE119, Shinetsu) and are designed to maintain contact with the conical feelers and the acrylic board and to make the columnar feelers touch an object. Each columnar feeler features a flange to fit into a counter bore portion in the fixing dome to protect the columnar feeler from horizontal displacement caused by shearing force.

2.1.2 Expressions for sensing element located on vertex

Dome brightness is inhomogeneous because the edge of the dome is illuminated and light converges on its parietal region. Since the optical axis coincides with the center line of the vertex, the apparent image of the contact area changes based on the sensing element's latitude. Although we must consider the above problems to formulate a series of equations for the three components of force, the most basic sensing element located on the vertex will be considered first.

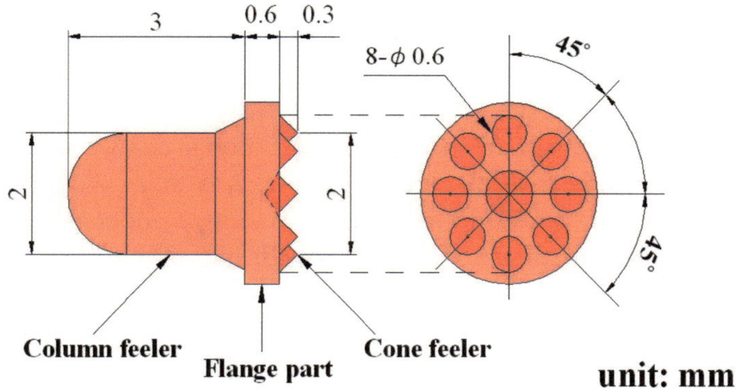

Fig. 2. Sensing element

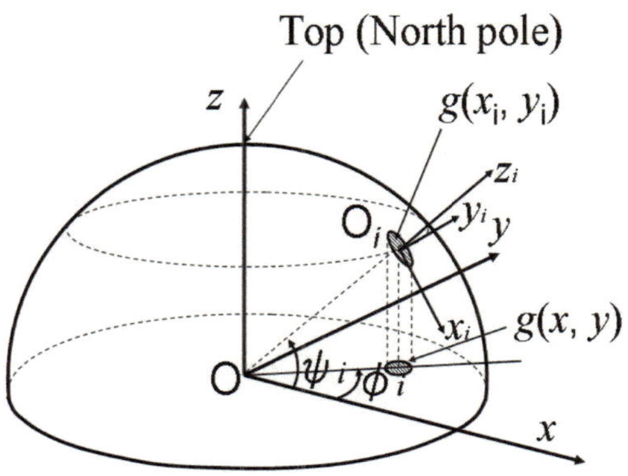

Fig. 3. Relationship between spherical and Cartesian coordinates

Coordinate O-xyz is adopted, as shown in Fig. 3. Based on previous studies (Ohka, 2005), since grayscale value $g(x,y)$ obtained from the image data is proportional to pressure $p(x,y)$ caused by contact between the acrylic dome and the conical feeler, normal force is calculated from integrated grayscale value G. Additionally, shearing force is proportional to the centroid displacement of the grayscale value. Therefore, the F_x, F_y, and F_z values are calculated using integrated grayscale value G and the horizontal displacement of the centroid of grayscale distribution $\mathbf{u} = u_x\mathbf{i} + u_y\mathbf{j}$ as follows:

$$F_x = f_x(u_x), \tag{1}$$

$$F_y = f_y(u_y), \tag{2}$$

$$F_z = -g(G), \tag{3}$$

where $\mathbf{i}$ and $\mathbf{j}$ are the orthogonal base vectors of the x- and y-axes of a Cartesian coordinate, respectively, and $f_x(x)$, $f_y(x)$, and $g(x)$ are approximate curves estimated in calibration experiments.

2.1.3 Expressions for sensing elements other than those located on vertex

For sensing elements other than those located on the vertex, each local coordinate O_i-$x_iy_iz_i$ is attached to the root of the element, where suffix i denotes element number. Each z_i-axis is aligned with the center line of the element and its direction is along the normal direction of the acrylic dome. The z_i-axis in local coordinate O_i-$x_iy_iz_i$ is taken along the center line of sensing element i so that its origin is located on the crossing point of the center line and the acrylic dome's surface and its direction coincides with the normal direction of the acrylic dome. If the vertex is likened to the North Pole, the directions of the x_i- and y_i-axes are north to south and west to east, respectively. Since the optical axis direction of the CCD camera coincides with the direction of the z-axis, information of every tactile element is obtained as an image projected into the O-xy plane. The obtained image data $g(x,y)$ should be transformed into modified image $g(x_i, y_i)$, which is assumed to be taken in the negative direction of the z_i-axis attached to each sensing element. The transform expression is derived from the coordinate transformation of the spherical coordinate to the Cartesian coordinate as follows:

$$g(x_i, y_i) = g(x, y) / \sin\varphi_i \tag{4}$$

Centroid displacements included in Eqs. (1) and (2), and $u_x(x,y)$ and $u_y(x,y)$ should be transformed into $u_x(x_i,y_i)$ and $u_y(x_i,y_i)$ as well. In the same way as Eq. (4), the transform expression is derived from the coordinate transformation of the spherical coordinate to the Cartesian coordinate as follows:

$$u_x(x_i, y_i) = \frac{u_x(x,y)\cos\phi_i + u_y(x,y)\sin\phi_i}{\sin\varphi_i}, \tag{5}$$

$$u_y(x_i, y_i) = u_x(x,y)\sin\phi_i + u_y(x,y)\cos\phi_i. \tag{6}$$

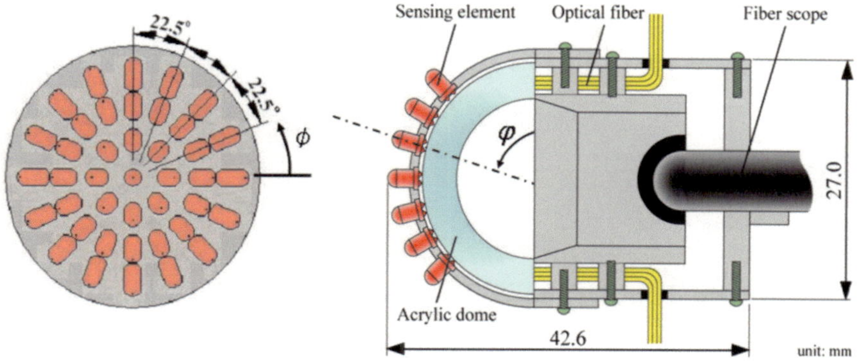

Fig. 4. Fingertip including three-axis tactile sensor

2.1.4 Design of optical three-axis tactile sensor

Since the tactile sensor essentially needs to be a lens system, it is difficult to make it thinner; thus, it should be designed as a type of integrated fingertip and hemispherical three-axis tactile sensor, as shown in Fig. 4 (Ohka et al., 2008). Forty-one sensing elements are concentrically arranged on the acrylic dome, which is illuminated along its edge by optical fibers connected to a light source (ELI-100S, Mitsubishi Rayon Co.) Image data consisting of bright spots caused by the feelers' collapse (MSGS-1350-III, Moritex Co.) are retrieved by an optical fiber scope connected to the CCD camera (C5985, Hamamatsu Photonics Co.)

2.2 Procedure of evaluation tests

2.2.1 Experimental apparatus

We developed a loading machine shown in Fig. 5 that includes an x-stage, a z-stage, rotary stages, and a force gauge (FGC-0.2B, NIDEC-SIMPO Co.) to detect the sensing characteristics of normal and shearing forces. The force gauge has a probe to measure force

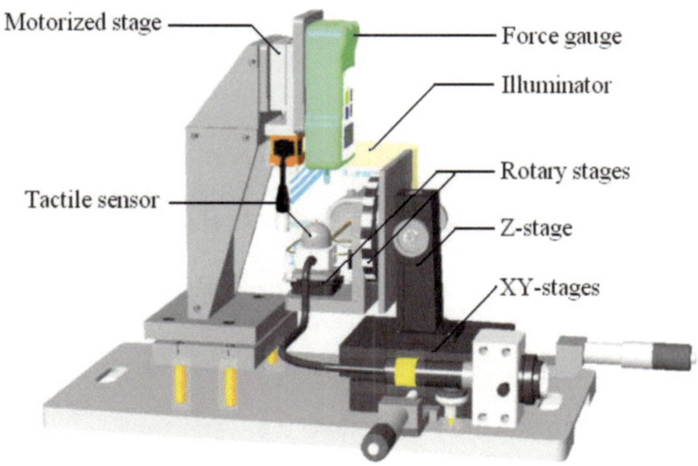

Fig. 5. Loading machine

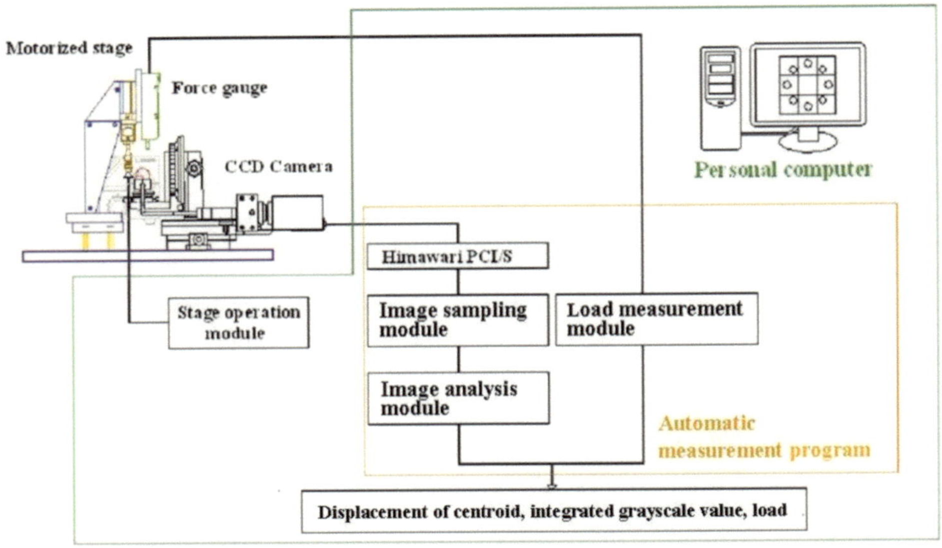

Fig. 6. Tactile data processing system

and can detect force ranging from 0 to 2 N with a resolution of 0.001 N. The positioning precisions of the *y*-, the *z*-, and rotary stages are 0.001 mm, 0.1 mm, and 0.1°, respectively. Output of the present tactile sensor is processed by the data processing system shown in Fig. 6. The system is composed of a tactile sensor, a loading machine, an image processing board (Himawari PCI/S, Library, Co.), and a computer. Image data acquired by the image processing board are processed by in-house software.
The image data acquired by the CCD camera are divided into 41 subregions, as shown in Fig. 7. The dividing procedure, digital filtering, integrated grayscale value and centroid displacement are processed on the image processing board. Since the image warps due to projection from a hemispherical surface, as shown in Fig. 7, software installed on the computer modifies the obtained data. The motorized stage and the force gauge are controlled by the software.

2.2.2 Procedure of sensing normal force test

Because the present tactile sensor can detect not only normal force but also shearing force, we must confirm the sensing capability of both forces. In normal-force testing, by applying a normal force to the tip of a sensing element using the z-stage after rotating the attitude of the tactile sensor, it is easy to test the specified sensing element using the rotary stage. Since the rotary stage's center of rotation coincides with the center of the present tactile sensor's hemispherical dome, testing any sensing element aligned along the hemisphere's meridian is easy.

2.2.3 Procedure of sensing shearing force test

When generating the shearing-force component, both the rotary and x-stages are adjusted to specify the force direction and sensing element. First, the rotary stage is operated to give force direction θ, as shown in Fig. 8. The *x*-stage is then adjusted to the applied tilted force

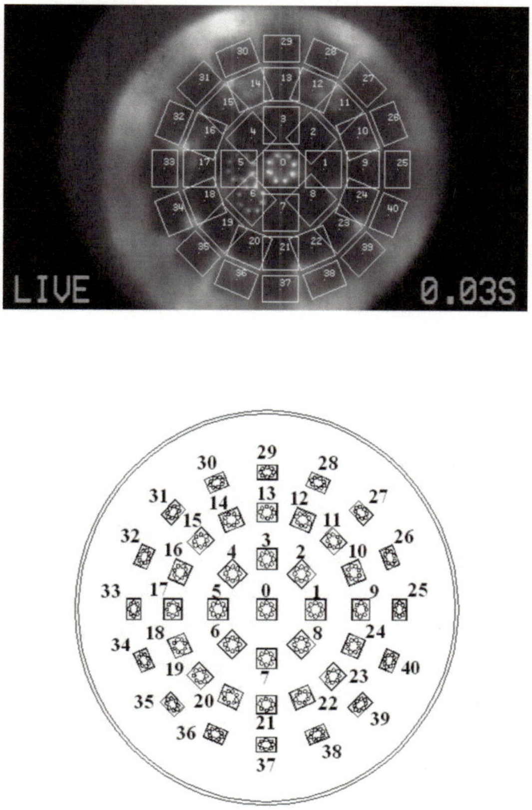

Fig. 7. Addresses of sensing elements

at the tip of the specified sensing element. Figure 8 shows that the sensing element located on the parietal region can be assigned based on the procedure described above. After that, a force is loaded onto the tip of the sensing element using the *z*-stage. Regarding the manner

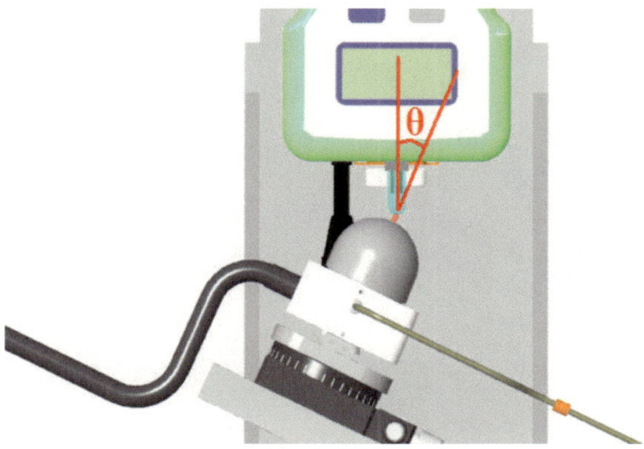

Fig. 8. Generation of shearing force component

of loading, since the force direction does not coincide with the axis of the sensing element, slippage between the probe and the tip of the sensing element occurs. To eliminate this problem, a spherical concave portion is formed on the probe surface to mate the concave portion with the hemispherical tip of the tactile element. Normal force F_N and shearing force F_S applied to the sensing elements are calculated using the following formulas, when force F is applied to the tip of the tactile element:

$$F_N = F\cos\theta \tag{7}$$

$$F_S = F\sin\theta \tag{8}$$

2.3 Sensing ability of optical three-axis tactile sensor

2.3.1 Sensing ability of normal force

To evaluate the sensing characteristics of sensing elements distributed on the hemispherical dome, we need to measure the variation within the integrated grayscale values generated by the sensor elements. Figure 9 shows examples of variation in the integrated grayscale value caused by increases in the normal force for sensors #00, #01, #05, #09, #17, #25, and #33. In these experiments, normal force is applied to a tip of each tactile element. As the figure indicates, the gradient of the relationship between the integrated grayscale value and applied force increases with an increase in φ; that is, sensitivity depends upon the latitude on the hemisphere. Dome brightness is inhomogeneous because the edge of the dome is illuminated and light converges on its parietal region. Brightness is represented as a function of latitude φ, and since sensitivity is uniquely determined by latitude, it is easy to modify the sensitivity according to φ.

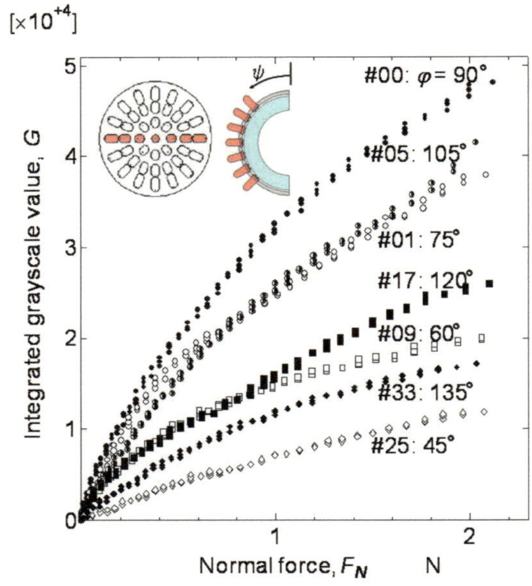

Fig. 9. Relationship between applied force and grayscale value

However, sensing elements located at the same latitude φ show different sensing characteristics. For example, the sensitivities of #09 and #17 should coincide since they have identical latitude; however, as Fig. 10 clearly indicates, they do not. The difference reflects the inhomogeneous brightness of the acrylic dome. Therefore, we need to obtain the sensitivity of every sensing element.

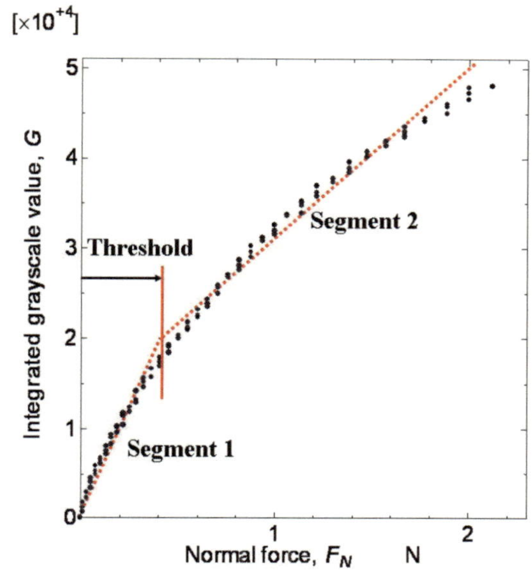

Fig. 10. Approximation using bi-linearity

As shown in Fig. 9, the relationship between the integrated grayscale value and applied normal force is not completely linear. Therefore, we adopt the bi-linear lines shown in Fig. 10 as function $g(x)$ in Eq. (3) to approximate these curves. Linear approximation adequately represents the relationship between force and integrated grayscale value for two tactile elements (#12 and #29) with high accuracy; for the other elements bi-linear approximation can represent the relationship with a rather high correlation factor ranging from 0.911 to 0.997.

To show that under the combined loading condition normal force component was independently obtained with Eq. (3), we applied inclined force to the tip of the tactile element to examine the relationship between the normal component of applied force and integrated grayscale value. Figure 11 displays the relationship for #00. Even if the inclination is varied from -30° to 30°, the relationship coincides within a deviation of 3.7%. Therefore, the relationship between the normal component of applied force and the integrated grayscale value is independent of inclination θ.

2.3.2 Sensing ability of shearing force

When force is applied to the tip of the sensing element located in the parietal region under several θ s, the relationships between the displacement of the centroid and the shearing-force component calculated by Eq. (5) are obtained, as shown in Fig. 12. Although the inclination of the applied force is varied in a range from 15° to 60°, the curves converge into a single one. Therefore, the applied shearing force is obtained independently from centroid displacement.

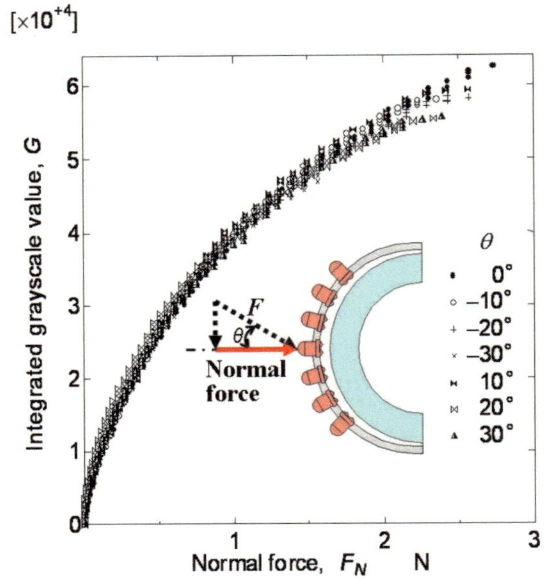

Fig. 11. Relationship between integrated grayscale value and applied normal force at several inclinations

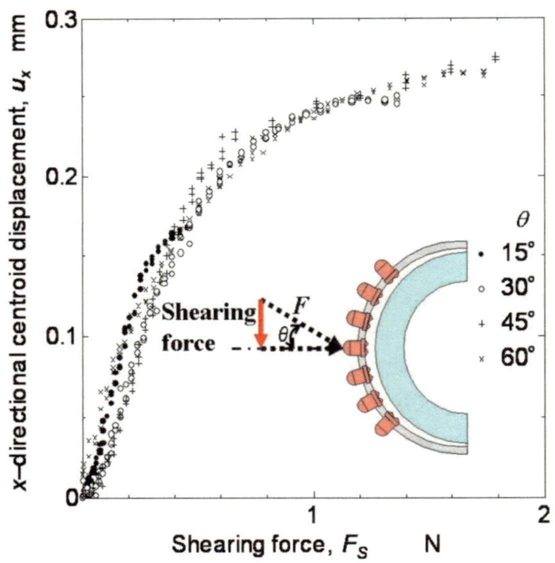

Fig. 12. Relationship between centroid displacement and applied shearing

When the tactile element accepts directional forces of 45°, 135°, 225°, and 315°, centroid trajectories are shown in Fig. 13 to examine shearing force detection under various directions except for the x- and y-directions. If the desired trajectories shown in Fig. 13 are compared to the experimental results, they almost trace identical desired trajectories. The present tactile sensor can detect various applied forces.

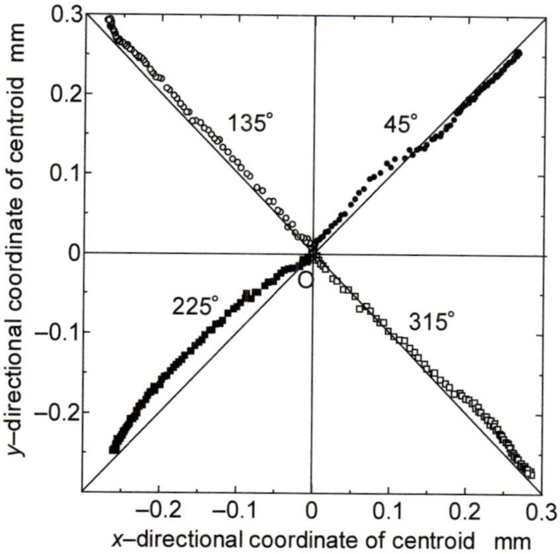

Fig. 13. Trajectory of centroid

3. Object handling based on tri-axial tactile data

3.1 Hand robot equipped with optical three-axis tactile sensors

We designed a two-fingered robotic hand as shown in Fig. 14 for general-purpose use in robotics (Ohka et al., 2009b, 2009c). The robotic hand includes links, fingertips equipped with the three-axis tactile sensor, and micro actuators (YR-KA01-A000, Yasukawa). Each micro actuator, which consists of an AC servo-motor, a harmonic drive, and an incremental encoder, was particularly developed for application to a multi-fingered hand. Since the tactile sensors must be fitted to a multi-fingered hand, we are developing a fingertip that includes the hemispherical three-axis tactile sensor shown in Fig. 4.

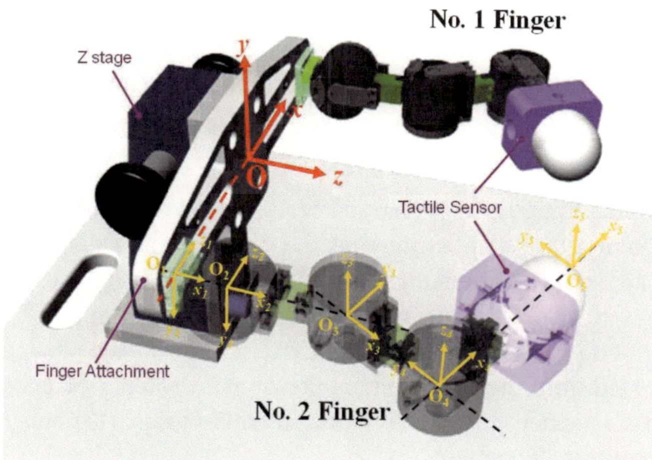

Fig. 14. Robotic hand equipped with three-axis tactile sensors

3.2 Kinematics of hand robot

As shown in Fig. 14, each robotic finger has three movable joints. The frame of the workspace is set on the bottom of the z-stage. The kinematics of the present hand is derived according to Denavit-Hartenberg notation shown in Fig. 14. The frame of the workspace is defined as O-*xyz*. The frames of O_i- $(x_i y_i z_i)$ (in the following, O-*xyz* is used instead of O_0- $x_0 y_0 z_0$) are attached on each joint, the basement of the z-stage, or the fingertip, as shown in Fig. 14. The velocities of the micro actuators ($\dot{\boldsymbol{\theta}} = \begin{pmatrix}\dot{\theta}_1 & \dot{\theta}_2 & \dot{\theta}_3\end{pmatrix}$) are calculated with

$$\dot{\boldsymbol{\theta}} = \mathbf{J}^{-1}(\boldsymbol{\theta})\dot{\mathbf{r}} \tag{9}$$

to satisfy specified velocity vector $\dot{\mathbf{r}}$ $(=\begin{pmatrix}\dot{x} & \dot{y} & \dot{z}\end{pmatrix})$, which is calculated from the planed trajectory. Jacobian $\mathbf{J}(\boldsymbol{\theta})$ is obtained by the kinematics of the robotic hand as follows:

$$\mathbf{J}(\boldsymbol{\theta}) = \begin{bmatrix} -R_{13}(l_2 + l_3 c_2 + l_4 c_{23}) & l_3(R_{11}s_3 + R_{12}c_3) + l_4 R_{12} & R_{12} l_4 \\ -R_{23}(l_2 + l_3 c_2 + l_4 c_{23}) & l_3(R_{21}s_3 + R_{22}c_3) + l_4 R_{22} & R_{22} l_4 \\ -R_{33}(l_2 + l_3 c_2 + l_4 c_{23}) & l_3(R_{31}s_3 + R_{32}c_3) + l_4 R_{32} & R_{32} l_4 \end{bmatrix}, \tag{10}$$

where

$$\begin{aligned} \begin{bmatrix} R_{11} & R_{12} & R_{13} \\ R_{21} & R_{22} & R_{23} \\ R_{31} & R_{32} & R_{33} \end{bmatrix} &= \begin{bmatrix} a_{11}c_{23} + a_{13}s_{23} & -a_{11}s_{23} + a_{13}c_{23} & -a_{12} \\ a_{21}c_{23} + a_{23}s_{23} & -a_{21}s_{23} + a_{23}c_{23} & -a_{22} \\ a_{31}c_{23} + a_{33}s_{23} & -a_{31}s_{23} + a_{33}c_{33} & -a_{32} \end{bmatrix}, \\ \begin{bmatrix} a_{11} & a_{12} & a_{13} \\ a_{21} & a_{22} & a_{23} \\ a_{31} & a_{32} & a_{33} \end{bmatrix} &= \begin{bmatrix} c\varphi_1 c_1 + s\varphi_1 s\varphi_2 s_1 & -c\varphi_1 s_1 + s\varphi_1 s\varphi_2 s_1 & s\varphi_1 c\varphi_2 \\ c\varphi_2 s\varphi_1 & c\varphi_2 s\varphi_1 & -s\varphi_2 \\ -s\varphi_1 c_1 + s\varphi_1 s\varphi_2 s_1 & s\varphi_1 s_1 + c\varphi_1 s\varphi_2 c\varphi_1 & c\varphi_1 c\varphi_2 \end{bmatrix} \\ c_i \equiv \cos\theta_i,\ s_i \equiv \sin\theta_i,\ &c\varphi_i \equiv \cos\varphi_i,\ s\varphi_i \equiv \sin\varphi_i, \\ c_{ij} \equiv \cos(\theta_i + \theta_j),\ s_{ij} &\equiv \sin(\theta_i + \theta_j),\ (i\,;\ j = 1, 2, 3). \end{aligned} \tag{11}$$

In the above equations, the rotations of the first frame around the x_0- and y_0-axes are denoted as φ_1 and φ_2, respectively. The distance between the origins of the *m*-th and *m*+1-th frames is denoted as l_m. The joint angles of the micro actuators on O_2- $x_2 y_2 z_2$, O_3- $x_3 y_3 z_3$ and O_4- $x_4 y_4 z_4$ are θ_1, θ_2, and θ_3, respectively.

Position control of the fingertip is performed based on resolved motion rate control. In this control method, joint angles are assumed at the first step, and displacement vector $\mathbf{r}_0$ is calculated with kinematics. Adjustment of joint angles is obtained by Eq. (9) and the difference between $\mathbf{r}_0$ and objective vector $\mathbf{r}_d$ to modify joint angle $\boldsymbol{\theta}_{t+1}$ at the next step. The modified joint angle is designated as the current angle in the next step, and the above procedure is repeated until the displacement vector at *k*-th step $\mathbf{r}_k$ coincides with objective vector $\mathbf{r}_d$ within a specified error. That is, the following Eqs. (12) and (13) are calculated until $|\mathbf{r}_d - \mathbf{r}_k|$ becomes small enough:

$$\dot{\mathbf{r}}_k = \mathbf{J}\dot{\boldsymbol{\theta}}_k \tag{12}$$

$$\boldsymbol{\theta}_{k+1} = \boldsymbol{\theta}_k - \mathbf{J}^{-1}\left(\mathbf{r}_d - \mathbf{r}_k\right) \tag{13}$$

3.3 Control algorithm

Our objective is to show that the robotic hand adapts its finger trajectory to the environment according to tri-axial tactile data. Hence, we make a simple control algorithm for the hand. In the algorithm, there is an assumption that finger trajectory provided beforehand to the hand is as simple as possible. The trajectory is modified to prevent normal force from exceeding a threshold and to stabilize slippage caused on the contact area according to tri-axial tactile data.

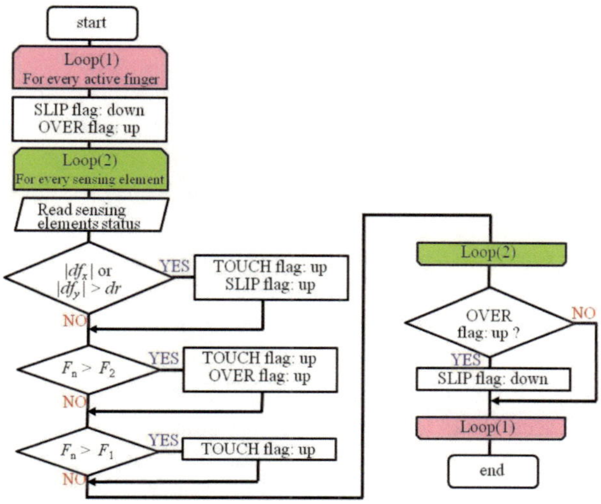

Fig. 15. Algorithm of flag analyzer

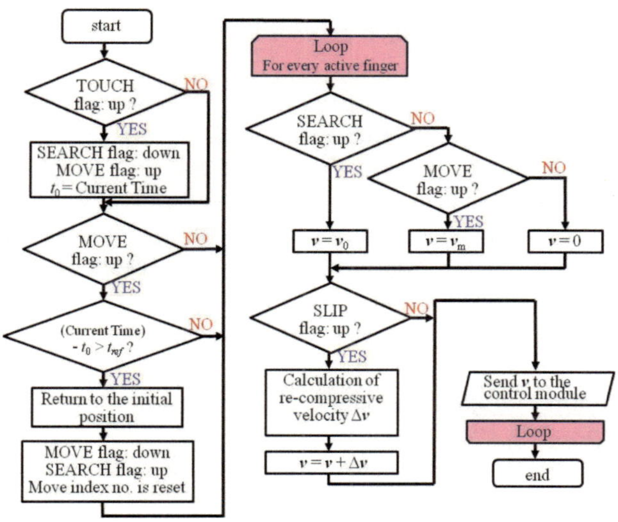

Fig. 16. Algorithm of finger speed estimator

The hand is controlled according to velocity control. First, hand status becomes "search mode" to make fingers approach an object with finger speed $\mathbf{v} = \mathbf{v}_0$. After the fingers touch the object, the hand status becomes "move mode" to manipulate the object with finger speed $\mathbf{v} = \mathbf{v}_m$. During both search and move modes, when the absolute time derivative of the shearing force of a sensing element exceeds a threshold dr, this system regards the sensing element as slippage. To prevent the hand from dropping the object, re-compressive velocity is defined as moving the fingertip along the counter direction of applied force.

However, if normal force of a sensing element exceeds a threshold F_2, the re-compressive velocity is cancelled to prevent the sensing element from breaking. The hand is controlled by a control module with applying total velocity obtained by adding the re-compressive velocity to current velocity.

In our system, the sensor control program and hand control program are executed in different computers because CPU time is efficiently consumed using a multi-task program method. These programs are synchronized with the following five flags.

SEARCH: Fingers search for an object with initial finger velocity $\mathbf{v}_0$ until normal force of a sensing element exceeds a threshold F_1 or Slip flag is raised.

MOVE: This flag is raised whenever the robotic hand manipulates an object.

TOUCH: This flag is raised whenever one of the fingers touches an object.

SLIP: This flag is raised whenever the time derivative of shearing force exceeds a threshold dr.

OVER: This flag is raised when normal force of a sensing element exceeds a threshold F_2.

These flags are decided according to tri-axial tactile data and finger motions. Since two modules, the flag analyzer and finger speed estimator, mainly play the role of object handling, these modules are shown in Figs. 15 and 16, respectively.

In the flag analyzer, TOUCH flag, SLIP flag and OVER flag are decided. The flag analyzer regards finger status as touching an object when normal force of a sensing element is exceeded or the absolute time derivative of the shearing force is exceeded (SLIP flag is raised). Whenever it regards finger status as touching an object, the TOUCH flag is raised. The OVER flag is raised when normal force of a sensing element exceeds F_1 to prohibit re-compressive motion.

In the finger speed estimator, the velocity of the fingertip is determined based on the five flag values and conserved whenever contact status is not changed. Since the cap-twisting problem requires touch-and-release motion, the MOVE and SEARCH flags are controlled according to the TOUCH flag and time spent. Whenever the SLIP flag is raised, a sensing element of the largest normal force is determined and the re-compressive velocity of the finger is determined as an inward normal line of the sensing element. The re-compressive velocity is added to the current velocity, and the resultant velocity is applied to the control module.

3.4 Evaluation experiment of object handling

3.4.1 Experimental apparatus and procedure

To examine the above algorithm, the robotic hand performed the bottle cap-closing task because this task requires a curved trajectory along the cap contour. Evaluation of cap closing is performed using an apparatus including a torque sensor. Figure 17 shows the apparatus composed of the two-fingered hand, the torque sensor (TCF-0.2N, Nippon

Tokushu Sokki, Co., Ltd.) and a PET bottle holder. A PET bottle is clamped with two twists of the PET holder, and its cap is turned by the robotic hand. The torque sensor measures torque with four strain gauges. Variation in gauge resistance is measured as voltage through a bridge circuit, and it is sent to a computer with an A/D converter to obtain the relationship between finger configuration and generated torque.

The experimental apparatus is shown in Fig. 17. A PET bottle is held by the holder. At first, two fingers approach the cap, and moving direction is changed to the tangential direction of the cap surface after grasping force exceeds 1 N. After the finger moves, keeping the direction within 10 mm, the fingers are withdrawn from the cap surface and returned to each home position from which they started moving. Consequently, the trajectory of the fingers is designed as shown in Fig. 17. During the task of closing the cap, variation in torque is monitored through the torque sensor to evaluate the task. Even if the trajectory is simple, we will show that it adapts to the cap contour in the following section.

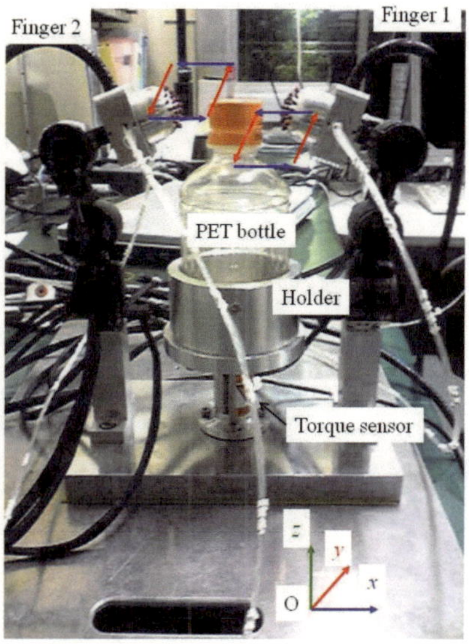

Fig. 17. Experimental apparatus for cap-twisting task

3.4.2 Relationship between grasping force and torque

The relationship between grasping force and torque while twisting the bottle cap is shown in Fig. 18 as an overview of the experiment. Since touch-and-release motion is continued four times, four groupings are found in Fig. 18. As shown in Fig. 18, compared to the first twisting motion, both grasping force and torque decrease considerably in the second twisting, and in the third and fourth twistings they increase compared to the former two twistings. Since the third and fourth twistings show almost the same variations in grasping force and torque, twisting seems to become constant. Therefore, after the third twisting, the cap seems to be closed. In the first twisting, we can observe the transition from light twisting to forceful twisting because torque increases in spite of constant grasping force. It is shown that the cap is turned without resistant torque at first. The reason for reducing grasping force and torque in the second twisting is the variation in contact position and status between the first and second twistings. Twisting on the cap was successfully completed as mentioned above.

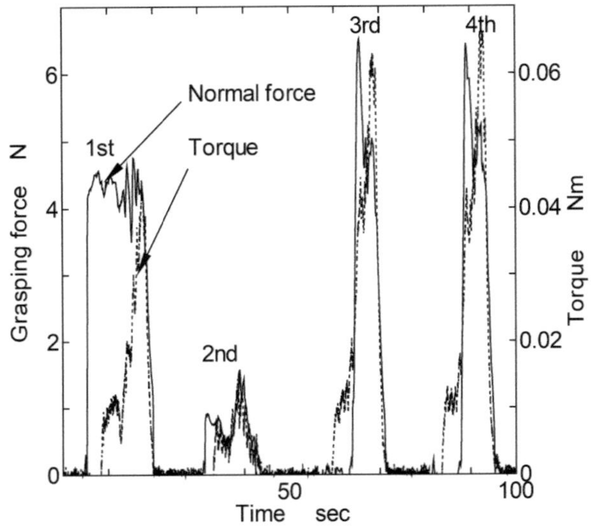

Fig. 18. Relationship between grasping force and torque

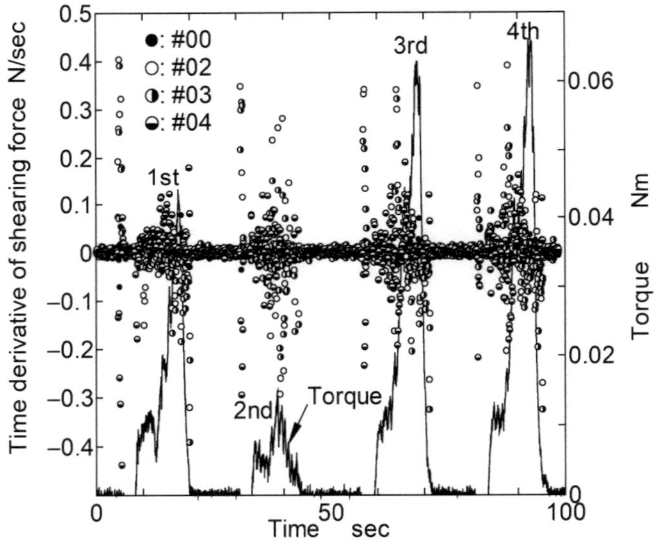

Fig. 19. Relationship between variations in time derivative of shearing force and torque

3.4.3 Relationship between time derivative of shearing force and torque

When the cap is twisted on completely, slippage between the robotic finger and the cap occurs. To examine this phenomenon, the relationship between the time derivative of the shearing force and torque is shown in Fig. 19. As can be seen, the time derivative of the shearing force shows periodic bumpy variation. This bumpy variation synchronizes with variation in torque. This means large tangential force induces the time derivative of the shearing force, which is caused by the trembling of the slipping sensor element.

To examine the cap-twisting, a comparison between the results of the first screwing and fourth twisting is performed with Figs. 20 and 21. In the first twisting, since the cap is loose, the marked time derivative of the shearing force does not occur in Fig. 20. On the other hand, in the fourth twisting, the marked time derivative of the shearing force does occur because of the securing of the cap (Fig. 21). Therefore, the robotic hand can twist on the bottle cap completely. Additionally, the time derivative of the shearing force can be adopted as a measure for twisting the cap.

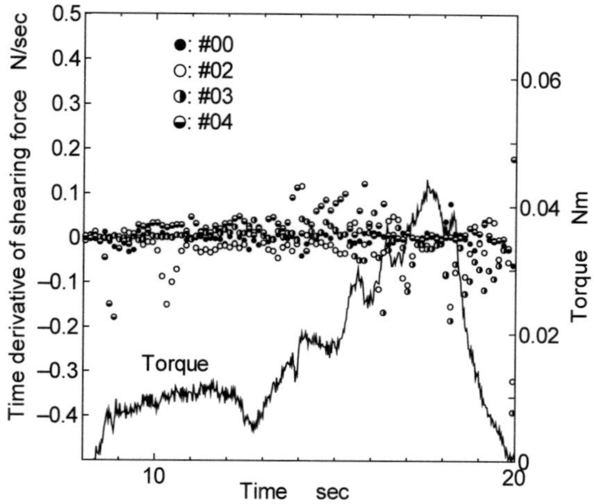

Fig. 20. Detailed relationship between variations in time derivative of shearing force and torque at first twisting

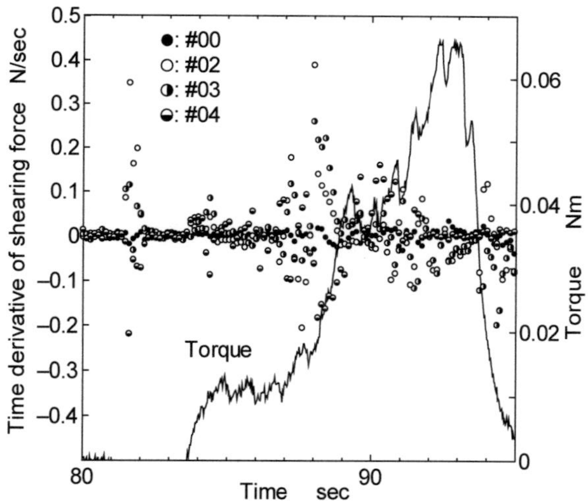

Fig. 21. Detailed relationship between variations in time derivative of shearing force and torque at fourth twisting

3.4.4 Trajectory of fingertip modified according to tri-axial tactile data

Trajectories of sensor element tips are shown in Figs. 22 and 23. If the result of Fig. 22 is compared with the result of Fig. 23, trajectories of Fig. 23 are closer to the cap contour. Modification of the trajectory is saturated after closing the cap. Although input finger trajectories were a rectangle roughly decided to touch and turn the cap as described in the previous section, a segment of the rectangle was changed from a straight line to a curved line to fit the cap contour.

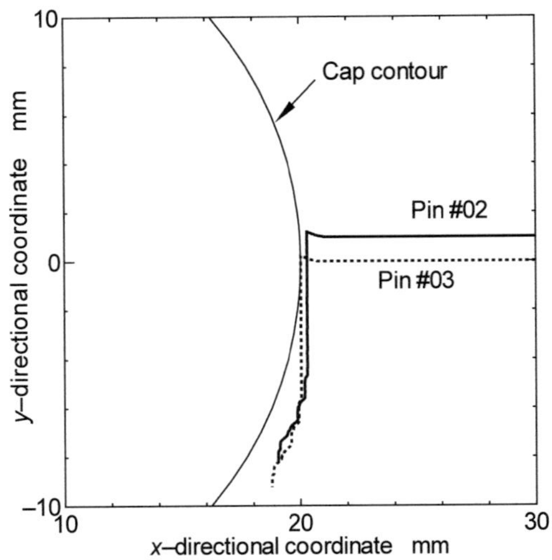

Fig. 22. Trajectories of sensor element before closing the cap

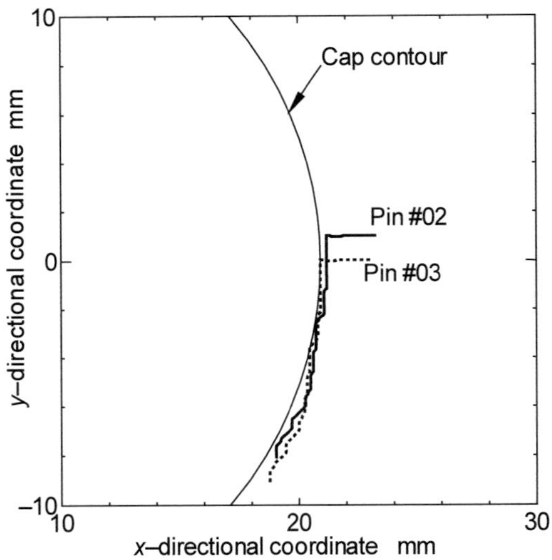

Fig. 23. Trajectories of sensor element after closing the cap

4. Conclusion

We developed a new three-axis tactile sensor to be mounted on multi-fingered hands, based on the principle of an optical waveguide-type tactile sensor comprised of an acrylic hemispherical dome, a light source, an array of rubber sensing elements, and a CCD camera. The sensing element of the present tactile sensor includes one columnar feeler and eight conical feelers. A three-axis force applied to the tip of the sensing element is detected by the contact areas of the conical feelers, which maintain contact with the acrylic dome. Normal and shearing forces are calculated from integration and centroid displacement of the grayscale value derived from the conical feeler's contacts.

To evaluate the present tactile sensor, we conducted a series of experiments using a *y-z* stage, rotational stages, and a force gauge. Although the relationship between the integrated grayscale value and normal force depended on the sensor's latitude on the hemispherical surface, it was easy to modify sensitivity based on the latitude. Sensitivity to normal and shearing forces was approximated with bi-linear curves. The results revealed that the relationship between the integrated grayscale value and normal force converges into a single curve despite the inclination of the applied force. This was also true for the relationship between centroid displacement and shearing force. Therefore, applied normal and shearing forces can be obtained independently from integrated grayscale values and centroid displacement, respectively. Also, the results for the present sensor had enough repeatability to confirm that the sensor is sufficiently sensitive to both normal and shearing forces.

Next, a robotic hand was composed of two robotic fingers to indicate that tri-axial tactile data generated the trajectory of the robotic fingers. Since the three-axis tactile sensor can detect higher order information compared to the other tactile sensors, the robotic hand's behavior is determined on the basis of tri-axial tactile data. Not only tri-axial force distribution directly obtained from the tactile sensor but also the time derivative of shearing force distribution is used for the hand-control program. If grasping force measured from normal force distribution is lower than a threshold, grasping force is increased. The time derivative is defined as slippage; if slippage arises, grasping force is enhanced to prevent fatal slippage between the finger and object. In the verification test, the robotic hand twists on a bottle cap completely. Although input finger trajectories were a rectangle roughly decided to touch and turn the cap, a segment of the rectangle was changed from a straight line to a curved line to fit the cap contour. Therefore, higher order tactile information can reduce the complexity of the control program.

We are continuing to develop the optical three-axis tactile sensor to enhance its capabilities such as sensing area, precision and sensible range of load. Furthermore, we will apply the hand to more practical tasks such as assemble-and-disassemble and peg-in-hole tasks in future work.

5. References

Borovac, B., Nagy, L., and Sabli, M., Contact Tasks Realization by sensing Contact Forces, Theory and Practice of Robots and Manipulators, *Proc. of 11th CISM-IFToNN Symposium, Springer Wien New York,* pp. 381-388, 1996.

Chigusa, H., Makino, Y. and Shinoda, H., Large Area Sensor Skin Based on Two-Dimensional Signal Transmission Technology, *Proc. World Haptics 2007,* Mar., Tsukuba, Japan, pp. 151-156, 2007.

Hackwood, S., Beni, G., Hornak, L. A., Wolfe, R., and Nelson, T. J., Torque-Sensitive Tactile Array for Robotics, *Int. J. Robotics Res.*, Vol. 2-2, pp. 46-50, 1983.

Hakozaki, M. and Shinoda, H., Digital Tactile Sensing Elements Communicating Through Conductive Skin Layers, *Proc. of 2002 IEEE Int. Conf. On Robotics and Automation,* pp. 3813-3817, 2002.

Harmon, L. D, Automated Tactile Sensing, *Int. J. Robotics Res.*, Vol. 1, No.2, pp. 3-32, 1982.

Hasegawa, Y., Shikida, M., Shimizu, T., Miyaji, T., Sakai, H., Sato, K., and Itoigawa, K., A Micromachined Active Tactile Sensor for Hardness Detection, *Sensors and Actuators (A Physical)*, Vol. 114, Issue 2-3, pp. 141-146, 2004.

Kamiyama, K., Vlack, K., Mizota, T., Kajimoto, H., Kawakami, N. and Tachi, S., Vision-Based Sensor for Real-Time Measuring of Surface Traction Fields, *IEEE Computer Graphics and Applications,* January/February, pp. 68-75, 2005

Kaneko, M., H. Maekawa, and K. Tanie, Active Tactile Sensing by Robotic Fingers Based on Minimum-External-Sensor-Realization, *Proc. of IEEE Int. Conf. on Robotics and Automation*, pp. 1289-1294, 1992.

Maekawa, H., Tanie, K., Komoriya, K., Kaneko M., Horiguchi, C., and Sugawara, T., Development of a Finger-shaped Tactile Sensor and Its Evaluation by Active Touch, *Proc. of the 1992 IEEE Int. Conf. on Robotics and Automation*, pp. 1327-1334, 1992.

Mott, H., Lee, M. H., and Nicholls, H. R., An Experimental Very-High-Resolution Tactile Sensor Array, *Proc. 4th Int. Conf. On Robot Vision and Sensory Control*, pp. 241-250, 1984.

Nicholls, H. R. & Lee, M. H., A Survey of Robot Tactile Sensing Technology, *Int. J. Robotics Res.*, Vol. 8-3, pp. 3-30, 1989.

Nicholls, H. R., Tactile Sensing Using an Optical Transduction Method, *Traditional and Non-traditional Robot Sensors (Edited by T. C. Henderson)*, Springer-Verlag, pp. 83-99, 1990.

Novak, J. L., Initial Design and Analysis of a Capacitive Sensor for Shear and Normal Force Measurement, *Proc. of 1989 IEEE Int. Conf. On Robotic and Automation*, pp. 137-145, 1989.

Ohka, M., Mitsuya, Y., Takeuchi, S., Ishihara, H. and Kamekawa, O., A Three-axis Optical Tactile Sensor (FEM Contact Analyses and Sensing Experiments Using a Large-sized Tactile Sensor), *Proc. of the 1995 IEEE Int. Conf. on Robotics and Automation*, pp. 817-824, 1995.

Ohka, M., Mitsuya, Y., Matsunaga, Y., and Takeuchi, S., Sensing Characteristics of an Optical Three-axis Tactile Sensor Under Combined Loading, *Robotica*, vol. 22, pp. 213-221, 2004.

Ohka, M., Mitsuya, Y., Higashioka, I., and Kabeshita, H., An Experimental Optical Three-axis Tactile Sensor for Micro-robots, *Robotica*, vol. 23, pp. 457-465, 2005.

Ohka, M, Kobayashi, H., Takata, J., and Mitsuya, An Experimental Optical Three-axis Tactile Sensor Featured with Hemispherical Surface, *Journal of Advanced Mechanical Design, Systems, and Manufacturing*, Vol. 2-5, pp. 860-873, 2008.

Ohka, M., Robotic Tactile Sensors, *Wiley Encyclopedia of Computer Science and Engineering*, , pp. 2454 - 2461 , 2009.

Ohka, M., Takata, J., Kobayashi, H., Suzuki, H., Morisawa, N., and Yussof, H. B., 60. Object Exploration and Manipulation Using a Robotic Finger Equipped with an Optical Three-axis Tactile Sensor, *Robotica*, vol. 27, pp. 763-770, 2009.

Ohka, M., Morisawa, N., and Yussof, H., B., Trajectory Generation of Robotic Fingers Based on Tri-axial Tactile Data for Cap Screwing Task, *Proc. of IEEE Inter. Conf. on Robotic and Automation*, pp. 883-888, 2009.

Shimojo, M., Namiki, A., Ishikawa, M., Makino, R. and Mabuchi, K., A Tactile Sensor Sheet Using Pressure Conductive Rubber with Electrical-wires Stitched Method, *IEEE Trans. Sensors*,Vol.5-4, pp.589-596, 2004.

Tanaka, M. Leveque, J., Tagami, H. Kikuchi, K. and Chonan, The "Haptic finger" - a New Device for Monitoring Skin Condition, *Skin Research and Techonology*, Vol. 9, pp. 131-136, 2003.

Tanie, K., Komoriya, K., Kaneko M., Tachi, S., and Fujiwara, A., A High-Resolution Tactile Sensor Array, *Robot Sensors Vol. 2: Tactile and Non-Vision*, Kempston, UK: IFS (Pubs), pp. 189-198, 1986.

Yamada, Y. & Cutkosky, R., Tactile Sensor with 3-Axis Force and Vibration Sensing Function and Its Application to Detect Rotational Slip, *Proc. of 1994 IEEE Int. Conf. On Robotics and Automation*, pp. 3550-3557, 1994.

Part 2

Applications

9

3D Terrain Sensing System using Laser Range Finder with Arm-Type Movable Unit

Toyomi Fujita and Yuya Kondo
Tohoku Institute of Technology
Japan

1. Introduction

A 3D configuration and terrain sensing is a very important function for a tracked vehicle robot to give precise information as possible for operators and to move working field efficiently. A Laser Range Finder (LRF) is widely used for the 3D sensing because it can detect wide area fast and can obtain 3D information easily. Some 3D sensing systems with the LRF have been presented in earlier studies (Hashimoto et al., 2008) (Ueda et al., 2006) (Ohno & Tadokoro, 2005). In those measurement systems, multiple LRF sensors are installed in different directions (Poppinga et al., 2008), or a LRF is mounted on a rotatable unit (Nuchter et al., 2005) (Nemoto et al., 2007). Those kinds of system still have the following problems:

a. The system is going to be complex in data acquisition because of the use of multiple LRFs for the former case,
b. It is difficult for both cases to do sensing more complex terrain such as valley, deep hole, or inside the gap because occlusions occur for such terrain in the sensing.

In order to solve these problems, we propose a novel kind of sensing system using an arm-type sensor movable unit which is an application of robot arm. In this sensing system, a LRF is installed at the end of the arm-type movable unit. The LRF can change position and orientation in movable area of the arm unit and face at a right angle according to a variety of configuration. This system is therefore capable of avoiding occlusions for such a complex terrain and sense more accurately. A previous study (Sheh et al., 2007) have showed a similar sensing system in which a range imager has been used to construct a terrain model of stepfields; The range imager was, however, fixed at the end of a pole. Our proposed system is more flexible because the sensor can be actuated by the arm-type movable unit.

We have designed and developed a prototype system of the arm-type sensor movable unit in addition to a tracked vehicle robot. In this chapter, Section 2 describes an overview of the developed tracked vehicle and sensing system as well as how to calculate 3D sensing position. Section 3 explains two major sensing methods in this system. Section 4 presents fundamental experiments which were employed to confirm a sensing ability of this system. Section 5 shows an example of 3D mapping for wide area by this system. Section 6 discusses these results.

2. System overview

The authors have designed and developed a prototype system of the arm-type movable unit. The unit has been mounted on a tracked vehicle robot with two crawlers that we have also developed. Fig. 1 shows the overview. The following sections describe each part of this system.

2.1 Tracked vehicle

We have developed a tracked vehicle robot toward rescue activities. Fig. 1 shows an overview of the robot system. The robot has two crawlers at the both sides. A crawler consists of rubber blocks, a chain, and three sprocket wheels. The rubber blocks are fixed on each attachment hole of the chain. One of the sprocket wheels is actuated by a DC motor to drive a crawler for each side. The size of the robot is 400[mm](length) × 330[mm](width) × 230[mm](height), when the sensor is descended on the upper surface of the robot.

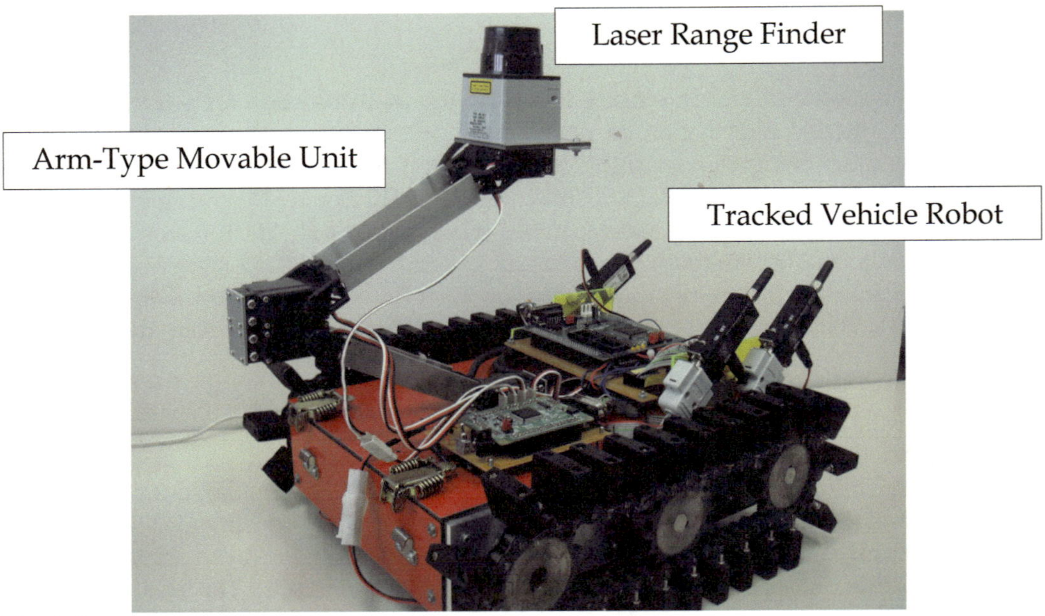

Fig. 1. System overview

2.2 Arm-type sensor movable unit

We have designed the arm-type sensor movable unit and developed a prototype system. This unit consists of two links having a length of 160[mm]. The links are connected by two servo motors as a joint in order to make the sensor horizontal orientation easily when folded. Another two joints are also attached to the both ends of the connecting links; one is connected to the sensor at the end and the other is mounted on the upper surface of the robot. The robot can lift the sensor up to a height of 340[mm] and change its position and orientation by rotating those joints.

2.3 Sensors

HOKUYO URG-04LX (Hokuyo Automatic Co. Ltd.) is used as the Laser Range Finder (LRF) in this system. This sensor can scan 240 degrees area and obtain distance data every 0.36

degree on a 2D plane. The robot is able to change the position and orientation of this sensor because it is equipped at the end of the arm-type movable unit.
In addition, we have installed an acceleration sensor around three orthogonal axes to detect tilt angle of the robot body and to control the orientation of the LRF to be flat corresponding to the tilt angle. The use of this sensor enables the arm-type movable unit to change the height of the LRF with keeping its orientation.

2.4 Control system

The control system of this robot system consists of two embedded micro computers: Renesas SH-2/7045F and H8/3052F for controlling the main robot and the arm-type sensor movable unit respectively. A Windows/XP host PC manages all controls of those units as well as scanned data of the sensor. The host PC sends movement commands to individual embedded micro computers for the robot and arm-type movable unit and request for sensor data acquisition to the sensor. The sensor can communicate directly with the host PC. All communications for those protocols are made by wireless serial communications using bluetooth-serial adapters: SENA Parani-SD100.

2.5 Calculation of 3D sensing position

In this system, the robot can obtain 3D sensing positions from distance data of the LRF. We gave coordinate systems to each joint of the arm-type unit and LRF as shown in Fig. 2. When the sensed distance by the LRF is d_s at a scan angle θ_s, the 3D measurement position vector $\mathbf{X}$ in the base coordinate system can be calculated by

$$\begin{pmatrix} \mathbf{X} \\ 1 \end{pmatrix} = {}^{0}\mathbf{P}_{1}\,{}^{1}\mathbf{P}_{2}\,{}^{2}\mathbf{P}_{3}\,{}^{3}\mathbf{P}_{4}\,{}^{4}\mathbf{P}_{5} \begin{pmatrix} \mathbf{X}_s \\ 1 \end{pmatrix} \tag{1}$$

where $\mathbf{X}_s$ shows a position vector of sensed point in the LRF coordinate system:

$$\mathbf{X}_s = d_s \left(\cos\theta_s \ , \ \sin\theta_s \ , \ 0 \right)^{\mathbf{T}} \tag{2}$$

${}^{i}\mathbf{P}_{i+1}$ (i=0,…,4) shows a homogeneous matrix that represents a transformation between two coordinate systems of joint-(i) and joint-(i+1) :

$${}^{i}\mathbf{P}_{i+1} = \begin{pmatrix} {}^{i}\mathbf{R}_{i+1} & {}^{i}\mathbf{T}_{i+1} \\ \mathbf{0}_3 & 1 \end{pmatrix} \tag{3}$$

where ${}^{i}\mathbf{R}_{i+1}$ shows a rotation matrix for the rotation angle θ_{i+1} around y_i axis,

$${}^{i}\mathbf{R}_{i+1} = \begin{pmatrix} \cos\theta_{i+1} & 0 & \sin\theta_{i+1} \\ 0 & 1 & 0 \\ -\sin\theta_{i+1} & 0 & \cos\theta_{i+1} \end{pmatrix} \tag{4}$$

for i= 0, …,4. ${}^{i}\mathbf{T}_{i+1}$ shows a translation vector on the link from joint-(i) to joint-(i+1) the length of which is ℓ_i :

$${}^{i}\mathbf{T}_{i+1} = \left(0 \ \ 0 \ \ \ell_i \right)^{\mathrm{T}} \tag{5}$$

for $i = 0, \ldots, 4$ ($\ell_0 = 0$). $\mathbf{0}_3$ shows a 3×1 zero vector.

The base position of the sensor is set to the position when the arms are folded: the joint angles θ_1, θ_2, θ_3, and θ_4 in Fig. 2 are 90, −90, −90, and 90 degrees respectively.

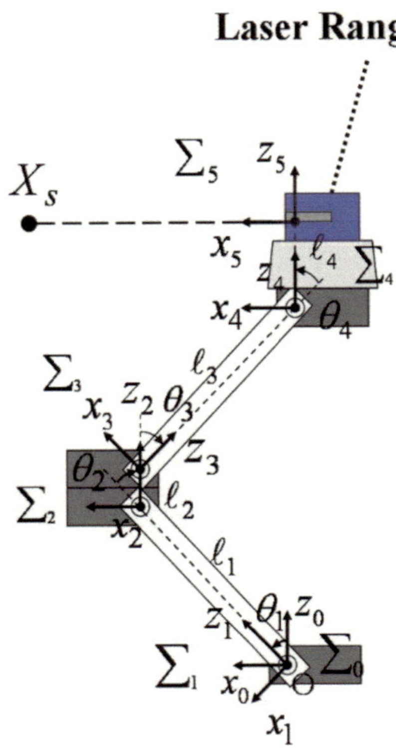

Fig. 2. Coordinate systems

3. Sensing method

The mechanism of this system enables the LRF to change position and orientation to face at a right angle corresponding to a variety of configuration. For example, this sensing system is able to do sensing deep bottom area without occlusions as shown in Fig. 3. Because the occlusion can be avoided by this mechanism even for complex terrain, the robot can measure a 3D configuration such as valley, gap, upward or downward stairs more accurately than conventional 3D sensing system with the LRF. In addition, a robot can do sensing more safely by this method because the robot does not have to stand at close to the border. It is important when the robot needs to work in an unknown site such as disaster area.

On the other hand, this arm-type movable unit can change the height of the LRF by keeping its orientation flat. In this way, 2D shape information in a horizontal plane is detected in each height with even distance. Consequently, the 3D shape of surrounding terrain can be obtained more efficiently by moving the LRF up vertically and keeping its orientation flat. We have installed acceleration sensors to detect tilt angle of the robot so that the robot performs this kind of sensing even when it is on uneven surface as shown in Fig. 4.

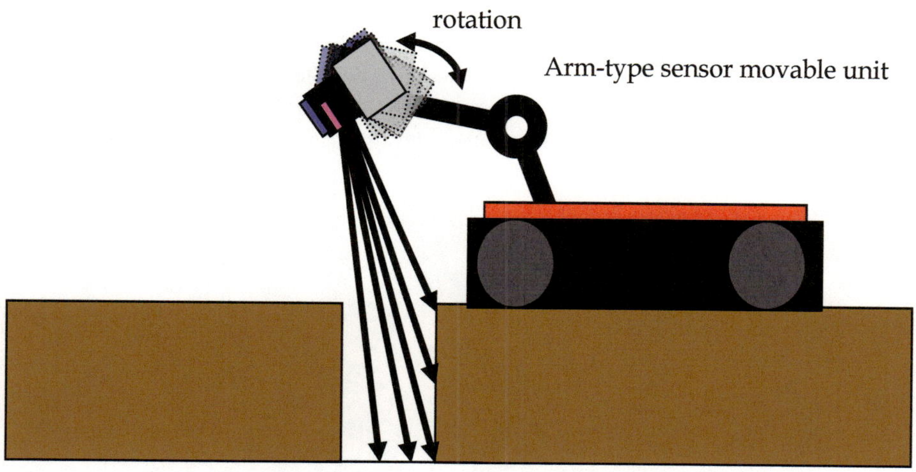

Fig. 3. Sensing of deep bottom area

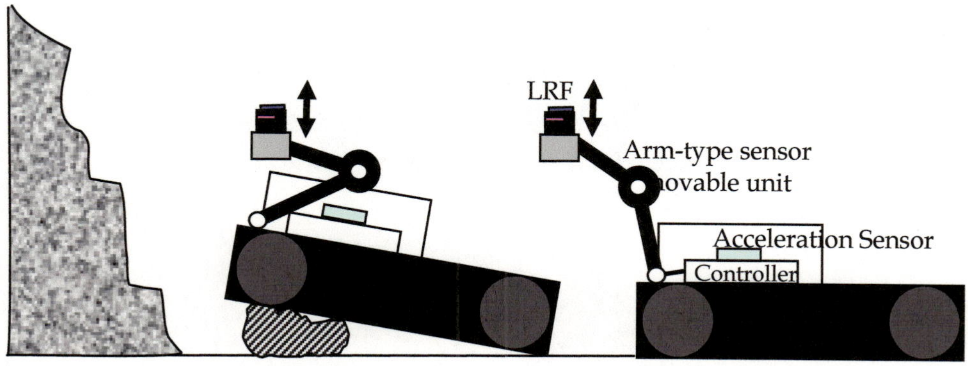

Fig. 4. Sensing by moving the LRF up vertically

We can switch these two kinds of sensing style, as shown in Fig. 3 and Fig. 4, depending on the kind of information that the robot needs to obtain.

4. Experiments

Several fundamental experiments were employed to confirm basic sensing ability of the sensing system for complex terrains: upward stairs, downward stairs, valley configuration, and side hole configuration under the robot. The results for these experiments showed that proposed system is useful for sensing and mapping of more complex terrain.

4.1 Upward stairs

We employed a measurement of upward stairs as an experiment for basic environment. Fig. 5 shows an overview of the experimental environment and Fig. 6 shows its schematic diagram. The stairs are located 1100[mm] ahead of the robot. Each stair is 80[mm] in height and depth. The robot stayed at one position and the LRF sensor was lifted vertically by the arm-type unit from the upper surface of robot to the height of 340[mm] with equal interval

of 50[mm]. Each scanning of the sensor was performed for each height. The robot was tilted 10 degrees to confirm the usefulness of the acceleration sensor in the robot. The robot detected its orientation by the sensor and controlled the height of the LRF according to the orientation.
Fig. 7 shows the measurement result; almost same configuration to actual environment was obtained in this sensing system.

Fig. 5. Overview of an experiment for measurement of upward stairs

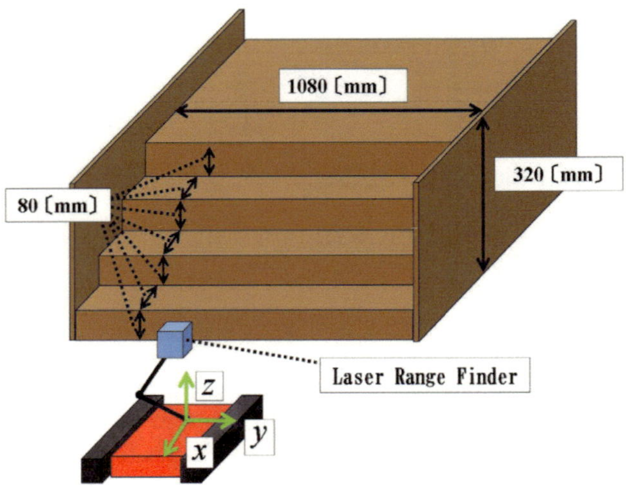

Fig. 6. Schematic diagram of experimental environment for measurement of upward stairs

4.2 Downward stairs

Fig. 8 shows an overview of the experimental setup and Fig. 9 shows its schematic diagram with reference points for 3D measurement of downward stairs. We picked some reference points from corner points for measurement error analysis. Each stair is 80[mm] in height

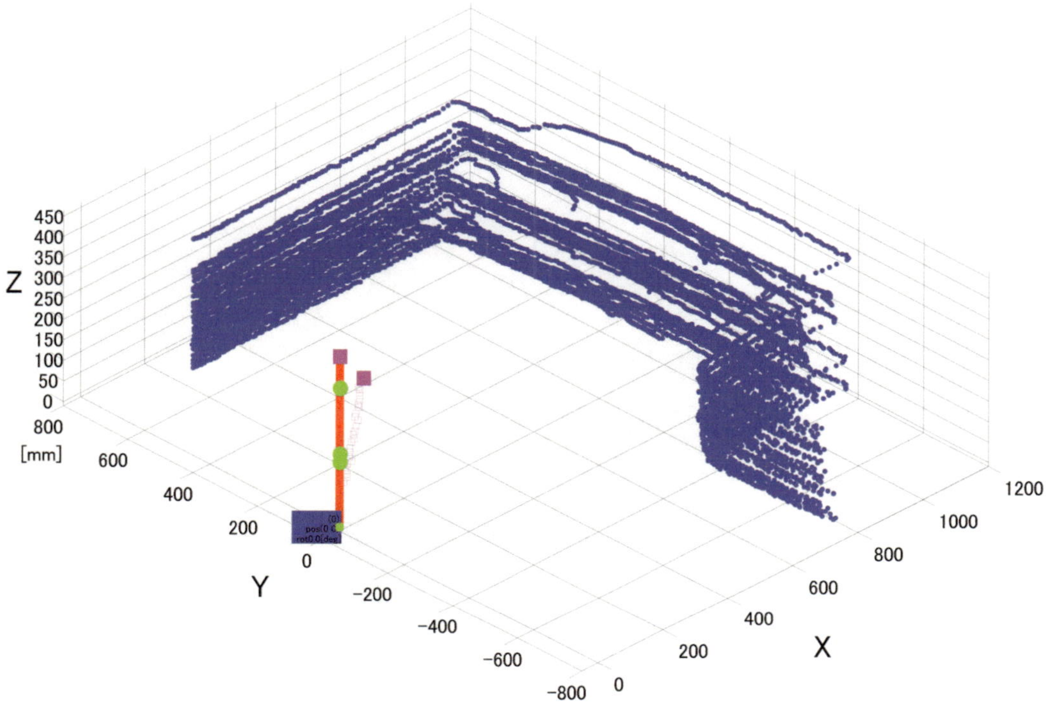

Fig. 7. Measured 3D shape of upward stairs

and depth. In this experiment, the robot stayed at one position, 330[mm] away from the top stair, and moved the arm-unit so that the sensor was located over the downward stairs. The sensor angle was changed by rotating the angle θ_4, shown in Fig. 2, with the same position of the end of the arm-unit by keeping the angles θ_1, θ_2, and θ_3. The rotation angle θ_4 was controlled remotely from 0 degree to 60 degrees every 1.8 degrees. The scanning of the LRF was performed for each sensor angle. The sensing data were then accumulated to make 3D map.

Fig. 8. Overview of an experiment for measurement of downward stairs

Fig. 10 shows the measurement result. We can see almost same configuration to actual environment. The measurement positions for reference points are also denoted in the figure. The results show that accurate position can be sensed by this system. Table 1 shows actual and measured distance with error ratio values on the reference points. This result also confirms valid sensing of this system because error ratio values of the distance were within 5.8% for all reference points.

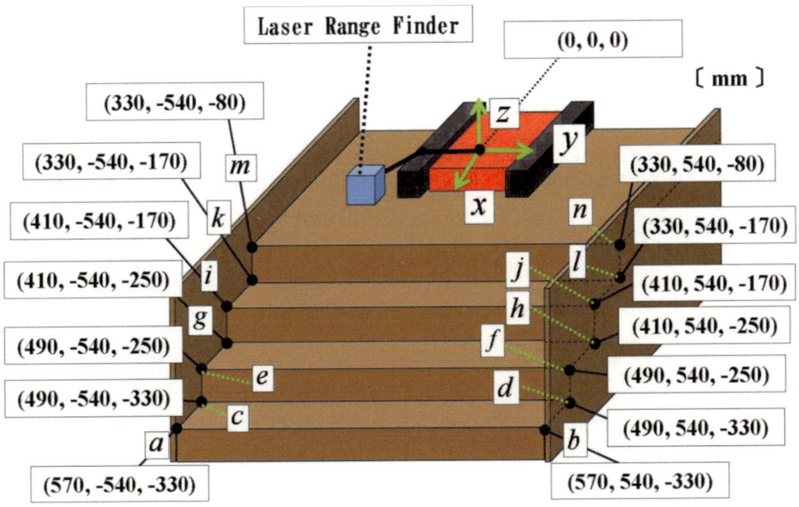

Fig. 9. Schematic diagram of experimental environment for downward stairs with reference points

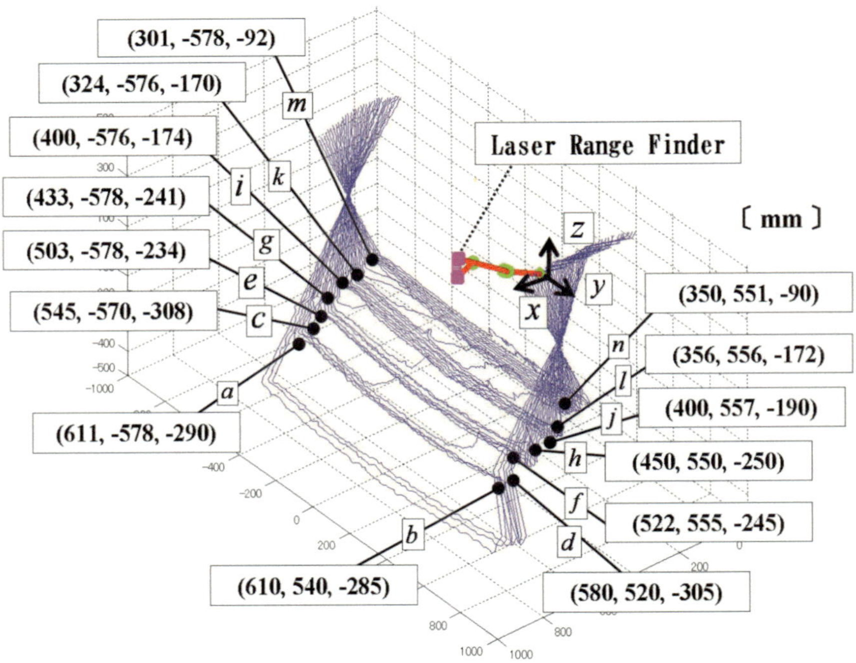

Fig. 10. Measured 3D shape of downward stairs with measurement position values for reference points (unit:[mm])

	distance [mm]			
point	actual	measured	error	error ratio [%]
a	851.7	889.7	38.0	4.5
b	851.7	863.1	11.4	1.3
c	800.4	846.6	46.3	5.8
d	800.4	836.6	36.2	4.5
e	770.8	801.2	30.3	3.9
f	770.8	800.3	29.5	3.8
g	722.6	761.4	38.7	5.4
h	722.6	753.3	30.7	4.2
i	699.0	722.5	23.5	3.4
j	699.0	711.6	12.6	1.8
k	655.3	682.4	27.1	4.1
l	655.3	682.2	27.0	4.1
m	637.9	658.1	20.3	3.2
n	637.9	658.9	21.1	3.3

Table 1. Measured distances and error ratios on reference points for downward stairs

4.3 Valley

A valley configuration was set up as an experimental environment as shown in Fig. 11. Fig. 12 shows its schematic diagram. The valley was 610[mm] deep and 320[mm] long. We gave reference points at each corner of the configuration to estimate actual error value on measurement points.

Fig. 11. Overview of an experiment for measurement of a valley configuration (left: front view, right: side view)

In the same way as previous experiment, the robot stayed at one position, 250[mm] away from the border, and the sensor was located over the valley by the arm-unit. The sensor angle only was changed and the other joint angles of the arm were kept to fix sensor position. The rotation angle of the sensor, θ_4, was varied from 0 degree to 90 degrees every 1.8 degrees. Each scanning was performed for each sensor angle.

Fig. 13 shows the measurement result. We can see very similar configuration to the actual valley. The measurement positions for reference points are also denoted in the figure. The

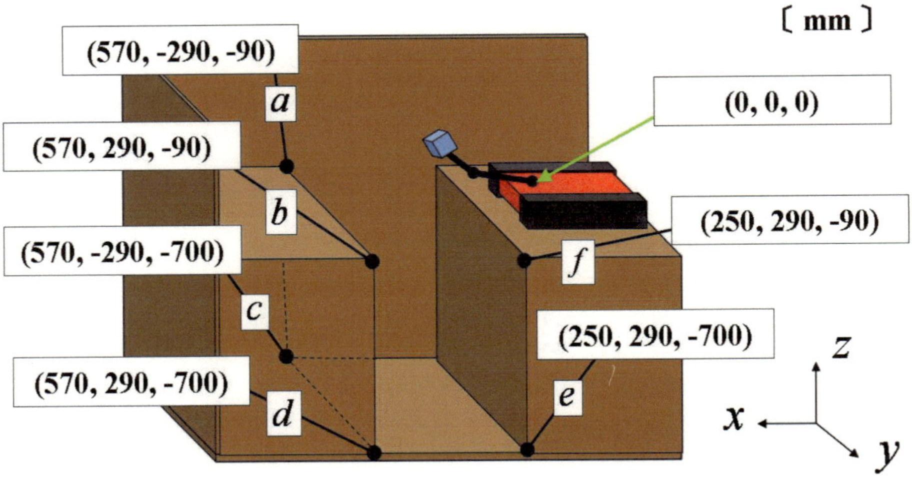

Fig. 12. Schematic diagram of experimental environment for a valley configuration with reference points

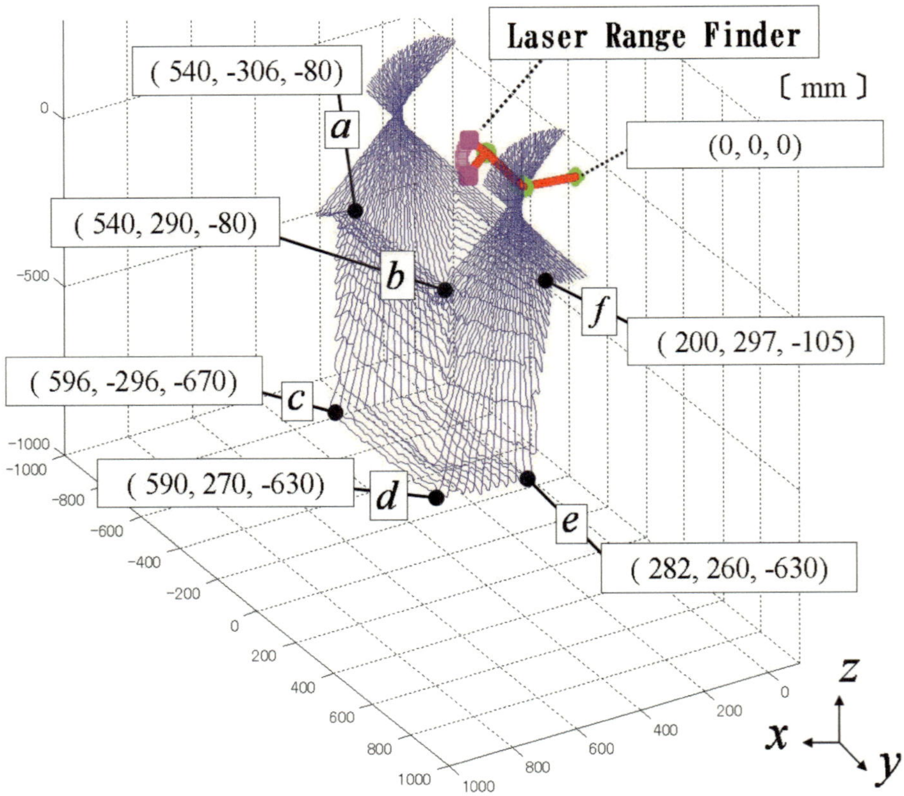

Fig. 13. Measured valley configuration with measurement position values for reference points (unit:[mm])

position values show that accurate position can be sensed by this sensing system. Table 2 shows actual and measured distance with error ratio values on the reference points. Even though the error ratio for the point e was higher, the most of the values are less than about 5% for the other points.

	distance [mm]			
point	actual	measured	error	error ratio [%]
a	645.8	625.8	20.0	3.1
b	645.8	618.1	27.7	4.3
c	948.2	944.3	3.8	0.4
d	948.2	904.4	43.8	4.6
e	797.9	737.6	60.3	7.6
f	393.3	373.1	20.2	5.1

Table 2. Measured distances and error ratios on reference points for a valley configuration

4.4 Side hole under robot

We employed an experiment of measurement for a side hole configuration under the robot. Fig. 14 shows an overview of the experiment and Fig. 15 shows a schematic diagram of its environment. The dimension of the hole was set to 880[mm](width) × 400[mm](height) × 600[mm](depth). Eight reference points were given at each corner of the hole to estimate actual errors.

Fig. 14. Overview of an experiment for measurement of a side hole configuration under the robot

The robot stayed at one position as previous experiments and the sensor was located in front of the hole by the arm-unit. Each rotation angles of joints except for the last joint was fixed to keep the sensor position. The sensor angle, θ_4, was only varied from 0 degree to 70 degrees every 1.8 degrees. Each scanning was performed for each sensor angle. Fig. 16 shows the measurement result. This result also showed almost same configuration to the actual environment. The measurement position values for reference points are also denoted in the figure. Table 3 shows actual and measured distance with error ratio values on the eight reference points. The error ratios demonstrated accurate sensing in this system; the maximum was 4.4% and average was 1.9% for all points.

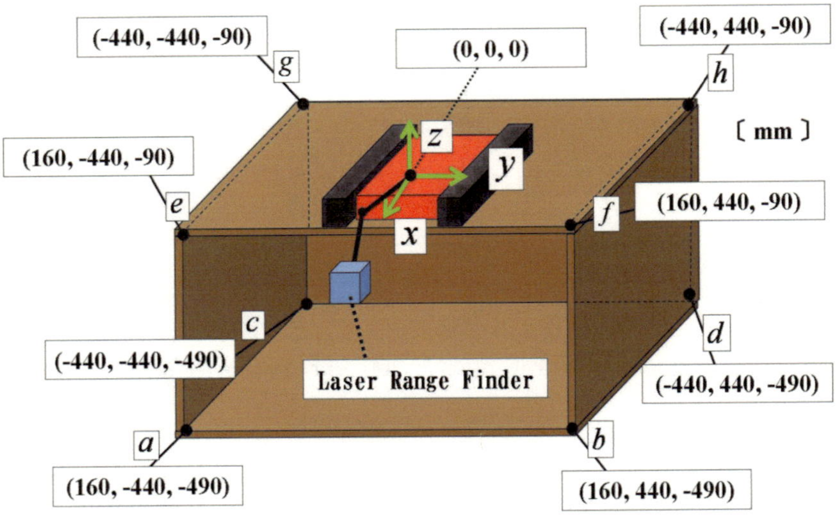

Fig. 15. Schematic diagram of experimental environment for a side hole configuration under the robot with reference points

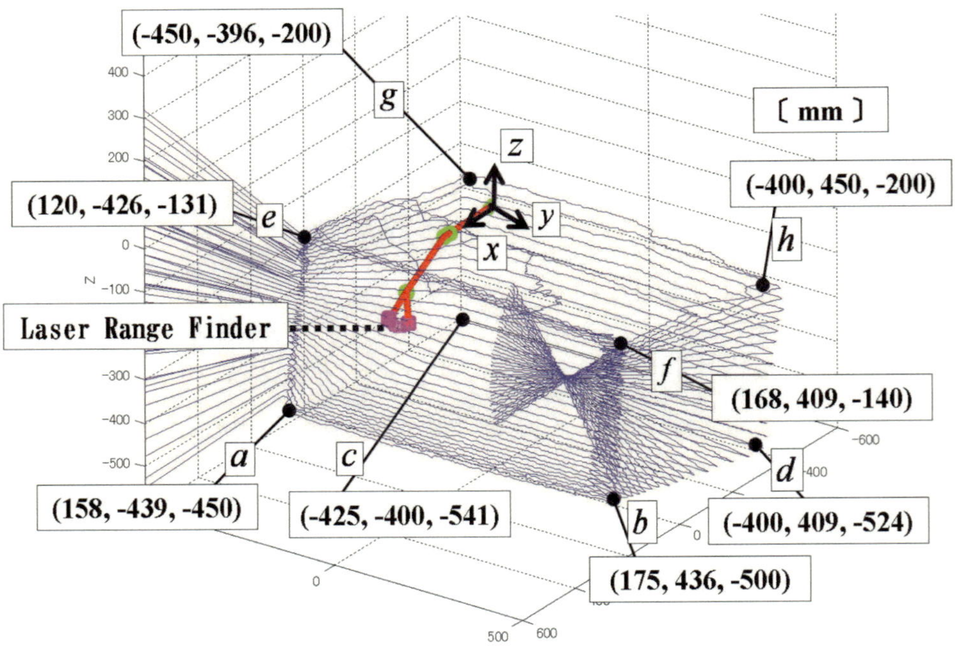

Fig. 16. Measured configuration of a side hole under the robot with measurement position values for reference points (unit:[mm])

	distance [mm]			
point	actual	measured	error	error ratio [%]
a	677.7	648.2	29.5	4.4
b	677.7	686.1	8.4	1.2
c	792.0	795.8	3.8	0.5
d	792.0	775.8	16.2	2.0
e	476.8	461.6	15.2	3.2
f	476.8	463.8	13.0	2.7
g	628.7	631.9	3.2	0.5
h	628.7	634.4	5.7	0.9

Table 3. Measured distances and error ratios on reference points for a side hole configuration under the robot

5. 3D mapping

A basic experiment of 3D mapping for wide area was employed by this sensing system. In the experiment, robot moved in a flat corridor shown in Fig. 17. The robot moved forward in the environment for every 40[cm] distance and made a 3D sensing on each location. The robot obtained 3D data by moving the LRF vertically from the upper surface of the robot to

Fig. 17. Experimental environment for 3D mapping

the height of 340[mm] in every 68[mm] for respective scanning at each sensing location. In order to build 3D map, all sensing data at the sensing locations were combined using odometry information of the robot. We put additional several obstacles in the environment to estimate how this system can detect these shapes and positions. The obstacles are put in the areas labeled by ① and ② as shown in Fig. 17.

Fig. 18 shows the result of 3D mapping. This result shows valid 3D shapes of the environment including added obstacles within the appropriate height. The areas of the obstacles are denoted by ellipse with each label. The built data for each sensing location were described by individual different color. Note that this result clearly shows the top surface detection for each obstacle. This sensing can be made by the mechanism of this system.

Fig. 19 shows the upper view of the built map in the left panel and actual map in the right panel. Obstacles were detected at almost correct location in the result.

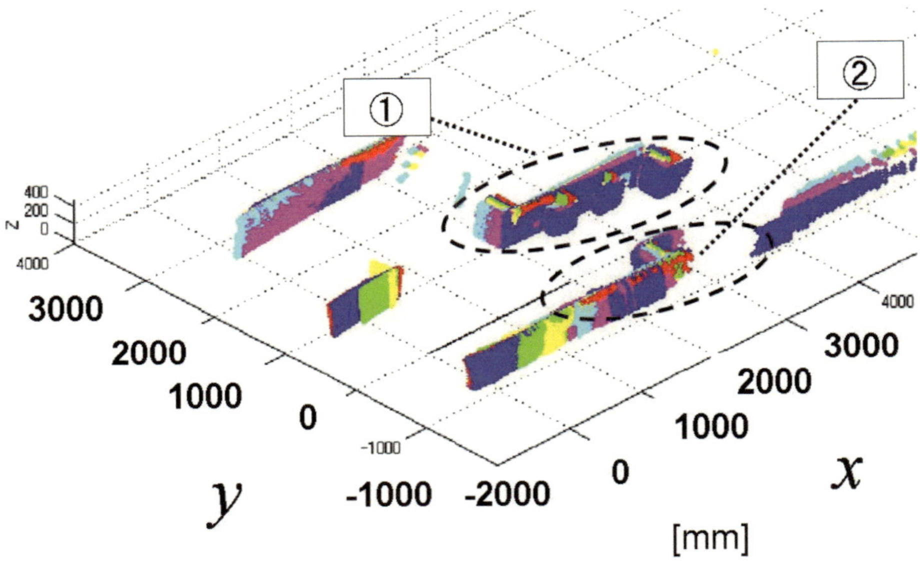

Fig. 18. Experimental result of 3D mapping

6. Discussions

We have employed fundamental experiments for sensing complex terrains: upward stairs, downward stairs, valley configuration, and side hole configuration under the robot. From Fig. 7, Fig. 10, Fig. 13, and Fig. 16, we can see that the almost same configuration was measured respectively. We therefore confirm that this sensing system has basic ability of 3D sensing and useful for more complex environment.

The result of sensing for upward stairs, as shown in Fig. 7, provided that the sensing by lifting the LRF vertically with equal interval was effective for getting whole 3D shape in the sensing area. We confirmed that the acceleration sensor was useful for this kind of sensing. This sensing method is also able to avoid a problem on accumulation point in conventional method which uses a rotating mechanism.

The result of sensing for downward stairs, as shown in Fig. 10 and Table 1, suggested that this system is possible to perform 3D mapping effectively even if the terrain has many

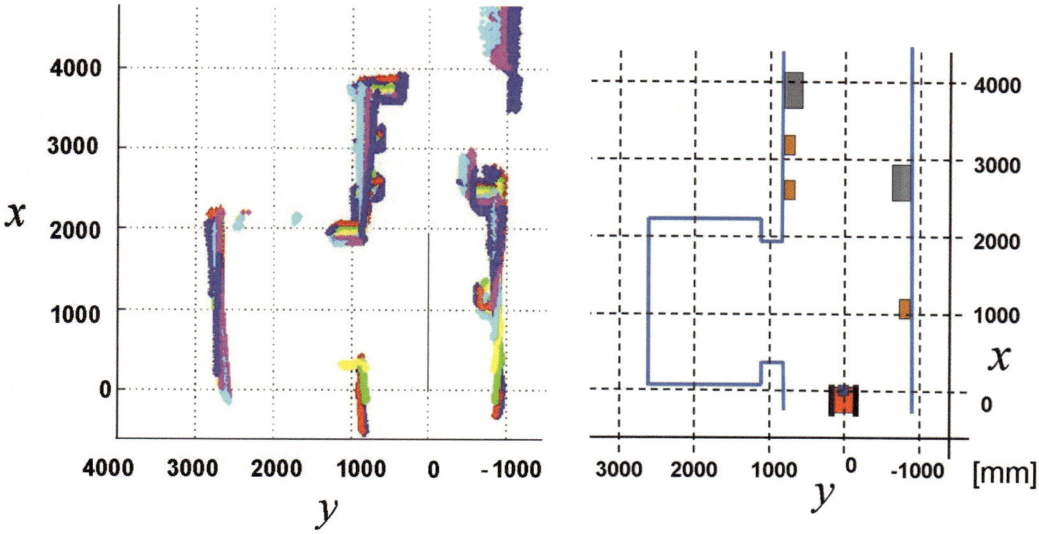

Fig. 19. Upper view of built map (left) and actual environment (right)

occlusions. The error ratio of distance was about 5% at a maximum. This error may be derived from mechanical errors of the unit in addition to original detection errors of the sensor device itself. It is necessary to develop the unit with mechanical stability. We however consider this error value is acceptable for a mapping for the purpose of movement or exploration by a tracked vehicle or a rescue robot.

The 3D shape of measurement result for a valley terrain, as shown in Fig. 13, indicated another advantage of the proposed sensing method. This sensing system is able to do sensing deep bottom area without occlusions. In addition, a robot can do it safely by this method because the robot does not have to stand at close to the border. We consider that the error ratio of 7.6% for the reference point e, shown in Table 2, occurred because the position was acute angle for the sensor. This error could be improved if the sensor is located properly so that it can face to the right position to the point. This sensing system can correspond to variety of terrain because the arm-type sensor movable unit can provide a lot of positions and orientations of the sensor.

The result of 3D measurement for a side hole under the robot also demonstrated further ability and strong advantage of the sensing system. Fig. 16 showed that this system enables us to obtain 3D information for such a shape which any conventional sensing system has never been able to measure. Moreover, the experimental result showed accurate sensing due to less error ratios, as shown in Table 3. This sensing system must be useful for 3D shape sensing specially in rough or rubble environments such as disaster area.

The experimental results for 3D mapping described in Section 5 indicated that this robot system was capable of building 3D map in wide area using odometry information. Fig. 18 showed almost actual shapes and positions of obstacles in the areas ① and ②. The sensing of top-surface of the obstacles also demonstrated one of advantages of this proposed system because such a sensing would be difficult for conventional method. Some errors however occurred in the far area from the beginning sensing location. We consider these errors may come from some odometry errors due to slip of tracks in the movement. More accurate mapping would be possible by solving this problem using external sensors with more sophisticated calculation method such as ICP (Nuchter et al., 2005) (Besl & Mckay, 2002).

7. Conclusions

This chapter proposed a novel 3D sensing system using arm-type sensor movable unit as an application of robot arm. This sensing system is able to obtain 3D configuration for complex environment such as valley which is difficult to get correct information by conventional methods. The experimental results showed that our method is also useful for safe 3D sensing in such a complex environment. This system is therefore adequate to get more information about 3D environment with respect to not only Laser Range Finder but also other sensors.

8. References

Besl, P. J. & Mckay, N. D. (1999). A method for registration of 3-d shapes, *IEEE Transactions on Pattern Analysis and Machine Intelligence,* 14(2), pp.239–256, August 2002

Hashimoto, M.; Matsui, Y. & Takahashi, K. (2008). Moving-object tracking with in-vehicle multi-laser range sensors, *Journal of Robotics and Mechatronics,* Vol.20, No.3, pp. 367-377

Hokuyo Automatic Co., Ltd., Available from http://www.hokuyo-aut.co.jp

Iocchi, L.; Pellegrini, S. & Tipaldi, G. (2007) Building multi-level planar maps integrating LRF, stereo vision and IMU sensors, *Proceedings of IEEE International Workshop on Safety, Security and Rescue Robotics 2007*

Nemoto, Z.; Takemura, H. & Mizoguchi, H. (2007). Development of Small-sized Omni-directional Laser Range Scanner and Its Application to 3D Background Difference, *Proceedings of IEEE 33rd Annual Conference Industrial Electronics Society(IECON 2007),* pp. 2284–2289

Nuchter, A.; Lingemann, K. & Hertzberg, J. (2005). Mapping of rescue environments with kurt3d, *Proceedings of IEEE International Workshop on Safety, Security and Rescue Robotics 2005,* pp. 158–163

Ohno, K. & Tadokoro, S. (2005). Dense 3D map building based on LRF data and color image fusion, *Proceedings of 2005 IEEE/RSJ International Conference on Intelligent Robots and Systems, 2005.(IROS 2005),* pp. 2792-2797

Poppinga, J.; Birk, A. & Pathak, K. (2008). Hough based terrain classification for realtime detection of drivable ground, *Journal of Field Robotics,* Vol. 25, No. (1-2), pp. 67–88

Sheh, R.; Kadous, M.; Sammut, C. & Hengst B. (2007) Extracting terrain features from range images for autonomous random stepfield traversal, *Proceedings of IEEE International Workshop on Safety, Security and Rescue Robotics 2007*

Ueda, T.; Kawata, H.; Tomizawa, T.; Ohya, A. & Yuta, S. (2006). Mobile SOKUIKI Sensor System-Accurate Range Data Mapping System with Sensor Motion, *Proceedings of the 2006 International Conference on Autonomous Robots and Agents,* pp. 309-304, December 2006

10

Design of a Bio-Inspired 3D Orientation Coordinate System and Application in Robotised Tele-Sonography

Courreges Fabien
Université de Limoges
France

1. Introduction

In designing a dedicated robotised telemanipulation system, the first approach should be to analyse the task targeted by such a teleoperation system. This analysis is essential to obtain cues for the robot mechanical, human-system interface, and the teleoperation control designs. In this chapter we will focus mainly on orientation-based tasks. That is to say, tasks consisting in orienting the remote robot's end-effector in 3D space. One major application considered here is the robotised telesonography medical examination. In this application a medical expert can pilot the orientation of an ultrasound (US) probe to scan a remote patient in real-time by means of a robot arm handling the probe. We have focused our approach on the telesonography application in order to analyse the task of setting the orientation of an object in space around a fixed centre of motion. For this analysis, several points of view have been taken into account: perceptual and psychophysical analysis, experimental tracking of the orientation applied by the hand, and the analysis of medical sonography practices recommendations. From these studies we have developed a new frame of three angles enabling the definition of an orientation. Indeed to define an orientation in 3D space (also said attitude), a representation system with at least three degrees of freedom or coordinates is required. This new frame was designed in such a way that its three degrees of freedom are decoupled with respect to the human psychophysical abilities. That is to say that each angle of this frame can be easily assessed and varied by hand without changing the value of the other angles of the frame. Hence the so-called hand-eye coordination can be improved with such a system of representation for interfaces design. We name this new system "*H-angles*" where the *H* recalls the Human-centred design of this system. We will also show that standard rotation coordinate systems such as the Euler and quaternions systems cannot offer such properties. Thereby our new frame of angles can lead to several applications in the field of telerobotics. Indeed we will provide cues indicating that the considerations used to design our new frame of angles are not limited to the context of the telesonography application. This chapter is devoted to present the foundations which led to the design of a new bio-inspired frame of angles for attitude description but we will also present one major application of this frame of angles such as the design of a mouse-based teleoperation interface to pilot the 3D orientation of the remote robot's hand-effector. This main application has arisen from the fact that the task of orienting an object in 3D space by means of a computing system requires the use of specific man-machine interfaces to be achieved fast and easily. Such interfaces often require the use of sophisticated and costly technologies to sense the orientation of the user's hand handling the interface. The fields of activity concerned are not limited to robot telemanipulation; we also find the computer-aided-design, the interaction with virtual reality scenes, and teleoperation of manufacturing machines. When the targeted applications are related to the welfare of the whole society, such as medical applications, the cost and availability of the system raises the problem of fair access to those high-tech devices, which is an ethical issue. The proposed system of angles enables the development of methods to perform advanced telemanipulation orientation tasks of a robot arm by means of low-cost interfaces and infrastructures (except probably the robot). Thus the most expensive element in such a teleoperation scheme will remain the robot. But for a networked-robot accessible to multiple users, we can imagine that the bundle of its cost could be divided up among the several users. In this chapter we will show a new method for using a standard wheeled IT mouse to pilot the 3D orientation of a robot's end-effector in an ergonomic fashion by means of the H-angles. In the context of the telesonography application, we will show how to use the aforementioned method to teleoperate the orientation of a remote medical ultrasound scanning robot with a mouse. The remaining of the chapter will be structured as follows: the second section coming next will provide our analysis in three parts to derive some cues and specifications for the design of a new frame of angles adapted to human psychophysical abilities. The third section is dedicated to our approach relying on the preceding cues to derive a new frame of angles for attitude description. It will also be shown that the new proposed system exhibits a much stronger improvement of decorrelation among its degrees of freedom (DOF) compared to the ZXZ Euler system. An analysis of the singularities of the new system is also proposed. The fourth section will address our first application of the new frame of angles that is to say the setting of 3D rotations with the IT mouse; this section will start with a review of the state-of-the-art techniques in the field and will end with experimental psychophysical results given in the context of the telesonography application. We show the large superiority of our frame of angles compared to the standard ZXZ Euler system. Last section concludes with an overview of further applications and research opportunities.

2. Design and analysis of a psychophysically adapted frame of angles for orientation description

The sensorimotor process of a human adult for achieving the task of orienting a rod by hand can be modelled according to the following simplified scheme from perception to action: This figure is a simplified scheme and may be incomplete but it reflects the present common trend of thought in the field of neuroscience concerning the information encoding and transformation from perception to action.

As it is reported in the neuroscience literature, the human brain can resort to several reference frames for perceptual modalities and action planning (Desmurget et al., 1998). Moreover, according to Goodale (Goodale et al., 1996), Human separable visual systems for perception and action imply that the structure of an object in a perceptual space may not be the same one in an interactive space which implies some coordinates frames transformations. This figure proposes the integration of multimodal information in the sensorimotor cortex to generate a movement plan into one common reference frame. This

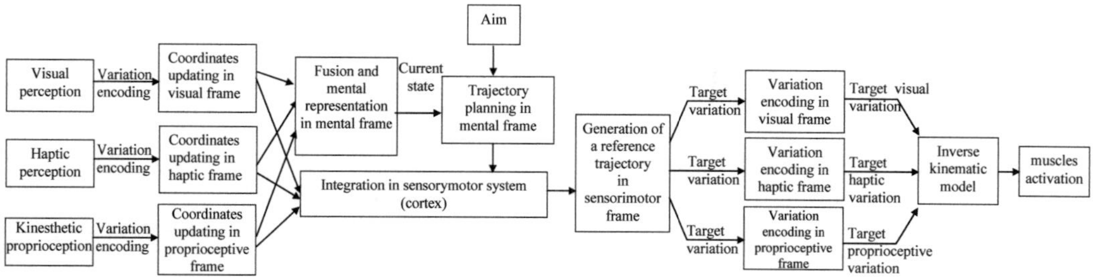

Fig. 1. Simplified Human perception to action process

concept comes from neurophysiological evidences reported by Cohen and Andersen (Y.E. Cohen & Andersen, 2002). Some research works (Paillard, 1987) also report the existence of two parallel information processing channels: cognitive and sensorimotor, which is reflected in figure 1. The idea of perception as action-dependent has been particularly emphasized by motor theories of perception, i.e. those approaches claiming that perceptual content depends in an essential way on the joint contribution of sensory and motor determinations (Sheerer, 1984). The theory underlies that action and perception are not independent cognitive domains and that perception is constitutively shaped by action. This idea is accounted in figure 1 by considering that motor variations are programmed in several frames of reference associated with each perceptual channel. Likewise, an inverse kinematics model learned by trials and errors in the infancy has been shown to be implemented by the central nervous system for the motor control (Miall & Wolpert, 1996). As depicted by figure 1, the task of handling a rod and making it rotate in space about a fixed centre of motion involves three perceptual modalities: visual, haptic, and kinaesthetic proprioception. The meaning of visual perception is unambiguous and this modality is essential for a precise motor control (Norman, 2002). The haptic modality involved here should be understood as "active touch" as defined by Gentaz (Gentaz et al., 2008): "*Haptic perception (or active touch) results from the stimulation of the mechanoreceptors in skin, muscles, tendons and joints generated by the manual exploration of an object in space… Haptic perception allows us, for example, to identify an object, or one of its features like its size, shape or weight, the position of its handle or the material of which it is made. A fundamental characteristic of the haptic system is that it depends on contact*". Haptics is a perceptual system, mediated by two afferent subsystems, cutaneous and kinaesthetic. Hence this perceptual system depends on spatio-temporal integration of the kinesthetics and tactile inputs to build a representation of the stimulus that most typically involves active manual exploration. The purely kinaesthetic proprioceptive perceptual system is a neurosensorial system providing the ability to sense kinaesthetic information pertaining to stimuli originating from within the body itself even if the subject is blindfolded. More precisely kinaesthetic proprioception is the subconscious sensation of body and limb movement with required effort along with unconscious perception of spatial orientation and position of body and limbs in relation to each other. Information of this perceptual system is obtained from non-visual and non-tactile sensory input such as muscle spindles and joint capsules or the sensory receptors activated during muscular activity and also the somato-vestibular system. Our aim in this section is to present our methodology to design an orientation frame comprehensible for both perception and action in performing a task of 3D orientation. We want a new frame of parameters whose values can be easily assessed from a perceived orientation, and easily set in orienting a rod by hand. Our approach was to seek for a system exhibiting three independent and decoupled coordinates when humans perform a planned trajectory in rotating a rod about a fixed centre of motion. For that purpose we have carried out an analysis in three parts given below. Before tackling this analysis we will provide some background and notations on orientation coordinate systems such as the quaternions and Euler angles.

2.1 Background on standard orientation coordinate system

We give in this section an insight on the most frequently used orientation representation systems in the field of human-machine interaction, namely quaternions and Euler systems.

2.1.1 The quaternions

The quaternions were discovered by Hamilton (Hamilton, 1843) who intended to extend the properties of the complex numbers to ease the description of rotations in 3D. A quaternion is a 4-tuple of real numbers related to the rotation angle and the rotation axis coordinates. Quaternions are free of mathematical singularities and enable simple and computationally efficient implementations for well-conditioned numerical algorithm to solve orientation problems. Quaternions constitute a strong formalization tool however it is not a so efficient mean to perform precise mental rotations. Quaternions find many applications especially in the field of computer graphics where they are convenient for animating rotation trajectories because they offer the possibility to parameterize smooth interpolation curves in SO(3) (the group of rotations in 3D space) (Shoemake, 1985).

2.1.2 Euler angles

Euler angles are intuitive to interpret and visualize and that's why that they are still widely used today. Such a factorization of the orientation aids in analyzing and describing the different postures of the human body. An important problem with using Euler angles is due to an apparent strength, it is a minimal representation (three numbers for three degrees of freedom). However all minimal parameterizations of SO(3) suffer from a coordinates singularity which results in a loss of a rotational degree of freedom in the representation also known as "gimbal lock". Any interpolation scheme based on treating the angles as a vector and using the convex sum will behave badly due to the inherent coupling that exists in the Euler angles near the singularity. Euler angles represent an orientation as a series of three sequential rotations from an initial frame. Each rotation is defined by an angle and a single axis of rotation chosen among the axes of the previously transformed frame. Consequently there are as many as twelve different sequences and each defines a different set of Euler angles. The naming of a set of Euler angles consists in giving the sequence of three successive rotation axes. For instance XYZ, ZXZ,... The sequences where each axis appears once and only once such as XYZ, XZY, YXZ, YZX, ZXY, ZYX are also named Cardan angles. In particular the angles of the sequence XYZ are also named *roll* (rotation about the x-axis), *pitch* (new y-axis) and *yaw* (new z-axis). The six remaining sequences are called proper Euler angles. In the present work it will be given a particular focus on the sequence ZXZ whose corresponding angles constitute the three-tuple noted (ψ,θ,φ). Angle ψ is called *precession* (first rotation about Z-axis), θ is the *nutation* (rotation about the new X-axis) and φ is named *self-rotation* (last rotation about the new Z-axis).

2.2 Neuroscience literature review

This section is dedicated to providing a comprehensive review of the neuroscience literature related to our purpose of identifying the 3D orientation encoding in the perceptual and sensory-motor systems. As indicated in figure 1 we have to investigate the orientation encoding in the following three perceptual systems: visual, haptic and proprioceptive. But we also have to consider the cognitive and the motor levels since Wang (Wang et al. 1998) argue that an interface design should not only accommodate the perceptual structure of the task and control structure of the input device, but also the structure of motor control systems. In the following the perceptual abilities (vision, haptic, proprioception) along with the mental cognition and motor control system will be indifferently denoted as *modalities*. We shall at first identify a common reference frame for all the modalities.

2.2.1 Common cross-modalities reference frame for orientations

As indicated previously, the reference frame may vary from one perceptual modality to another (Desmurget et al. 1998). Furthermore numerous studies have reported that for each modality its reference frame can be plastic and adapted to the task to be performed leading to conclude that several encodings of the same object coexist simultaneously. Importantly, the framework of multiple interacting reference frames is considered to be a general principle in the way the brain transforms combines and compares spatial representations (Y.E. Cohen & Andersen, 2002). In particular the reference frame can swap to be either egocentric (intrinsic or attached to the body) or allocentric (extrinsic to the body). This duality has been observed for the haptic modality (Volcic & Kappers 2008), the visual perception (Gentaz & Ballaz, 2000), the kinaesthetic proprioception (Darling & Hondzinski, 1999), the mental representation (Burgess, 2006) and the motor planning (Fisher et al., 2007; Soechting & Flanders, 1995). It is now a common opinion that both egocentric and allocentric reference frames coexist to locate the position and orientation of a target. In most of the research work it was found that whatever the modality, when the studied subjects have a natural vertical stance, the allocentric reference frame is gravitational or geocentric. It means that one axis of this allocentric reference frame is aligned with the gravitational vertical which is a strong reference in human sensorimotor capability (Darling et al. 2008). The allocentric reference frame seems to be common to each modality whereas this is not the case for the egocentric frame. It was also found for each modality that because of the so called "oblique effect" phenomenon the 3D reference frame forms an orthogonal trihedron. On a wide variety of tasks, when the test stimuli are oriented obliquely humans perform more poorly than when oriented in an horizontal or vertical direction. This anisotropic performance has been termed the "oblique effect" (Essock, 1980). This phenomenon was extensively studied in the case of visual perception (Cecala & Garner, 1986; Gentaz & Tschopp, 2002) and was brought to light also in the 3D case (Aznar-casanova et al. 2008). The review from Gentaz (Gentaz et al., 2008) suggests the presence of an oblique effect also in the haptic system and somato-vestibular system (Van Hof & Lagers-van Haselen, 1994) and the haptic processing of 3D orientations is clearly anisotropic as in 2D.

In the experiments reported by Gentaz the haptic oblique effect is observable in 3D when considering a plane-by-plane analysis, where the orientation of the horizontal and vertical axes in the frontal and sagittal planes, as well as the lateral and sagittal axes in the horizontal plane, are more accurately reproduced than the diagonal orientations even in the absence of any planar structure during the orientation reproduction phase. The oblique effect is also present at the cognitive level (Olson & Hildyard 1977) and is termed "oblique effect of class 2" (Essock, 1980). The same phenomenon has been reported to occur in the kinesthetic perceptual system (Baud-Bovy & Viviani, 2004) and for the motor control (Smyrnis et al., 2007). According to Gentaz (Gentaz, 2005) the vertical axis is privileged

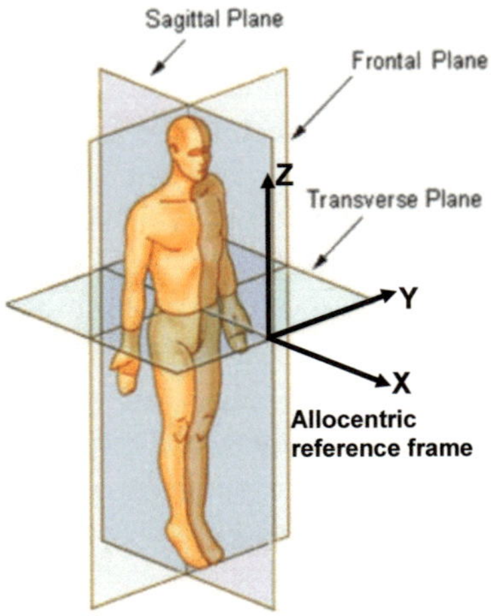

Fig. 2. Body planes and allocentric reference frame (picture modified from an initial public domain image of the body planes).

because it gives the gravitation direction and the horizontal axis is also privileged because it corresponds to the visual horizon. The combination of these two axes forms the frontal plane. A third axis is necessary to complete the reference frame and we will follow Baud-Bovy and Gentaz (Baud-Bovy & Gentaz, 2006a) who argue that the orientation is internally coded with respect to the sagittal and frontal planes. The third axis in the sagittal plane gives the gaze direction when the body is in straight vertical position (see figure 2). It should also be noticed that when the body is in normal vertical position, the allocentric and egocentric frames of most of the modalities are congruent. From now on, as it was found to be common to all modalities, it will be considered that the orientations in space are given with respect to the allocentric reference frame as described in figure 2.

2.2.2 Common cross-modalities orientation coordinate system

From (Howard, 1982) the orientation of a line in 2D should be coded with an angle with respect to a reference axis in the visual system. When considering the orientation of a rod in 3D space, two independents parameters at least are necessary to define an orientation and it seems from Howard that angular parameters are psychophysically preferred. It can be suggested from the analysis in the previous section about the common allocentric reference frame that the orientation encoding system should be spherical. For instance, the set of angles elevation-azimuth could well be adapted to encode the orientation of a rod in the allocentric reference frame of figure 2. Indeed the vertical axis constitutes a reference for the elevation angle and the azimuth angle can be seen as a proximity indicator of an oriented handled rod with respect to the sagittal and frontal planes. It should be noticed that the sets of spherical angles can carry different names but all systems made up of two independent spherical angles are isomorphic. We find for instance for the first spherical angle the naming: elevation, nutation, pitch,... and for the second angle : precession, yaw, azimuth,... Soechting and Ross (Soechting & Ross, 1984) have early demonstrated psychophysically that the spherical system of angles elevation-yaw, is preferred in static conditions for the kinaesthetic proprioceptive perception of the arm orientation. Soechting et al. have concluded that the same coordinate system is also utilized in dynamic conditions (Soechting et al. 1986). According to Darling and Miller (Darling & Miller, 1995), perceived orientations of the forearm in the kinaesthetic proprioception modality are preferably coded in spherical coordinates (elevation-yaw) with respect to an ego-centric body-centered reference frame. This frame coincides with the allocentric gravitational frame when the body trunk is in natural vertical position. The spherical system in orientation encoding is also supported by Baud-Bovy and Gentaz for the haptic perceptual system (Baud-Bovy & Gentaz, 2006b). Another interesting study in the context of the ultrasound scanning application is about visual perception of the orientation of a plane surface in a 3D space. Gibson (Gibson, 1950) has early proposed that the visual orientation of a surface in space is internally coded in spherical slant-tilt form, which was supported by Stevens psychophysical experiments (Stevens, 1983). The slant-tilt angles system is a spherical orientation encoding of the vector normal to the plane. This angles system is exactly the same as the elevation-azimuth system. From the previous discussion it can be stated that the orientation of a rod in 3D space is coded in spherical coordinates made up of two angles. But the orientation of a complete frame of three axes requires at least three parameters. For the consistency of the coordinate system the third parameter should preferably be an angle. To our knowledge very few psychophysical or neurophysiological studies have been carried out to identify a full set of three angles, coding the orientation of a frame in space. In the proprioceptive kinaesthetic context, Darling and Gilchrist (Darling & Gilchrist, 1991) confirm the finding of Soechting and Ross (Soechting & Ross, 1984) that the angles elevation and yaw are parts of the preferred DOF system for hand orientation. They also suggest from their experimental results that the roll angle in the ZXY Cardan system could constitute the third preferred DOF to define a complete orientation of the hand. This suggestion was contradicted by Baud-Bovy and Viviani (Baud-Bovy & Viviani, 1998) who have shown that the last angle in the ZXY Cardan system is strongly correlated with both first angles of that system and also with the reaching length. This result lets think that the six sets of Cardan angles are improper to code the orientation of a frame in a biomimetic way.

2.2.3 Discussion for orientation coding system design

This literature review enables to establish that a psychophysically and sensorimotor adapted coordinate system to encode the orientation of a rod in 3D space should be made-up of a set of two spherical angles with respect to an allocentric gravitational reference frame. As a matter of fact the quaternions of Hamilton whereas elegant and efficient in interpolating orientations doesn't seem to be the most appropriate system of orientation coding to fit with the psychophysical human abilities. In return even the most recent researches in the field are unable to identify a third necessary degree of freedom to define completely the orientation of a frame of three axes. This failure is probably due to the fact that this third DOF may be dependent on the task to perform and the kinematics postures of the acting arm and wrist during this task. Indeed the singularity arising in a minimal-coding system may be incompatible with the task to perform. For the task of handling a rod, a natural axis is given by the direction of the rod itself. Hence a spherical coordinate system such as (nutation, precession) should be used to code the orientation of the rod. Concerning the singularities it should be noticed that when orienting a rod the spherical coordinate system exhibits intrinsic singularities. Indeed when the nutation reaches 0 or π radians, the precession angle is undetermined which may lead to discontinuities in this angle. However, since there seems to be a consensus in favour of the spherical coordinate system in the field of the neurosciences it should be considered that those singularities truly reflect the human functioning mode and a biomimetic 3D orientation coding coordinate system should also probably exhibit such kind of singularity.

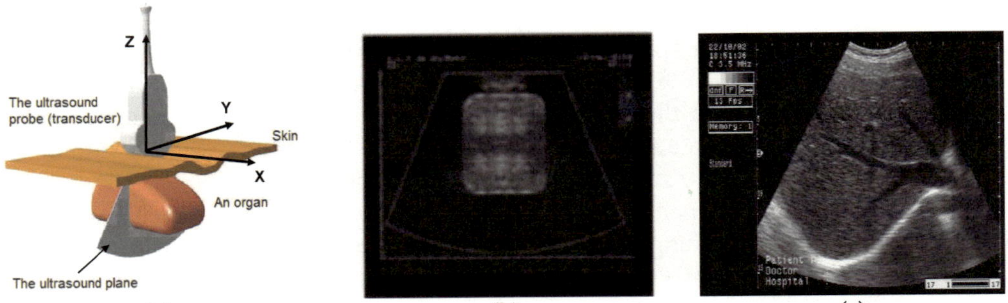

(a) (b) (c)

Fig. 3. (a) Ultrasound plane generated by the transducer and (b) corresponding generated image during a medical ultrasound examination (computer generated image). In (c) we have a real ultrasound slice of the hepativ vein.

When considering the application of ultrasound scanning, the rod is in fact the transducer generating an ultrasound (US) plane (figure 3) and a 3-axis frame should be attached to the US-plane to define its 3D orientation in space. Indeed the full 3D orientation description is important in the case of an ultrasound plane since the notions of right and left of the plane have a meaning in such an application: when the transducer is rotated around its own axis with an angle of π radians, it generates an US-plane which geometrically speaking remains the same plane in space, but the image obtained does not remain the same, right and left are inverted. It will be considered in the remaining of this chapter that the axes of the frame attached to the plane are arranged as given in figure 3. Axis Z is chosen to correspond with the longitudinal axis of the transducer. Among the Euler angles systems excluding the Cardan systems, only the sets ZXZ and ZYZ can offer spherical angles to code the direction of vector Z. Those two sets are perfectly equivalent and only the ZXZ system will draw our attention in the forthcoming sections. Next section will provide a deeper experimental analysis of this system with respect to the sonography application.

2.3 Experimental correlation analysis

This section's aim is to study the coupling within the angles of the ZXZ Euler system defining the orientation of the US plane when a medical expert performs a sonography examination on a real patient. This frame is preferably used in the robotized tele-echography context (Courreges et al. 2005; Gourdon et al. 1999; Vilchis et al. 2003) because it was found to be the one **among existing standard frames** that best suits the required mobilities during an US examination according to medical specialists (Gourdon et. al 1999). It is recalled that the DOF of the ZXZ Euler system is the triplet of angles: (ψ, θ, φ), where ψ is the precession, θ is the nutation and φ the self-rotation. We have set-up an experimental protocol in order to assess the dependencies of the degrees of freedom of the ZXZ Euler system and analyze the task to be performed by a medical tele-sonography robot. For that purpose we have captured the 6 DOFs movements of a real US specialist performing an abdominal examination of a healthy patient. The acquisition duration is about 5 minutes. During this examination the ultrasound (US) probe trajectories have been captured and recorded using a 6D magnetic localization sensor « Flock Of Bird » settled on the US probe

(Fig. 6). This kind of examination is frequently performed in routine and especially in emergency situations. The trajectories applied to the US probe by any expert would be roughly the same since these gestures come from the learning of recommended medical practices and not only from individual experience and is also subject to the human hand kinematics limitations (Tempkin, 2008). To better identify the correlation between angles and to be independent of the angles range and rollover we have considered the angles velocities.

2.3.1 Correlation in the angles

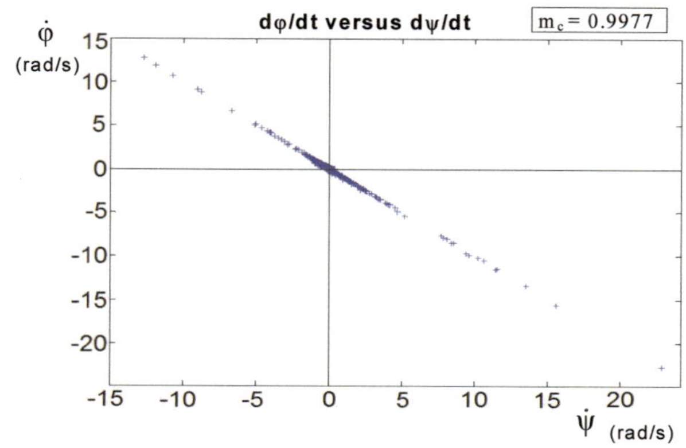

(a)

variables	mc
$\dot{\psi}, \dot{\theta}$	0.0041
$\dot{\varphi}, \dot{\theta}$	0.0164
$\dot{\psi}, \dot{\varphi}$	0.9977
average	0.3394

(b)

Fig. 4. (a) Phase plot of $\dot{\varphi}$ versus $\dot{\psi}$ and (b) correlation measures m_c.

To emphasize the dependencies among the Euler angles for this kind of application, we have analyzed the phase plots of each angle derivative versus the other ones (Courreges et al. 2008b). We can easily obtain a correlation measure by considering the absolute value of the Pearson correlation coefficient (J. Cohen et al., 2002). Let us name this measure m_c which is null for uncorrelated signals and is equal to 1 when the signals are linearly dependent. Figure 4b reports the correlation measures. From the plots obtained and correlation measures one can conclude that $\dot{\psi}$ and $\dot{\theta}$ are uncorrelated, $\dot{\varphi}$ and $\dot{\theta}$ are also uncorrelated, but $\dot{\psi}$ and $\dot{\varphi}$ are strongly correlated (see also figure 4a). Consequently it is clear that the ZXZ Euler system is not perfectly suited for this application, as it can't provide decoupled DOF to describe the US scanning task.

2.3.2 Data analysis

These experimental data show that the spherical coordinates (ψ, θ) are uncorrelated DOF which is in agreement with the previous neuroscience literature review. The previous data also clearly reveal that the ZXZ Euler system exhibits a strong correlation between the precession and self rotation angles. In other words applying a variation on angle ψ should induce a near proportional variation on angle φ according to figure 4(a). Since this proportionality applies whatever the value of θ, one can notice that the correlation of the angles is not related to the singularity of this Euler system (when θ=0). Thereby we conclude that the standard ZXZ Euler system is not the most appropriate system to represent the human privileged rotations directions when handling a rod. This analysis shows the need for the definition of a new non standard frame capable of providing decoupled DOF for this kind of task. Since according to figure 4, ψ and φ angles are strongly correlated, a principal component analysis (PCA) (Joliffe, 2002) of the phase plots of the moves expressed in the ZXZ Euler system should provide us with decorrelated DOF. Indeed we can define a new coordinate system by using the Karhunen-Loève transform (Loève, 1978) which provides a very good decorrelation of the DOF. Let us name (α,θ,β) this DOF triplet. According to the PCA, θ is the same as the Euler nutation. Variables α and β are linear combination of the Euler angles ψ and φ (see equation 1). Whereas this transformation is simple, this PCA based system doesn't provide meaningful variables: α and β are not intuitive for the hand-eye coordination. Moreover this transformation is optimal only for the particular conditions chosen for this experiment and may not be appropriate in other circumstances.

$$\begin{cases} \alpha = (\psi - \varphi)/\sqrt{2} \\ \beta = (\psi + \varphi)/\sqrt{2} \end{cases} \tag{1}$$

2.4 Sonography practice analysis

To obtain enough information to build a complete orientation coordinate system we studied the practice of sonography. More specifically we have analysed the way a 3D rotation is decomposed into simpler moves for pedagogical purpose in teaching the technique of medical US scanning. An US transducer works by generating a planar wave of ultrasounds. Waves reflected by the tissues are measured by the probe along with their time of flight, which enables to build a map of the density of the tissues (fig. 3c). Hence a medical expert has to think to rotate a plane in a 3D space to visualize the desired slice of the patient's body. In fact sonographers are used to describe their scan orientation by reference to three basis rotations (Tempkin, 2008; Block, 2004): probe angulation, probe rocking (fig.5) and self rotation. And in standard medical practice the examination is executed in two phases combining these three basis rotations: first, choosing an initial incidence for the ultrasound plane combining probe angulation and probe rocking so as to perform a narrow sweep of the scanned organ. This first move is intended to grossly identify lesions or cysts. Second phase consists in rotating the US plane around the probe axis so as to identify small structures as tumors or traumas and precisely locate their extent. A bio-inspired orientation frame should exhibit this same combination of movements. Consequently it was found that the professional field of medical sonography gives practical guidelines to maneuver the orientation of a probe in 3D space. Whereas the conclusions of this analysis are related to the specific field of sonography it is interesting to notice that the pragmatical rules for this task are consistent with the previous neuroscience conclusions. Indeed the recommended movements of probe angulation and probe rocking correspond exactly to the plane slant-tilt rotations as indicated in section 2.2.1. Moreover this analysis provides a complete and intuitive combination of hand movements to set the full 3D orientation of a frame in space since this combination was practiced and taught for ages in the field of sonography. The next section takes advantage of this analysis to propose a new frame of orientation description with a set of three angles satisfying the human preferences when someone performs an intentional orientation tracking. It is reasonable to think that this new frame could be satisfying not only for the fields of sonography-related applications but also for any other tasks implying the rotations of a rod about a fixed centre of motion.

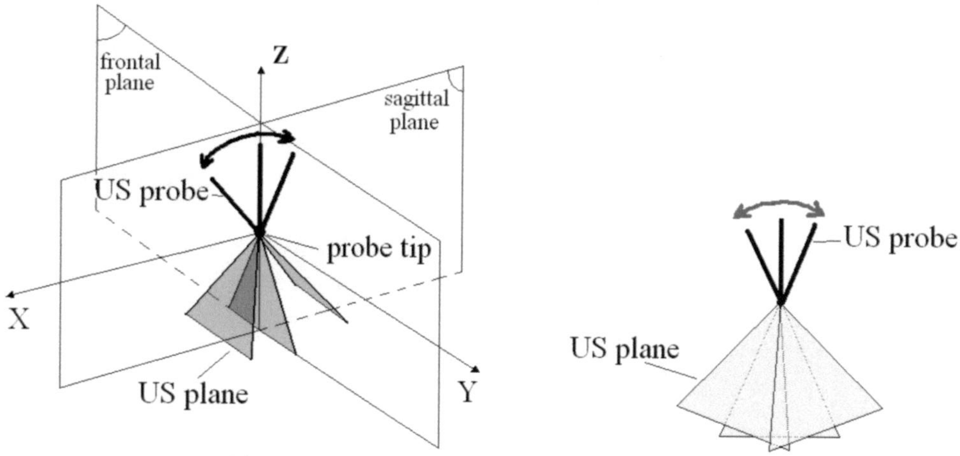

(a) (b)

Fig. 5. Two basic moves in medical US scanning. (a) probe angulation for organ sweeping (in this illustration, moves of the US probe remain in the sagittal plane). (b) probe rocking used to extend the scanning plane.

3. H-angles, a new attitude coordinate system

We will develop here our methodology to design a new frame of angles satisfying the criteria of the previous analyses. We call this frame *H-angles*. We will provide the transformations from the H-angles to the rotation matrix and inversely from the rotation matrix to the H-angles. We will also conduct an exhaustive analysis of the singularities of the rotation matrix. This section will be concluded with the excellent decorrelation results brought by our new system compared to the Euler system. Notice that in the following the angles unit should be understood in radians.

3.1 Rotations combinations

From previous analysis we propose a new frame of angles which we name H-angles and denoted as (ψ_n, θ_n, φ_n) parameterising an orientation obtained by a sequence of two consecutives rotations as for medical practice. Let's give some notations: let R_0 = (O, X0, Y0, Z0) be the fixed main reference frame with centre O, axis (X0, Y0, Z0) and basis B_0 = $(\vec{x_0}, \vec{y_0}, \vec{z_0})$. The basis obtained by the first transform on basis B_0 is denoted by $B_1 = (\vec{x_1}, \vec{y_1}, \vec{z_1})$. The framework with basis B_1 and origin O is noted R_1. The first movement is a complex rotation. According to the previous conventions the moving vector z gives the direction of the handled rod and vector x is normal to the moving plane corresponding to the US plane in sonography. This first rotation has two main functions:

- defining vector $\vec{z_1}$ by its nutation $\theta_n \in [0; \pi]$ and precession $\psi_n \in]-\pi; \pi]$, which is consistent with the neuroscience requirements depicted in §2.2.1;
- forcing vector $\vec{y_1}$ to stay in the plane $(\vec{z_1}\, O\, \vec{y_0})$ so as to constrain the first move to be only a combination of probe angulation and probe rocking as for medical practice (§2.4). This constraint implies $\vec{x_1} \cdot \vec{y_0} = 0$.

Given the previous constraints the orientation of basis B_1 is not totally determined and the sign of the dot product $\vec{x_1} \cdot \vec{x_0}$ must be defined according to the kind of application. In order to obtain a transformation with the minimum rotation angle (for minimum rotation effort) we have chosen to set: sign($\vec{x_1} \cdot \vec{x_0}$) = sign($\cos(\theta_n)$). Notice that the sign of $\cos(\theta_n)$ indicates in which hemisphere vector $\vec{z_1}$ is. Hence for the typical application of rotating a rod about a fixed centre of motion with its workspace located in the North hemisphere of the orientation space, we have sign($\vec{x_1} \cdot \vec{x_0}$) ≥ 0 indicating angle $\widehat{(\vec{x_1}, \vec{x_0})}$ is acute. Figure 6 provides a graphical overview of this first move, where the origin's definition of ψ_n angle has been chosen in analogy with the precession of the ZXZ Euler angles.

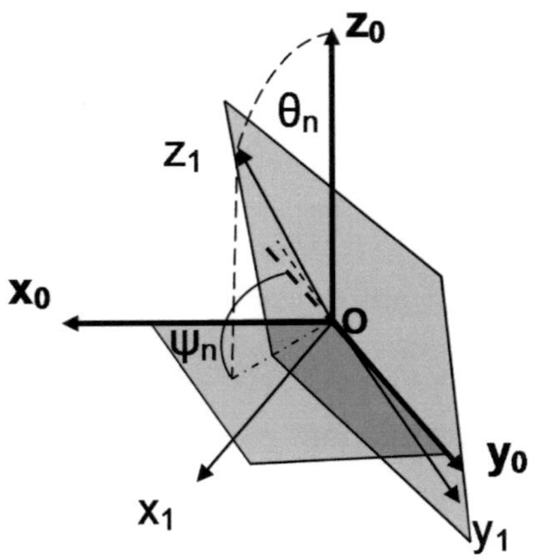

Fig. 6. First movement from B_0 to B_1. Vectors z_1, y_0 and y_1 are in the same plane.

The second transform is a simple rotation about vector $\vec{z_1}$ of angle $\varphi_n \in]-\pi; \pi]$ which we name "self rotation". On setting the same value for the precession and nutation angles in the ZXZ Euler svstem and in the new H-angles proposed system, we obtain the same position for vector $\vec{z_1}$. Hence the difference resides in the self rotation angle and the directions of vectors x and y. This modification of the self-rotation can be seen as an anticipation on the hand movement considering θ_n and ψ_n as inputs of this anticipator.

3.2 Rotations matrices

As indicated in the previous section the proposed orientation description system is decomposed into two sequential rotations. For each of these two rotations it is possible to write its rotation matrix and then multiply the matrices so as to express the global rotation matrix M. For the first rotation we denote M1 the rotation matrix. Let's define (z_x, z_y, z_z) the components of vector $\vec{z_1}$ in basis B0. We have:

$$\begin{cases} z_x = \sin\theta_n \sin\psi_n \\ z_y = -\sin\theta_n \cos\psi_n \\ z_z = \cos\theta_n \end{cases} \tag{2}$$

Components of vectors $\vec{x_1}$ and $\vec{y_1}$ can then be expressed as a function of (z_x, z_y, z_z) and we derive an expression of matrix M1 as function of (z_x, z_y, z_z):

$$M1 = \begin{bmatrix} \dfrac{z_z}{\sqrt{z_x^2 + z_z^2}} & \dfrac{-z_x z_y}{\sqrt{z_x^2 + z_z^2}} & z_x \\ 0 & \sqrt{z_x^2 + z_z^2} & z_y \\ \dfrac{-z_x}{\sqrt{z_x^2 + z_z^2}} & \dfrac{-z_z z_y}{\sqrt{z_x^2 + z_z^2}} & z_z \end{bmatrix} \qquad (3)$$

For the second rotation, we note M2 the standard rotation matrix operating a rotation about vector $\vec{z_1}$ in the frame R_1 with magnitude φ_n. The global rotation matrix M in the frame R_0 can then be computed as M = M1.M2 which leads to the following expression of M as a function of the H-angles:

$$M = \begin{bmatrix} \dfrac{\cos\theta_n \cos\varphi_n + \frac{1}{2}\sin^2\theta_n \sin(2\psi_n)\sin\varphi_n}{\sqrt{1-\sin^2\theta_n \cos^2\psi_n}} & \dfrac{-\cos\theta_n \sin\varphi_n + \frac{1}{2}\sin^2\theta_n \sin(2\psi_n)\cos\varphi_n}{\sqrt{1-\sin^2\theta_n \cos^2\psi_n}} & \sin\theta_n \sin\psi_n \\ \sin\varphi_n \sqrt{1-\sin^2\theta_n \cos^2\psi_n} & \cos\varphi_n \sqrt{1-\sin^2\theta_n \cos^2\psi_n} & -\sin\theta_n \cos\psi_n \\ \dfrac{\sin\theta_n(-\sin\psi_n \cos\varphi_n + \cos\theta_n \cos\psi_n \sin\varphi_n)}{\sqrt{1-\sin^2\theta_n \cos^2\psi_n}} & \dfrac{\sin\theta_n(\sin\psi_n \sin\varphi_n + \cos\theta_n \cos\psi_n \cos\varphi_n)}{\sqrt{1-\sin^2\theta_n \cos^2\psi_n}} & \cos\theta_n \end{bmatrix} \qquad (4)$$

As a first analysis we can see that matrix M is not defined when $\theta_n = \pi/2$ and $\psi_n = 0$ or π. When these conditions are met, vectors $\vec{x_1}$ and $\vec{y_1}$ are undetermined. Hence M can be rewritten as function of the components of vector $\vec{x_1}$ in basis B0 since we have $\vec{x_1} = (x_x, 0, x_z)$ and $z_x=z_z=0$ and $z_y = -\cos(\psi_n) = \pm 1$. We find:

$$M = \begin{bmatrix} x_x \cos\varphi_n - x_z \cos\psi_n \sin\varphi_n & -x_z \cos\psi_n \cos\varphi_n - x_x \sin\varphi_n & 0 \\ 0 & 0 & -\cos\psi_n \\ x_z \cos\varphi_n + x_x \cos\psi_n \sin\varphi_n & -x_z \sin\varphi_n + x_x \cos\psi_n \cos\varphi_n & 0 \end{bmatrix} \qquad (5)$$

The values of x_x and x_z can be context dependent. In practical applications of rotating a rod about a fixed centre of motion, the case $\theta_n = \pi/2$ is a limit hardly reachable. It can be found severable possible reasons to this:

- the application itself exhibits bounds that avoid reaching such a limit for θ_n such as the robotised tele-sonography application (Courreges et al., 2008a);
- or more simply the centre of motion may be on a plane surface and this surface avoids the hand from reaching this limit nutation.

3.3 Expression of the H-angles ($\psi_n, \theta_n, \varphi_n$) from the rotation matrix M components; singularities analysis

3.3.1 Extraction of the H-angles from the matrix outside singularities

The component of matrix M at line i and column j is noted m_{ij}. We find from equation (4):

$$\theta_n = \arccos(m_{33}) \tag{6}$$

Outside singular configurations it can be found:

$$\psi_n = \mathrm{a}\tan 2(m_{13}, -m_{23}) \tag{7}$$

$$\varphi_n = \mathrm{a}\tan 2(m_{21}, m_{22}) \tag{8}$$

Where "atan2" is an algorithmic function able to compute the arc tangent from two arguments so as to determine the quadrant of the angle on the trigonometric circle.

3.3.2 Singularities analysis

Two types of singularities can be identified and are studied hereafter.

1. When $m_{13}=0$ and $m_{23}=0$ simultaneously.

This singularity is obtained for $\theta_n=0$ or $\theta_n=\pi$. This situation implies: $z_x=z_y=0$ and angle ψ_n is undetermined. In rewriting the rotation matrix M (equation 9 hereafter), one can see that M is independent of ψ_n as z_z depends only on θ_n. Hence any value can be set for ψ_n without hindering the orientation. In our tele-sonography application we have given ψ_n a value of 0 when this configuration is met (Courreges et al. 2008b).

$$\mathrm{M} = \begin{bmatrix} sign(z_z)\cos\varphi_n & -sign(z_z)\sin\varphi_n & 0 \\ \sin\varphi_n & \cos\varphi_n & 0 \\ 0 & 0 & z_z \end{bmatrix} \tag{9}$$

Those singularities seem to be cumbersome as they correspond to some psychophysically preferred directions: the vertical axis. However it ensues from the discussion in section 2.2.2 that these singularities may be well integrated in the human internal orientations encoding and indeed the forthcoming results with the H-angles will enforce this conclusion. Moreover it can be noticed that those singularities are very different from the singularities of the ZXZ Euler system. Indeed in such Euler system its singularities can disrupt two of its degrees of freedom namely the precession and self-rotation; whereas with the H-angles only the precession angle is affected.

2. When $m_{21}=0$ and $m_{22}=0$.

When this singularity is met angle φ_n is undetermined. This situation implies both z_z and z_x are null simultaneously, hence $\theta_n = \pi/2$ and $\psi_n = 0$ or π and we meet the case where vector $\vec{x}_1$ is undetermined. When matrix M is rewritten according to equation (5), one can find:

$$\varphi_\mathrm{n} = \mathrm{atan2}\left(-\mathrm{x_x m_{12}} - \mathrm{x_z m_{32}}, \mathrm{x_x m_{11}} + \mathrm{x_z m_{31}}\right) \tag{10}$$

Components x_x and x_z can be chosen freely according to the context but have to satisfy: $x_x^2 + x_z^2 = 1$.

3.4 Decorrelation results

We have computed the velocities of the new H-angles attitude system for the same medical trajectory than in section 2.3 with the ZXZ Euler angles (Courreges et al. 2008b). As one could expect from the definition of the new system, there are no changes on $\dot{\psi}_n$ versus $\dot{\theta}_n$

compare to the plot $\dot{\psi}$ versus $\dot{\theta}$ of the ZXZ Euler angles velocities, hence they are still uncorrelated in our new system. We have obtained a good decorrelation between $\dot{\varphi}_n$ and $\dot{\psi}_n$ with a low coefficient m_c = 0.0116 which is a great improvement compared to the Euler system. The correlation m_c = 0.1552 of the variables $\dot{\varphi}_n$ and $\dot{\theta}_n$ has been raised in a relative important way compare to the homologous variables in the Euler system. However this value still remains low enough to consider the angles uncorrelated. To quantify the decorrelation improvement we can compute the average correlation coefficient for each system of angles. Our new system exhibits an average correlation $\hat{m}_{cn}$ =0.057 whereas the ZXZ Euler angles system exhibits an average correlation $\hat{m}_{cE}$ =0.339 (figure 4b). Consequently our new system provides a decorrelation improvement of more than 83% with respect to the average correlation measure. For comparison purpose, the average correlation measure of the PCA based system given in equation (1) is $\hat{m}_{cKL}$ =0.0302 which is much closer to our new system than to the Euler system.

4. Application: ergonomic mouse based interface for 3D orientation in robotised tele-sonography

We will show in this section how to exploit our new frame of angles to render the use of the standard IT mouse feasible to pilot efficiently the 3D orientation of a rod. We have tested this technique in the context of the particular application of robotised telesonography. In a first subsection we will draw some design requirements for 3D rotations techniques with a mouse. The following subsection will provide a short overview of existing techniques to set a 3D orientation with a mouse. This subsection is focused on reporting the evaluation and comparison of the various techniques considered with respect to the design principles promulgated in the preceding subsection. Next subsection will present our approach in exploiting the new frame of angles. The fourth subsection will describe the chosen psychophysical experimental protocol along with quantified results.

4.1 Design recommendations for 3D rotations techniques with a mouse

From their experience Bade et al. (Bade et al., 2005) have promulgated a number of four general principles as crucial for predictable and pleasing rotation techniques:

1. *Similar actions should provoke similar reactions:* the same mouse movement should not result in varying rotations.
2. *Direction of rotation should match the direction of 2d pointing device movement.*
3. *3d rotation should be transitive:* the rotation technique must not have hysteresis. In other words to one pointing location with the interface should correspond one and only one 3D orientation whatever the trajectory ending to that location.
4. *The control-to-display ratio should be customizable:* tuneable parameters must be available to find the best compromise between speed and accuracy according to the task and user preferences and is therefore crucial for performance and user satisfaction.

We also add a fifth principle:

5. *The input interface should allow the setting of an orientation by an integrated manipulation:* Hinckley showed (Hinckley et al., 1997) that the mental model of rotation is an integral manipulation in opposition with separable manipulation as defined by Jacob (Jacob R.J.K. et al., 1994). From a practical point of view the input interface should be designed to enable a simultaneous variation of each degree of freedom of the rotation.

4.2 Overview of 3D rotations techniques with a mouse and evaluations

This field of research has not much evolved this last decade and the works related to the use of the computer mouse to set an orientation in 3D are in applications to the fundamental research topic known as "2D interface for 3D orientation". Hence the following review reports some techniques where the mouse is considered as an input device with only two DOFs and which omit the input of the mouse wheel. The most well known and popular techniques because they are preferred (Chen et al., 1988) are based on the virtual trackball principle. It consists in surrounding the object to rotate by a virtual sphere fixed with the object (but the sphere may not be always displayed). The object is rotated by operating the virtual sphere with the mouse pointer. The common principle of these techniques to generate a rotation consists in letting the user select two locations with the mouse pointer. The first position is validated by a mouse click and remains constant until the next click; the second position can be moving. Those two points are then mapped to the virtual sphere and the projected points on the sphere enable to define an arc on a great circle. The angle of rotation is chosen as the aperture angle of the arc viewed from the sphere centre; and the axis of rotation is chosen perpendicular to the plane formed by the centre of the sphere and the arc. The virtual trackball-like techniques are preferred among other existing techniques with 2D input devices because they enable perform faster for both rotations and inspection tasks (Jacob I. & Oliver, 1995). From Henriksen (Henriksen et al., 2004): "*Virtual trackballs allow rotation along several dimensions simultaneously and integrate controller and the object controlled, as in direct manipulation. The main drawback of virtual trackballs is a lack of thorough mathematical description of the projection from mouse movement onto a rotation.*" This class of techniques comprise the techniques known as the Virtual sphere of Chen (Chen et al., 1988), the Arcball of Shoemake (Shoemake, 1992), the Bell's virtual trackball (Henriksen et al., 2004), the two-axis valuator (Chen et al., 1988) and the two-axis valuator with fixed up-vector (Bade et al., 2005). These techniques essentially differ in their plane-to-sphere projection. Not much experimental comparisons and evaluations of these techniques have been proposed in the literature. Bade et al. (Bade et al., 2005) have presented a tabular comparison of these techniques (excluding Chen's Virtual Sphere) with respect to the first four principles reported in the preceding section 4.1. We propose hereafter an extended comparison table (table 1 below).

Techniques / Design principles	Chens' Virtual sphere	Shoemake's Arcball	Bell's virtual trackball	Two-axis valuator	Two axis valuator with fixed up-vector
Principle 1	-	-	-	+	-
Principle 2	+	+	+	+/-	-
Principle 3	-	+	-	-	+
Principle 4	+	-	-	+	+
Principle 5	-	-	-	-	-

Table 1. Comparison of state of the art rotation techniques with respect to the five design principles.

For Chen's Virtual sphere we set principle 4 as satisfied since the ratio between the sphere radius and the radius of the mouse's workspace on the table can be tuned which will affect the compromise velocity/precision. Shoemake's Arcball renounces to this principle to be able to satisfy principle 3 (avoid hysteresis) (Hinckley et al., 1997). We set the principle 5 to be unsatisfied for each technique despite the preceding quote of Henriksen because once a first point is selected with the mouse the possible rotation axes are constraint to lie in a bounded space defined by the position selected. Hence every rotation can't be performed within a single smooth hand movement. This is supported by Hinckley (Hinckley et al., 1997) who argues that practically both the Virtual Sphere and the Arcball techniques require the user to achieve some orientations by composing multiple rotations each initiated by a cursor repositioning and mouse click which breaks the movement smoothness. The study from Hinckley did not provide evidence that the Arcball performs any better than the Virtual Sphere for both accuracy and completion time. The main usability problem with the virtual trackballs compare to free hand input devices was that users were unsure about the difference between being inside and outside the virtual sphere. The experiments of Bade et al. (Bade et al., 2005) combining inspection and rotations tasks revealed that users significantly perform faster with the two-axis valuator technique which was perceived as more predictable and comfortable for task completion than other trackball techniques. Bade et al. also suggest that these results were expected as the two-axis valuator fulfils most of the design principles. In these experiments the Shoemake's Arcball arrives in second position outstripping the Bell's virtual trackball and the two-axis valuator with fixed up-vector. A strong drawback of these techniques comes from their lack in satisfying principle 5 which make these techniques much slower than compared to the natural rotation of object by hand in 3D free space (Hinckley et al., 1997; Pan, 2008). The proposed method presented hereafter enables to satisfy all four principles within a large continuous range of orientations.

4.3 Angles coding in setting a rod attitude with a mouse

To ease the hand-eye co-ordination of the operator, for interface design, it is necessary to take care that the operator can easily assess the orientation changes when moving by hand the input device (Wolpert & Ghahramani, 2000). That is why we have chosen a biomimetic approach for the orientation control with a mouse. As discussed previously, the attitude of a US probe is defined by a sequence of two movements where the first movement enables to set the nutation and precession of the probe axis. This observation leads to state that humans have the sensorimotor ability to easily control the nutation and precession of a rod. In fact defining the precession and nutation of a constant length rod is the same task as placing a point, representing the top end of that rod on a sphere. This sphere radius is the length of the rod, namely in our application, the US probe length, and the sphere centre is the probe bottom tip. In a telesonography application only the north hemisphere is to be considered. It is possible to make a mental bijective transform from the orientation hemisphere to the mouse plane. However such a projection is not unique (Kennedy & Kopp, 2001). To make a proper choice of a particular sphere-to-plan projection it is necessary to account for the human sensorimotor abilities so as to maintain decoupled DOF with respect to the nutation and precession. We have chosen the class of projections that preserves the precession angle, namely the azimuthal projections, which are projections of the sphere on a tangent plane. The chosen tangent point is the North Pole, which defines the so important vertical axis (see §.2.2.1), because such transform generates less distortion around the tangent point. This choice also allows to visualize the mouse control of the probe from an overhead view (fig. 7b), where the origin is the tangent point between the orientation sphere and the plane. This transform guarantees the hand-eye coordination since it allows establishing one to one decoupled relations between the orientation DOF of probe nutation-precession and the visually and kinesthetically perceived polar coordinates of the mouse. Indeed the precession is growing and linearly dependent on the polar angle of the mouse and the nutation is a growing function of the distance from the mouse to the origin. To totally determine the projection, we have to set the perspective point. It has been reported that, because it is cognitively preferred, the path adopted by hand when moving on a plane from an initial position to a target position is a fairly straight-line (Sergio & Scott, 1998; Desmurget et al., 1999). Consequently a sphere to map projection that preserves orthodromy should be preferred (the shortest path between two points on the sphere - which is a great circle - should map to a straight line on the plane). Such a projection is a gnomonic projection where the projection center is at the center of the sphere (see figure 7a). Despite the drawback of the chosen sphere-to-plane transform implying sending a nutation of $\pi/2$ radian to infinity, it is however well suited to tele-sonography application for routine examination. Indeed we have shown in a previous work that the nutation remains lower than $\pi/4$ radians during 95% of the examination time in routine abdominal US scanning (Courreges et al., 2008a). To rotate the probe around its own axis and define the self rotation angle, the mouse scrolling wheel is used.

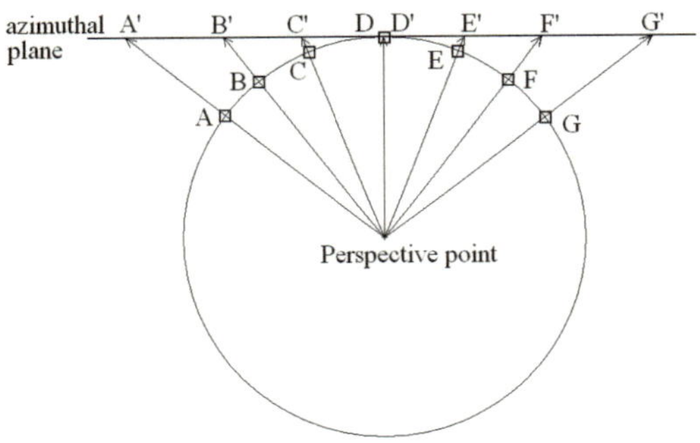

(a)

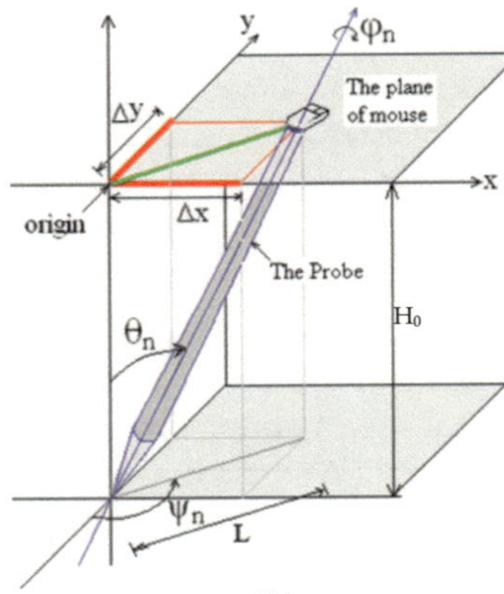

(b)

Fig. 7. (a) Gnomonic projection. Each point of the North hemisphere is projected on the azimuthal plane along a radius originating from the sphere center. In (b), use of a computer mouse as telerobotic control interface to set the frame of H-angles.

This point can be a limitation since a computer mouse wheel moves generally in discreet steps. Hence a compromise has to be adopted when choosing the increment factor to convert the wheel increment to angle increment. A great factor allows driving fast but reduces the accuracy whereas a small factor exhibits the contrary. This factor has to be chosen according to the mouse wheel's total number of increments and by considering the application needs. The representation proposed in figure 7b makes the virtual mouse controlled probe behave as if it were of variable length. The bottom tip of this virtual variable length probe is fixed, and top corresponds to the mouse position on the plan as is shown in the illustration figure 7b. The presented experiment revealed that operators could easily adapt to a variable length virtual probe (in the range employed for the nutation in this experiment) since it doesn't affect the attitude of the probe. To operate the simulated robot, the first step is to fix an origin for the mouse pointer. This origin position would correspond to a null calibration of every angle, where the probe is in vertical position. The mouse controls directly the angles in the chosen frame of angles: (ψ_n,θ_n,φ_n) for the new bio-inspired H-angles system envisaged and (ψ,θ,φ) for the Euler angles. The following equations use the new angles but are the same with the Euler angles. Let's define the following notations used in figure 7b:

Δx : displacement along X axis of the mouse from its origin position.

Δy : displacement along Y axis of the mouse from its origin position.

L: length of the projection of the mouse controlled probe in the XY plan.

H_0: minimal length of the virtual variable length mouse controlled probe. H_O is a tunable parameter enabling to set the control sensibility on angle θ_n such that the preceding design principle 4 (control-to-display ratio) can be satisfied.

ΔW_{inc} : mouse wheel increments variation (in radians) from the origin calibration position.

K_{i_a}: conversion factor from mouse wheel increments to angle in radians. This factor is tunable by the user and contributes to satisfy the design principle 4.

Using the previous definitions and according to figure 7b, the expressions of the orientation angles as a function of the mouse inputs are:

$$\theta_n = \operatorname{atan}\left(\frac{L}{H_0}\right) \quad (11)$$

$$\psi_n = \pi/2 + \operatorname{atan2}\left(\Delta x, \Delta y\right) \quad (12)$$

$$\varphi_n = K_{i\_a}.\ \Delta W_{inc} \quad (13)$$

One can notice that when making small incremental displacement of the mouse, the mental transform from a sphere to a plan is lighten since a spherical surface can be well locally approximated to a planar surface if the constant H_O is taken large enough compared to the variations of L. By construction and the exploiting of the mouse wheel our approach enables **fulfilling all five design principles** (section 4.1). In particular principle 5 is enforced by the fact that there is no need to proceed to multiple mouse button clicks to change an orientation. However in its current form our system does not allow to reach all orientations. A nutation of $\pi/2$ radians can't be attained. Typically we restricted the nutation to lie within the range [0; $\pi/4$] radians.

4.4 Experimental assessment protocol

The experimental setup is made to resemble the actual tele-echography setting that would be used in real conditions when using a mouse as interface as depicted in previous section. Consequently the setup is made up of a PC workstation displaying in 3D a simulated tele-echography robot handling a bright green probe and which end-effector orientation is controllable by the computer mouse (fig. 9b). The Robot Simulator has been built within Windows XP environment using OpenGL and Microsoft Visual C++. It accurately simulates the design and mobility of an actual OTELO tele-echography robot (Delgorge et al., 2005)(see figure 8). The view chosen for experimentation (shown on fig. 8a) is in accordance with the actual scenario in a tele-echography operation (Canero et al. 2005). In figure 8a, the red rings on the right side of the screen, which look like a target, work as a guide to move the mouse. Its centre is chosen as the origin for the mouse. And the farthest ring defines the region of maximum bending in nutation of the probe. These circles where useful for the users during their learning phase only. Human-Machine interfaces (HMI) are generally assessed with static targets, which give no information on their dynamic capabilities. Hence **we have imagined an original** way of interface evaluation consisting for the subjects to track the moves of an opaque red dummy probe which is overlaid on screen and animated from a previously recorded datafile during a real abdominal US examination. Some parts of the robot have been given transparency to make it easier to visualize both the probes simultaneously. We only have considered the orientations in this experiment; hence both the dummy probe and the simulated robot probe are fixed in translation.

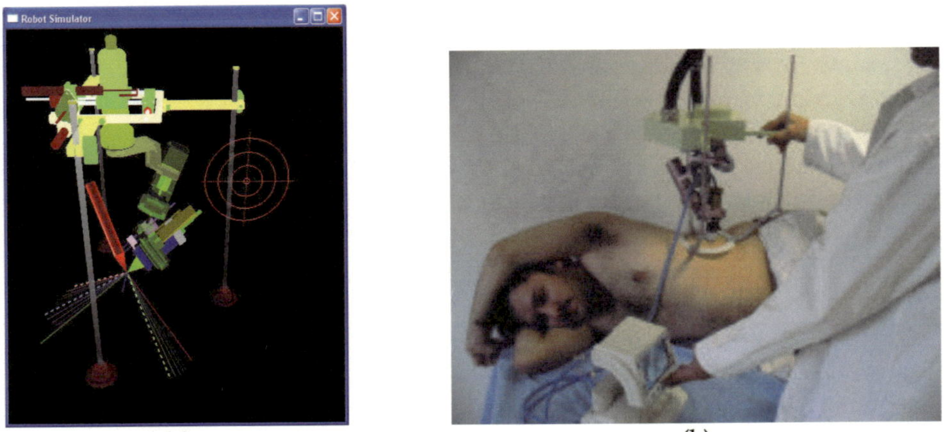

(a) (b)

Fig. 8. (a) The OTELO tele-echography robot simulated in OpenGL within a virtual-reality simulator for psychophysical assessment. (b) Actual OTELO robot in action.

A three-axis framework with differently coloured axis was also attached to this dummy probe and displayed for a better visualization of its orientation. Better telepresence could be achieved with a HMD (Head Mounted Display) for depth perception. However this would annihilate the interest in using a computer mouse for proposing simple low cost control interface, so we preferred using a standard 2D screen displaying 3D graphics. This teleoperation is simulated with no time-delay to avoid interfering effects on the assessment of the new H-angles frame. It should be noticed that this protocol induces a cognitive load on the test users that is heavier than on a medical specialist who would perform a real tele-echography examination by means of the mouse as input device. This is due to the fact that in our protocol users have a few prior knowledge of the trajectory to be tracked whereas the medical expert imposes his desired trajectory that he is used and trained to plan to navigate through the human body with a US image as feedback. In other words in real practice the movements are intentional and performed in a know environment with known landmarks whereas in the proposed experimental protocol the trajectory to track is imposed. Nevertheless it is not desirable to try filling this workload gap by providing the test users with a trajectory representation in the angles space. Indeed this trick would unpredictably lighten the cognitive load compare to the medical expert who has to make mental rotation in 3D space which is known to be a heavy mental load (Shepard & Metzler, 1971). Six different non-medical test users were solicited to carry out the experiment. They were all used to mouse manipulation and computer interaction. Each untrained user was shown the animation once just to get accustomed to the trajectory. Then he had an unlimited training session to understand how to control the robot orientation by the mouse and to have a preview of the trajectory to track. No more than five minutes of training was sufficient for every test user. The medical reference trajectory duration is three minutes long. Each test user had three trials to track this trajectory by using the H-angles coordinate system associated to the mouse, and next they had three other trials using the standard ZXZ Euler system, for comparison purpose. The session of orientation matching with the Euler system is intended to assess the performance improvement provided with the H-angles system. For each smallest possible turn of the mouse wheel we have set an increase of 10° for angle φ_n. This causes a limitation to the accuracy. However, whereas a smaller increment in the angle would have increased the precision, this would have make the robot probe difficult to rotate fast enough, to be able to track the animated probe rotations. We needed to strike a balance between, good rotation speed and higher precision, so as to obtain the optimum results. With the Euler system it was noticed that allowing faster rotations gave better results.

4.5 Psychophysical results

The orientation tracking error is computed as the minimum rotation angle between the frameworks of the controlled probe and dummy probe which is known to be a good metric in the rotations space. Let us notice this angle as Ω.

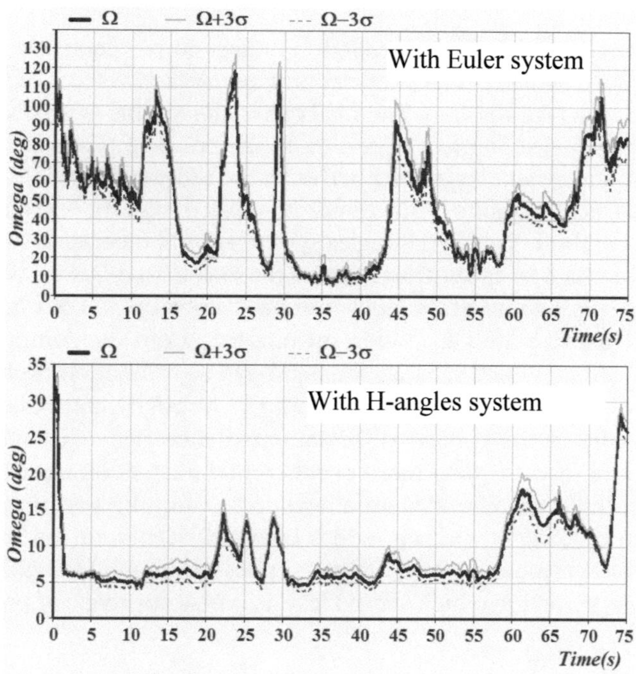

Fig. 9. Observed variations in average Ω values with bounding curves at Ω plus or minus three times the standard deviation σ.

Figure 9 reports the average of Ω orientation error among the users versus time of trajectory tracking. First plot is for the mouse used to set the Euler angles and second plot for the mouse used to set the H-angles. Plots of figure 9 reveal practically an indisputable superiority of our new system compared to standard Euler system. With our new system the tracking error remains most of the time lower than 10°, whereas with the Euler system the error rarely drops below 10°. Whatever the experimentation time considered, the error with the new system is at least two times lower than with the Euler system. From testimony of the test users the new system acts as if the self-rotation were anticipated. Whereas with the Euler system the trackings were confusing mainly because of the singularity of this system tending to produce fast variations of the X and Y axis when the nutation is close to zero. Notice that the presented results integrate the human response time lag. The simulator measures the difference in the framework angles in real time but the user takes some time to perceive and react to the animated moves. Hence there is always a lag in the controlled probe's movements compared to the animated probe. This lag is not a constant in time, but perhaps is a function of various other unaccounted parameters. For example the probe velocity, the visual angle and so on. The overall average values of Ω obtained for the two systems can also be used to compare the degree of effectiveness of the two systems. For the Euler system average Ω value was found 46.28° while for our new attitude coordinate system it was observed to be 8.69°. The values of standard deviation can be understood as the inconsistency, in being able to accurately orient the probe. Its averaged values over time where observed as 1.85° for Euler system and 0.347° with the new system.

5. Conclusion

We designed a new coordinate system called H-angles; to parameterize the attitude of an object in 3D space such as the Human central nervous system would do when rotating the object about a fixed centre of rotation. The final cue to derive this system was obtained from the analysis of the medical sonography practice. In the practical case considered we showed experimentally that our system largely outperforms the Euler systems in the decorrelation of the DOFs and in practical usability of the mouse as input device for 3D rotations. The design considerations lead to think that the H-angles system should theoretically maintain its good properties in a large range of applications where the task is to rotate an object about a fixed point. Some more experimental evaluations will have to be carried out to verify this claim. We have exploited the H-angles to design an interface from the computer mouse to the attitude parameterisation, which satisfies the hand-eye co-ordination needs for the purpose of poly-articulated robot orientation telecontrol through computer network. Our psychophysical results in the context of a simulated robotised tele-sonography are very promising and should lead to some more experimental evaluation in comparison with the virtual trackballs techniques. Our system allows imagining the performing of 6D mouse-based teleoperation by using switching modes between orientation and translation control with a standard wheeled mouse. Some further application of the H-angles system could be for hand orientation prediction which should lead to a new approach of predictive control.

6. References

Aznar-casanova, J. A., Torrents, A., & Alves, N. T. (2008). The role of vertical disparities in the oblique effect. *Psychol. Neurosci.*, Vol. 1, No. 2, (2008), pp. 167-175

Bade, R., Ritter, F. & Preim, B. (2005). Usability comparison of mouse-based interaction techniques for predictable 3d rotation. *SmartGraphics*, (2005), pp. 138-150

Baud-Bovy, G., & Viviani, P. (1998). Pointing to kinaesthetic targets in space. *J. Neuroscience*, Vol.18, (1998), pp. 1528–1545

Baud-Bovy, G., & Viviani, P. (2004). Amplitude and direction errors in kinesthetic pointing. *Exp. Brain Res.*, Vol. 157, (2004), pp. 197–214

Baud-Bovy, G., & Gentaz, E. (2006a). The hapic reproduction of orientations in three dimensional space. *Springer-Verlag Exp. Brain Res.*, Vol. 172, (2006), pp. 283-300

Baud-Bovy, G., & Gentaz E., (2006b) The haptic perception of orientations in the frontal plane and in space. *Exp Brain Res*, Vol. 172, (2006), pp. 283–300

Block, B. (2004). *The Practice of Ultrasound, a step by step guide to abdominal scanning*. Georg Thieme Verlag, ISBN 3-13-138361-5, Stuttgart, Germany

Burgess, N. (2006). Spatial memory: how egocentric and allocentric combine. *Trends in Cognitive Sciences*, Vol. 10, No. 12, (Dec. 2006), pp. 551-557

Canero, C., Thomos, N. ,Triantafyllidis, G. A., Litos, G. C. & Strintzis, M. G. (2005). Mobile tele-echography: user interface design. *IEEE Trans. Inform. Technol. Biomed.*, Vol. 9, No. 1, (2005), pp. 44–49

Cecala, A.J., & Garner, W.R. (1986). Internal frame of reference as a determinant of the oblique effect. *J. Exp. Psychol. Hum. Percept. Perform.*, Vol. 12, (1986), pp. 314–323

Chen, M., Mountford, S.J. & Sellen, A. (1988). A Study in Interactive 3-D Rotation Using 2-D Control Devices. *Computer Graphics*, Vol. 22, No. 4, (Aug. 1988), pp. 121-129

Cohen, J., Cohen, P., West S.G., & Aiken L.S., (2002). *Applied multiple regression/correlation analysis for the behavioral sciences* (3rd edition), Lawrence Erlbaum Assoc Inc, ISBN 0805822232, Mahwah, New Jersey (USA)

Cohen, Y. E., & Andersen, R. A. (2002). A common reference frame for movement plans in the posterior parietal cortex. *Nature Review Neuroscience*, Vol. 3, (2002), pp. 553-562

Courreges, F., Vieyres, P., Poisson G., Novales, C. (2005). Real-Time Singularity Controller for a Tele-Operated Medical Ecography Robot. *IEEE/RSJ IROS*, Edmonton, Alberta, Canada, Aug. 5th, 2005

Courreges, F., Poisson, G., & Vieyres, P. (2008a). Robotized Tele-Echography. In : *Teleradiology*, Sajeesh Kumar, Elizabeth Krupinshi (Eds.), pp. 139-153, Springer-Verlag, ISBN 978-3-540-78870-6, Berlin

Courreges, F., Vieyres, P., Poisson, G. (2008b). DOF Analysis of the Ultrasonography Technique for Improving Ergonomy in Tele-Echography. *IEEE International Conference on Robotics and Biomimetics, Robio'08*, ISBN 978-1-4244-2678-2, Bangkok, Thailand, pp. 2184-2189, Feb. 2009

Darling, W.G., & Gilchrist, L. (1991). Is there a preferred coordinate system for perception of hand orientation in 3-dimensional space? *Exp Brain Res*, Vol. 85, (1991), pp. 405-416

Darling, W. G., & Miller G. F. (1995). Perception of arm orientation in three-dimensional space. *Exp. Brain Res*, Vol. 102, (1995), pp. 495–502

Darling, W.G., & Hondzinski, J.M. (1999). Kinesthetic perceptions of earth- and body-fixed axes. *Exp Brain Res*, Vol. 126, (1999), pp. 417–430

Darling, W.G., Viaene, A.N., Peterson, C.R., & Schmiedeler, J.P. (2008) Perception of hand motion direction uses a gravitational reference. *Springer-Verlag Journal of Exp. Brain Res.* Vol. 186, (2008), pp. 237-248

Delgorge, C., Courreges, F., Al Bassit, L., Novales, C., Rosenberger, C., Smith-Guerin, N., Brù, C., Gilabert, R., Vannoni, M., Poisson, G., & Vieyres P. (2005). A Tele-operated mobile ultrasound scanner using a light weight robot. *IEEE Trans. on*

Innovation Technology in Biomedicine Special Issue, Vol. 9, No. 1, (March 2005), pp. 50-58

Desmurget, M., Pélisson, D., Rossetti, Y., & Prablanc, C. (1998). From eye to hand: planning goal-directed movements. *Neurosc. and Biobehav. Rev.*, Vol. 22, (1998), pp. 761-788

Desmurget, M., Prablanc, C., Jordan, M., & Jeannerod, M. (1999). Are reaching movements planned to be straight and invariant in the extrinsic space? Kinematic comparison between compliant and unconstrained motions. *Quarterly Journal of Experimental Psychology*, Vol 52, (1999), pp. 981-1020

Dorst, L., Doran, Ch., & Lasenby, J. (Eds.) (2002). *Applications of Geometric Algebra in Computer Science and Engineering*, Birkhäuser, ISBN 0-8176-4267-6, New York, USA

Essock, E.A. (1980). The oblique effect of stimulus identification considered with respect to two classes of oblique effects. *Perception*, Vol. 9, (1980), pp. 37-46

Gentaz, E., & Ballaz, C. (2000). La perception visuelle des orientations et l'effet de l'oblique. *Ann. Psychol.*, Vol. 100, (2002), pp. 715–744.

Gentaz, E., & Tschopp, C. (2002). The oblique effect in the visual perception of orientations. In : *Advances in Psychology Research*, Shovov S. (Ed.), Vol. 11, pp. 137-163, Nova Sciences Publishers, ISBN 1-59033-186-9, New-York

Gentaz, E. (2005). Explorer pour percevoir l'espace avec la main. In : *Agir dans l'espace*, Thinus-Blanc, C. (Ed), MSH, pp. 33-56, Paris

Gentaz, E., Baud-Bovy, G., & Luyat, M. (2008). The haptic perception of spatial orientations. *Experimental Brain Research*, Vol. 187, (2008), pp. 331-348

Gibson, J.J. (1950). The perception of visual surfaces. *Am. J. Psychol.*, Vol. 63, (1950), pp. 367-384

Goodale, M.A., Jakobson, L.S., & Servos, P. (1996). The visual pathways mediating perception and prehension, In : *Hand and Brain*, ed. by Wing, A. M., Haggard, P. & Flanagan, J. R., pp. 15-31, Academic Press, New York

Gourdon, A., Poignet, Ph., Poisson, G., Vieyres, P., Marché, P. (1999). A new robotic mechanism for medical application. *IEEE/ASM Int. Conf. On Adv. Intel. Mechatronics*, pp 33-38, Atlanta, USA, September 1999

Hamilton, W. R. (1843). On a new Species of Imaginaries Quantities connected with a theory of Quaternions. *Proc. of the Royal Irish Academy*, Vol. 2, (Nov. 13, 1843), pp. 424-434

Hinckley, K., Tulio, J., Pausch, R., Proflitt, D., & Kassell, N. (1997). Usability Analysis of 3D Rotation Techniques. *Proceedings of ACM Symp. User Interface Software and Technology (pp. 1-10)*, New York : Association for computing machinery, 1997

Henriksen, K., Sporring, J., & Hornbaek, K. (2004). Virtual trackballs revisited. *IEEE Transactions on Visualization and Computer Graphics*, Vol. 10, No.2, (2004), pp. 206-216

Howard, I. P. (1982). *Human visual orientation*, Wiley, ISBN 978-0471279464, New York

Jacob, R.J.K, Sibert, L.E., Mcfarlane, D.C., & Mullen M. P. (Jr) (1994). Integrality and Separability of Input Devices". *ACM Transactions on Computer-Human Interaction*, Vol. 1, No. 1, (March 1994), pp. 3-26

Jacob, I., & Oliver, J., (1995). Evaluation of Techniques for Specifying 3D Rotations with a 2D Input Device. *Proc. Human Computer Interaction Symp. '95*, (1995), pp. 63-76

Jolliffe, I.T. (2002). *Principal Component Analysis* (2nd edition), Springer series in Statistics, ISBN 9780387954424, New-York

Kennedy, M., & Kopp, S. (2001). *Understanding Map Projections*. Esri Press, ISBN 9781589480032, Redlands (USA)

Loève, M. (1978). *Probability theory Vol. II,(Graduate Texts in Mathematics)* (4th edition), Vol. 46, Spinger Verlag, ISBN 0-387-90262-7, New-York (USA)

Miall, R.C., & Wolpert, D.M. (1996). Forward Models for Physiological Motor Control. *Neural Networks*, Vol. 9, No8, (1996), pp. 1265-1279

Norman, J. (2002). Two visual systems and two theories of perception: An attempt to reconcile the constructivist and ecological approaches. *Behavioral and brain sciences*, Vol. 25, No. 1, (2002), pp. 73-144

Olson, D. R., & Hildyard, A. (1977). On the mental representation of oblique orientation. *Canadian Journal of Psychology*, Vol. 31, No. 1, (March 1977), pp. 3-13, ISSN 0008-4255

Paillard, J. (1987). Cognitive versus sensorimotor encoding of spatial information. In: *Cognitive processes and spatial orientation in animal and man, Vol. II, Neurophysiology and developmental aspects*, Ellen, P., & Thinus-Blanc, C. (Eds.), pp. 43-77, Martinus Nijhoff, ISBN 90-247-3448-7, Dordrecht, Netherland

Pan, Q., (2008). Techniques d'interactions mixtes isotonique et élastique pour la sélection 2D et la navigation / manipulation 3D. *PhD thesis of the Lille 1 University (France)*, defended the 19 décembre 2008, theme instrumentation and advanced analysis

Sergio, L.E., & Scott, S.H. (1998). Hand and joints paths during reaching movements with and without vision". Journal of Exp. Brain. Res., Vol. 122, (1998), pp. 157-164

Sheerer, E. (1984). Motor Theories of Cognitive Structure: A Historical Review. In : *Cognition and Motor Processes*, Prinz W., & Sanders A. F. (Eds), SpringerVerlag, ISBN 978-0387128559, Berlin

Shepard, R. & Metzler. J. (1971). Mental rotation of three dimensional objects. *Science*, Vol. 171, No. 972, (1971), pp. 701-703.

Smyrnis, N., Mantas, A., & Evdokimidis, I. (2007). The "motor oblique effect": perceptual direction discrimination and pointing to memorized visual targets share the same preference for cardinal orientations. *J. Neurophysiol.*, Vol. 97, (2007), pp. 1068–1077

Soechting, J.F., & Ross, B. (1984). Psychophysical determination of coordinate representation of human arm orientation. *Journal of Neuroscience*, Vol. 13, (1984), pp. 595-604

Soechting, J.F., Lacquaniti, E., & Terzuolo, C.A. (1986). Coordination of arm movements in three-dimensional space. Sensorimotor mapping during drawing movement. *Neuroscience*, Vol. 17, No. 2, (Feb. 1986), pp. 295-311

Soechting, J. F., & Flanders, M. (1995). Psychophysical approaches to motor control. *Current Opinion in Neurobiology*, Vol. 5, (1995), pp. 742-748

Shoemake, K. (1985). Animating rotation with quaternion curves. *Computer Graphics (Proceedings of SIGGRAPH 85)*, Vol. 19 (1985),pp. 245–254

Shoemake, K., (1992). ARCBALL: A User Interface for Specifying Three-Dimensional Orientation Using a Mouse. *Graphics Interface*, (1992), pp. 151-156

Stevens, K. A. (1983). Slant-tilt: the visual encoding of surface orientation. *Biological Cybernetics*, Vol. 46, (1983), pp. 183–195

Tempkin, B.B. (2008). *Ultrasound Scanning, Principles and Protocols* (edition), Saunders, Elsevier publishers, 3rd ed., ISBN 0721606361, Philadelphia, USA

Van Hof, M. W., & Lagers-van Haselen, G. C. (1994). The oblique effect in the human somatic sensory system. *Acta Neurobiol. Experimentalis*, Vol. 54, (1994), pp. 259-262

Vilchis, A., Troccaz, J., Cinquin, P., Masuda, K., Pellissier, F. (2003). A new robot architecture for tele-echography. *IEEE Trans. Rob. and Autom., Special issue on Medic. Rob.*, Vol. 19, No. 5, (October 2003), pp. 922-926

Volcic, R., Kappers, A. M. L. (2008). Allocentric and egocentric reference frames in the processing of three-dimensional haptic space. *Experimental Brain Research*, Vol. 188, (2008), pp. 199-213.

Wang, Y., Mackenzie, C., Summers, V.A., & Booth, K.S. (1998). The structure of object Transportation and Orientation in Human-Computer Interaction. *Proceedings of ACM CHI'98*, pp. 312-319

Wolpert, D.M., & Ghahramani, Z. (2000). Computational principles of movement neuroscience. *Nature America, Neurosci.*, Vol. 3, (2000), pp. 1212-1217

11

Object Location in Closed Environments for Robots Using an Iconographic Base

M. Peña-Cabrera[1], I. Lopez-Juarez[2], R. Ríos-Cabrera[2]
M. Castelán[2] and K. Ordaz-Hernandez[2]
[1]*Universidad Nacional Autonoma de México*
[2]*Centro de Investigación y de Estudios Avanzados del IPN (CINVESTAV)*
Mexico

1. Introduction

In order to have an efficient use of digital image processing and pattern recognition techniques, it is necessary to understand the human visual system behaviour. The human visual system comprises the human eye and some areas of the human brain which performs neurological processing information. Human eye and brain together convert the optical information in a visual scene perception, the eye functions as the human visual system camera, and his function is to transform light electromagnetic waves to small electrical signals carrying the information to the brain to achieve data analysis and build a structured high resolution image.

For a machine vision, this process represents a very complex task to implement; adequate elements have to exist like: environment perception, information processing and primitive base knowledge generation to achieve specific actions for decision making based on interpretation of such information. Artificial vision then might be defined as: machine implementation with capabilities of visual perception of the surrounding environment, extraction of region of interest, analysis, scene interpretation and decision making. In this regard some authors like Haralick & Shapiro have defined machine vision as the science which study and develop the theoretical bases and algorithms to obtain information of the real world from one or some images (Haralick&Shapiro, 1992). Another notable definition is proposed by Pajares et al. that define it as a machine capability to see its rounded world, to understand the structure and properties of a 3D world based in the analysis of one or more 2D images.

The system described in this article resembles the above capability in automated industrial applications using manufacturing machinery or robots, which in some moment of the industrial process need to obtain its position and location within their working space, so to accomplish their labour tasks efficiently. A real example is locating and positioning working pieces within a manufacture area during assembly, painting, sorting or storage operations. The goal is to identify the working piece and make reference to a visual scene, then with a geometrical model, the working piece position is calculated and so the necessary path to reach the pieces using a robot-arm as described by de Lope (de Lope, et al., 1997) typically in an eye-in-hand configuration.

A frequent requirement that we have found during manipulative tasks using mobile robots or multiple collaborative robot arms is to accurately locate objects within the work space or

in the case of mobile robots, the robot itself. In this chapter we present the design and implementation of an artificial vision system capable of obtaining the position of an object (working piece), tool or end-effector device of a robot manipulator, or the camera's position during eye-in-hand configuration for robot arms within an enclosed environment by the way of "wall marks" recognition.
An iconographic base formed by icon symbols containing enough information for the object is employed in order to obtain the robot's spatial position (end-effector location, for instance). All the information is obtained in one camera shot. The systems comprises a digital camera with pan-tilt movements attached to the object to get its location, the object might be a working piece, the end-effector of a robot-arm or any other mobile object. The system obtains the location and position of the camera by way of performing an observation around the walls, the system finds "wall marks" which indicates where the object is and its position. The camera gets enough information to obtain sufficient parameters to allow calculations from the exact position of an object; the information allows the graphical representation of the closed environment and the camera location.
Pattern recognition techniques are widely used in computer vision with excellent results in industrial inspection and automation applications; however, some of the algorithms are not for practical use in real time applications. Using a visual system based on the recognition of symbols makes the possibility to design and implement a visual system to get working pieces locations by way of using an iconographic descriptive symbol set with real time interpretation, and making possible to build a basic geometrical symbol recognition method, which interpretation, might indicate information of location. By making analysis of depicted symbols, the object camera within an enclosed environment might know where it is, what its direction, and what its inclination is.
The used symbols represent movements, paths in 2D and potentially 3D and allow establishing references to fixed zones within the scene; the symbols are extracted from general scene and from its analysis, the location information is obtained.

2. Scope

The scope of this research is to design a methodology implementing an artificial vision system capable to obtain in real time the exact position of an object within an enclosed working environment by using symbols and icon recognition. The system comprises a camera located in the object, working piece or end-effector segment of a robot-arm. The customized software communicates with the camera to get visual information of its environment in the same place where the object, working piece or end-effector segment is located. The parameter acquisition is carried out by image analysis and the exact location of the camera is obtained. Enough information is obtained as well to have a graphical representation of the environment. Applications for autonomous agents and robot manipulators are natural for this methodology.
The possibility of using this methodology involves using closed environments, the walls are decorated with symbols (icons) in specific areas, the symbols contain by itself graphical information about where they are located; in this sense, by image processing and interpretation is possible to know the location of the object by looking at the icons.

3. Design and implementation

The design and implementation of the system, as well as the experimental platform is integrated with the following operation modules.

Software and Hardware (camera and dedicated algorithms). - The camera is a digital solid state wireless camera with pan/tilt movement support, used to look for the appropriate icon in a specific working area. The wireless communication with the control computer is achieved via a wireless access point.
Iconographic set. - For environment representation, the icons are basic geometric forms implemented within the environment (walls).
Dummy enclosed environment for experiment. - is a specific environment to carry out test experimentation with the system, this dummy model, has practical dimensions to acquire enough data to match the digital camera specifications and satisfy the application requirements.
Image acquisition. - It is the module in the system to capture the image and produce a standard format to be implemented in an image model.
Region of Interest (ROI) algorithm.- It detects the icon presence within the scene, and extracts it from scene.
Detection, interpretation and camera location algorithm.- This algorithm performs data processing of an image structure of the extracted icon, and together with a geometrical model obtain the camera coordinates within the scene.

3.1 Software

Visual Basic 6.0 and Matlab 7.0 platforms were used to implement the system, software application development was basically developed in Visual Basic 6.0, interaction with Matlab 7.0 were carried out by way of "Matlab Automation Server" component. The main purpose of the system is to obtain the coordinates of a video camera, within a pre-establish coordinate reference system in a known working space. The system has to have recalibration capabilities in order to work in different working space dimensions for different applications. Figure 1 shows the software modules.

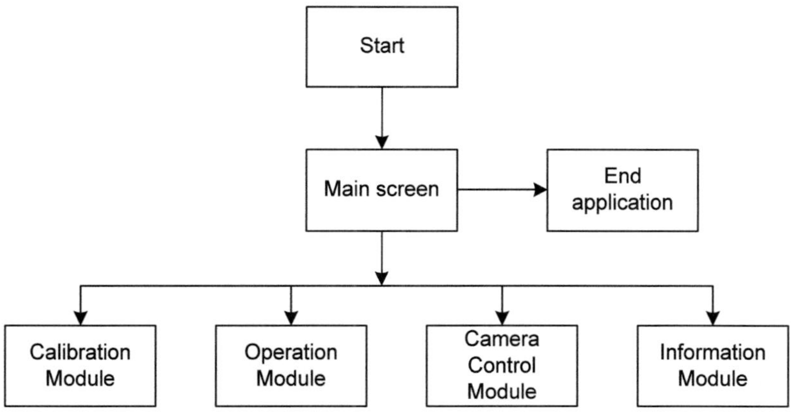

Fig. 1. General configuration of software modules.

3.2 Hardware

Hardware components of the developed system are: a digital camera, a connectivity network card and a PC computer. The used camera is a "Veo Wireless Observer", it has communication with computer through a wireless local network using a wireless access point and a wireless network card "3Com® 11 Mbps Wireless LAN PCI Adapter" PCI with protocol 802.11b of IEEE and using a point to point configuration (figure 2).

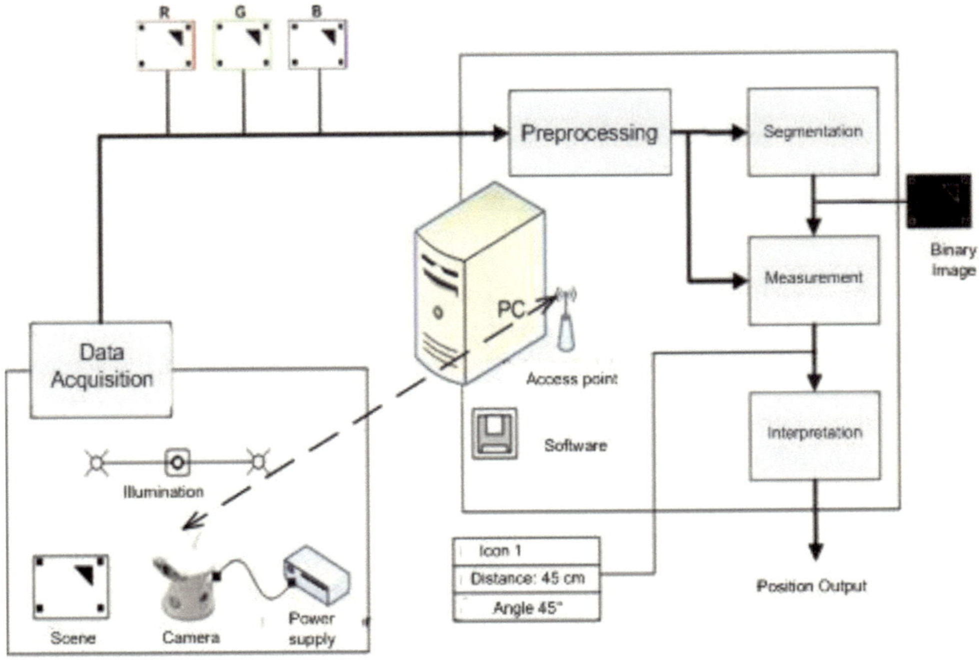

Fig. 2. General configuration of the system.

3.3 Iconographic base

Simple figures were used to implement the iconographic base, combination of these figures have to get enough features to obtain the location information, a symbol set was created to conform what we have called "iconographic base", this symbols were physically painted in the wall of the closed environment for different zones within the working place. For testing purposes this was implemented in a dummy model of practical dimensions. Four "working icons" were proposed and each icon has to be located in the specific area of each working

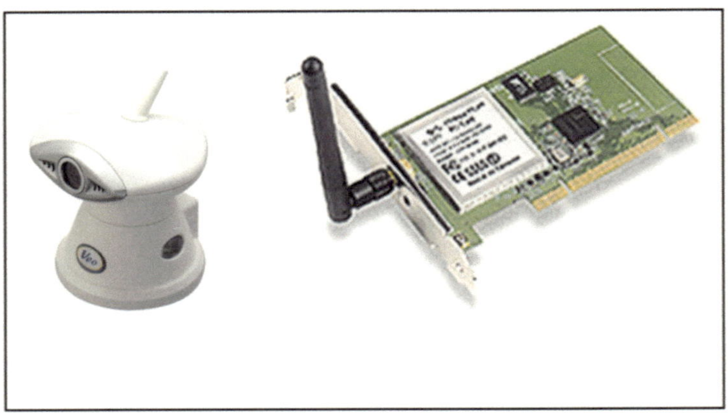

Fig. 3. Camera and network card.

area zone in order to get correlated information with physical real world location. Each "working icon" is conformed by four small squares, their centroid conform the corner points of a bigger imaginary square. The position of the icons allow subsequent analysis of displacements due to the perspective factors, by using geometrical models, according with this, it is possible to get the distance between the camera lens and the centroid of imaginary square as well as the angle between optical axis of the camera and the normal line crossing the imaginary square centroid. Dimensions for different square, were obtained by "try and error" in order to reach optimum resolution and view area for a particular used camera.

3.4 Specifications for the design of the dummy model environment

For the design of the model environment, a camera resolution of 320 x 240 pixels was used in order to achieve an image size suitable for fast processing and acquisition. Several test shots to look for ranges of distance between the camera lens and the centre of the icon to get an acceptable work experiment were carried out, resulting a maximum distance of 60 [cm] and a minimum distance of 20 [cm] for the experiment, for distances outside of this range was impossible to obtain the necessary parameters to obtain the required distance and angle of vision to get a real situation of experimental measurements of objects locations within the dummy model enclosed environment. The maximum value of angle of vision to guarantee the experiment was 55 °. Once established these parameters it can be an operational region for each icon, in which the system will be able to obtain the position of the camera, this region is applicable for each of the four icons and will be called the Operating Region of the working space (Figure 4a and 4b).

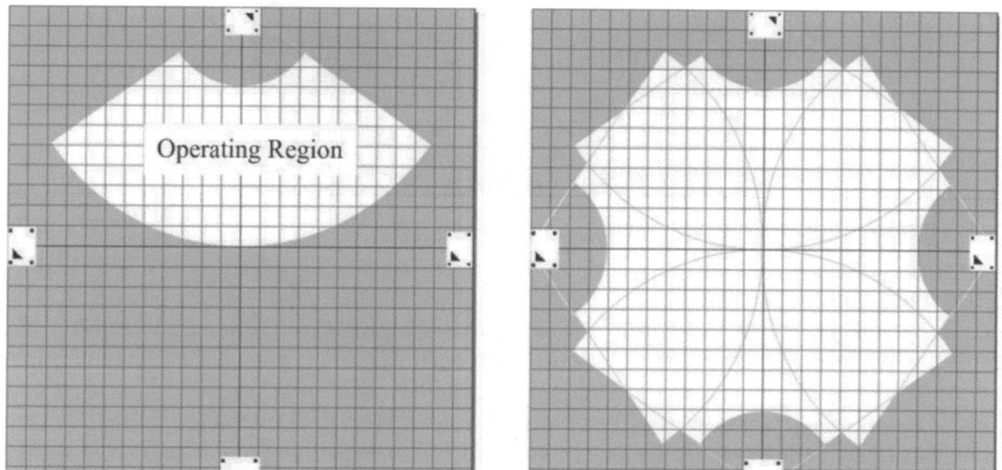

(a) (b)

Fig. 4. (a) Icon's Operating Region, (b) Operating Region of the Working Space.

Each icon frame was designed as depicted in Figure 5a and they were located in each wall as indicated in Figure 5b.

In addition to the above parameters, it was also necessary to obtain the minimum height in order to avoid external scenes images to the working environment being acquired , which will act in the form of noise causing failures in image processing, the height was established as 50 [cm]. The resulting dimensions proposed for the construction of model tests were: 50cm x 120cm x 120cm, as shown in figure 6.

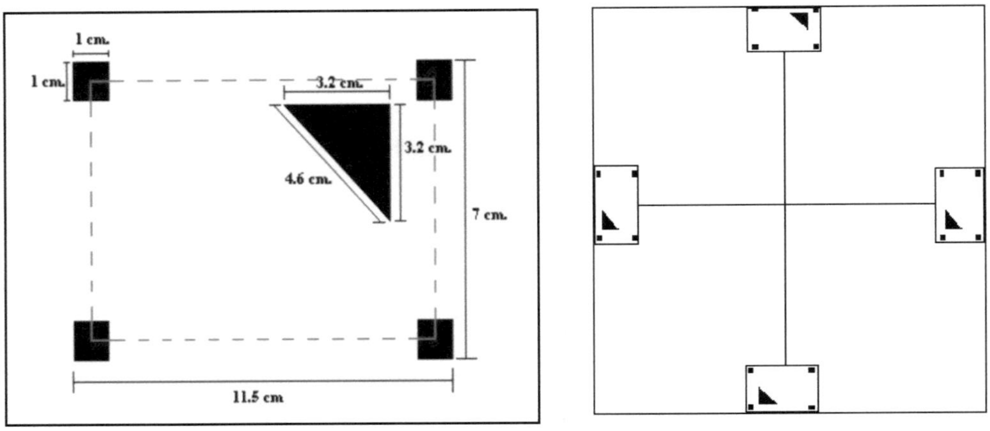

Fig. 5. (a)icon's frame design, (b) icon's wall location

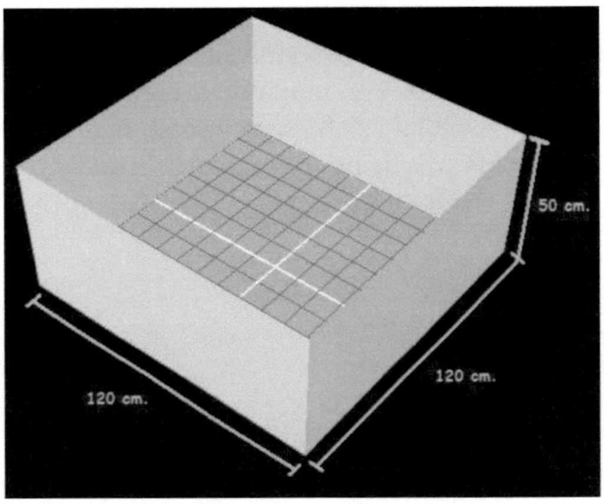

Fig. 6. Dummy model environment for experimental test.

3.5 Working icon detection

In order to detect just one of the four working icons at a time in a particular scene, we used a "minimum and maximum distances" defined for each operation region, so if they are out of an specific range the icon is not considered for analysis. For "icon" detection, we first select a "merit number" so to know which region of interest has been associated for each "icon", this is the "P2A" number, which is a known factor between perimeter and area used in digital image processing on binary images, in our case we define a black object (icon zone) on a white background.

3.6 Icon identification

The main objective in having the icon set is to get enough parameters to find the camera-icon distance, vision angle and the region being used within all enclosed environment.
Three algorithms were designed to perform each task, they are:

a. Algorithm to get camera-icon distance (CIDA),
b. Algorithm to get the angle of vision (AVA),
c. and algorithm to find which of the four working icons is being used within the current image in the application and is called icon identification algorithm (IIA).

One important requirement is to have some of the working icons as a complete image within the image frame; otherwise, the camera had to be moved to find it in order to have enough confidence on data values before calculations were made.
Icon identification algorithm (IIA). It uses a binary image and tag process, so that to get the difference among different elements comprising the working icons, it is necessary to obtain the "centroid" of each element to calculate distances as given by equation (1) with a reference element (triangle) and with a different orientation for each working region in order to obtain the identification of the working icon (see figure 5 and 7.)

$$d_{\overline{AB}} = \sqrt{(X_A - X_B)^2 + (Y_A - Y_B)^2} \quad (1)$$

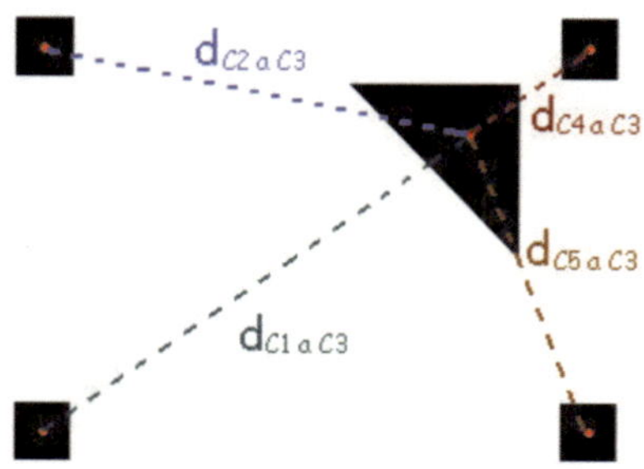

Fig. 7. Centroid distances representation.

Once all distances are obtained, the algorithm finds which is the smaller distance so to get the criteria to determine which "working icons" has been founded and which "working region" (geometrical square) is being used within the environment. This information tells the system where the camera is because the "working icon" center, has been draw in the very border of each geometrical working region in the environment walls as shown in figure 8.

3.7 Calculation of distance to the icon

The algorithm to get the camera-icon distance (CIDA), is useful to obtain the distance between the camera lens and the center of the icon, which make use of the following formula:

$$d = {}^{K}\!/_{iconheight} \quad (2)$$

where:
d = distance
k = proportional constant
icon height = icon height in pixels
The procedure is to get the distance from the analysis of the image perspective, which relate the magnitudes of objects with the distance being considered, such as the figures appear to be smaller with a greater distance and vice versa, this relationship is given by equation (2).
To make use of the equation, the first step is proceed to a labelling process of icon elements (see figure 9), to identify them within the icon working area.
The centroid of each element is calculated in order to qualify for two points: one point corresponds to a midpoint between the centroid of element 2 and the centroid of element 4,

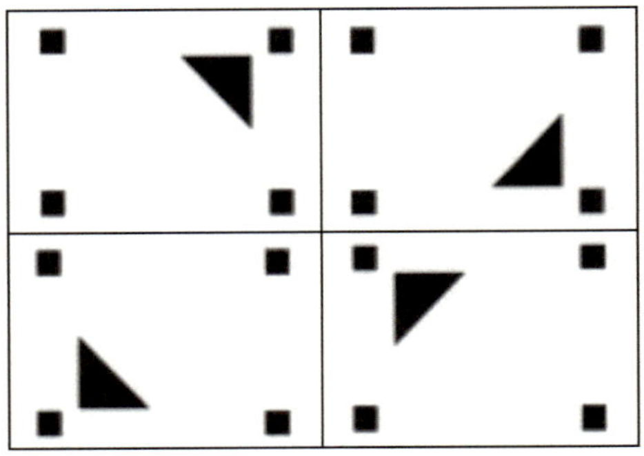

Fig. 8. Working icons for different working regions

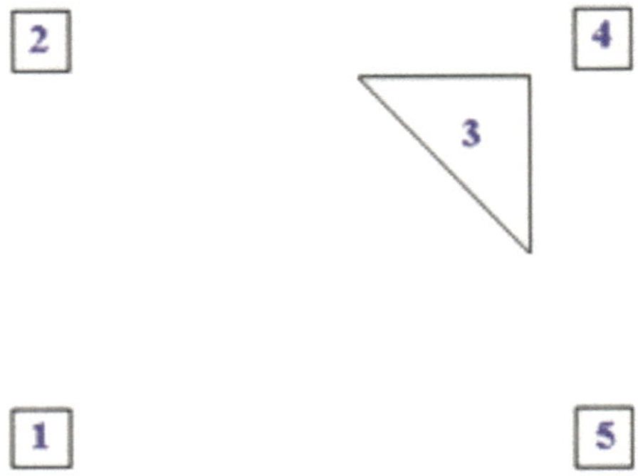

Fig. 9. Icon elements identification on image.

the second corresponds to the midpoint between centroid of element 1 and centroid of element 5. It important to notice that for the calculation of the distance, the centroid of element 3 (triangle) is not used, and calculation of previous defined midpoints points, equations 2 and 3 are used, once obtained these points, it is calculated the magnitude than exists among them (called "apparent height") as illustrated in Figure 10.

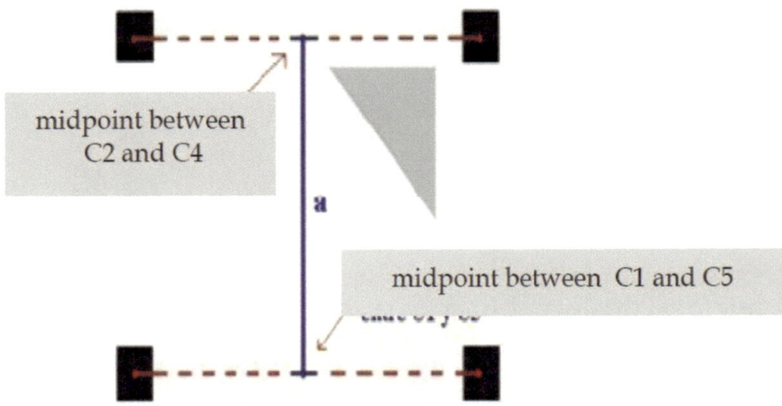

Fig. 10. Original Icon image

The distance a is given by equation 7 as follows:

$$a = \sqrt{(PM2_x - PM1_x)^2 + (PM2_y - PM1_y)^2} \quad (7)$$

where:
a = icon height (apparent height)
PM1x, PM1y corresponds to the x-y coordinates of the mid point centroid between centroid of element 2 and centroid of element 4.
PM2x, PM2y corresponds to the x-y coordinates of the mid point centroid between centroid of element 1 and centroid of element 5.
They are calculated as:

$$PM1x = (xC2 + xC4) / 2 \quad (3)$$

$$PM1y = (yC2 + yC4) / 2 \quad (4)$$

and

$$PM2x = (xC1 + xC5) / 2 \quad (5)$$

$$PM2y = (yC1 + yC5) / 2 \quad (6)$$

where
xC2 and yC2 are coordinates of centroid 2
xC4 and yC4 are coordinates of centroid 4
xC1 and yC1 are coordinates of centroid 1
xC5 and yC5 are coordinates of centroid 5
A similar procedure is made when an image is obtained from another point in the scene with different image perspective as shown in Figure 11.

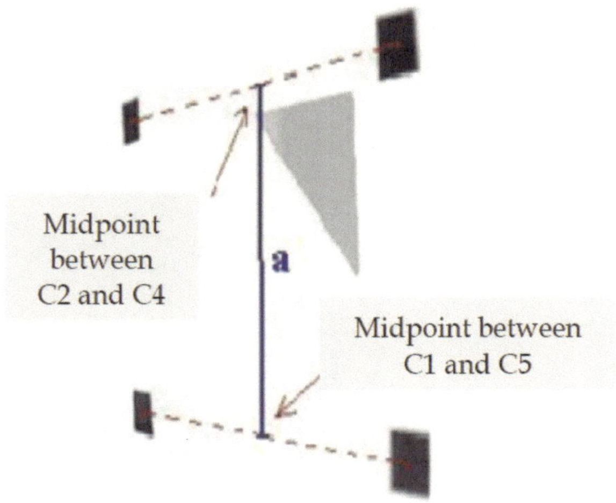

Fig. 11. Icon image at 55° with reference to the perpendicular line

At this point, we have only obtained a parameter of the equation (2), the next step is to obtain the proportionality constant (k), which depends on the lens characteristics and each visual scene, therefore the best practical way to get it is by means of laboratory tests.
Tests for obtaining "k" were carried out in the following way:

1. Place the camera at a known distance “d” to obtain a focused and central positioned image of the icon.
2. With the acquired image get the parameter "apparent height".
3. Values "d" and "apparent high" were replaced in the equation (2) to obtain a proportionality constant "k_i".
4. Repeat steps 1 to 3 for 5 different distances from the operational region of a specific icon (see table 1).
5. We calculated the average of the "ki" in order to find a constant "k" allowing us to get the value of "d" at any point within the operating region of each icon. The obtained values of K_i obtained in laboratory tests are provided in Table 1.

3.8 Camera position

The following actions were performed to obtain a final position of the camera: First, a reference position is found by a system initialization (HOME), then camera will begin to make a PAN movement to find some available “working -icon” and checking validated distances, once the icon is found, the system moves the camera to get the icon in the image centre of current image, calculates the criteria distances explained before to know which icon is within the scene.

Once the icon is in the centre, the system calculates the distance between camera and icon and the vision angle θ (angle between optical camera axis and normal line to icon centroid). The distance and the angle are shown in Figure 12 and the corresponding values obtained in laboratory tests are given in Table 1.

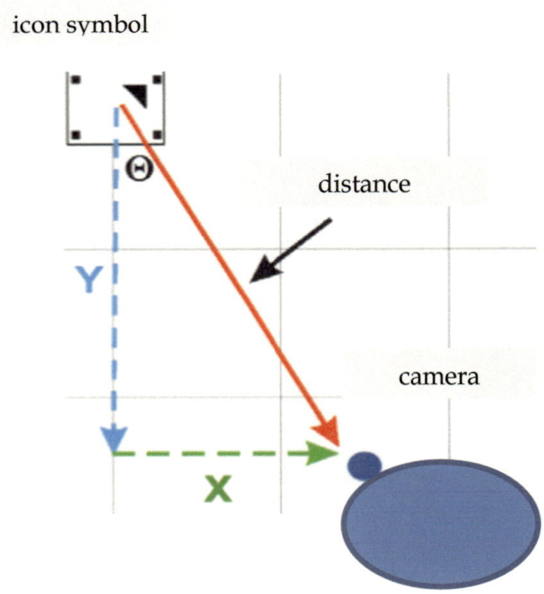

Fig. 12. Position vector

With this information a geometrical model is used to get the (x,y) coordinate of the camera in the working region being used and a graphical representation of the camera and the environment interaction is obtained. In order to get this situation, three parameters are obtained and used to get the final position of the camera:

a. Icon identification (minimum, maximal distance),
b. camera-icon distance and
c. Angle of vision θ,

Distance [cm]	K_i	*Icon height* [pixels]
20	2985.8	149
30	2990.3	100
40	3007.9	75
50	3000.3	60
60	3001.7	50

$_{Average}K = 2997.2$

Table 1. Values obtained with laboratory measurements.

4. Experimental results

Experimental tests were made to obtain the performance of the system in real conditions, to this purpose an enclosed squared environment was built, the iconographic symbols set as described before were and painted on the four different walls of the environment, which represents four working icons areas.

4.1 Experimental method

Experimental tests showed very good results with real time performance of the system. Once the system was implemented, and the practical operation checked, the precision of the system was verified. In order to achieve this task, we used different working regions for each working icon. For experiment purposes, the area of the enclosed environment for each icon was divided in three zones with 20, 40 and 55cms distances from each icon as shown in figure 13

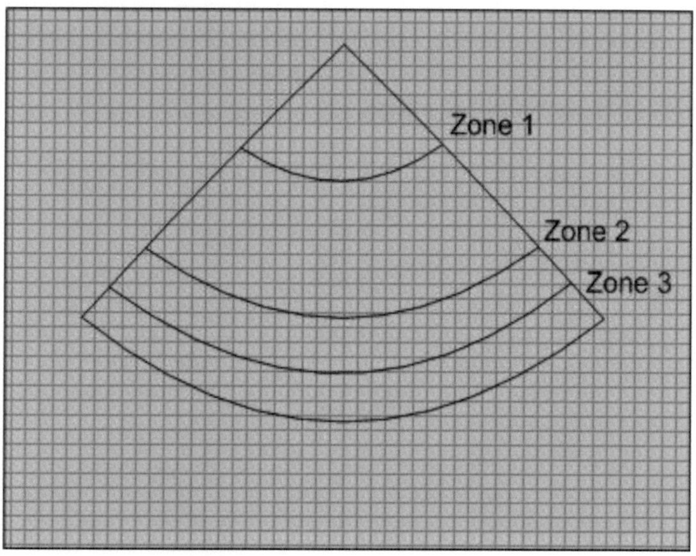

Fig. 13. Divided zones for each working icon.

The difference between desired and real locations was measured and the results are showed in Table 2. Eight points were selected in a random manner for each zone and real desired physical coordinates were obtained. The camera was positioned on each selected point and the system calculated the positions to compare its results (table 1).
The experiment was made for all points in all different regions for all different icons a graphical representation was made with the obtained values to get a better feedback of the system performance, Figures 14 and 15 shows a graphics for two different icon working regions.
Testing of the complete system with software and hardware integrated was done by selecting ten random points inside of the workspace, then the camera along with a driver support which performs the pan/tilt movements was located in real points and compare the response given by the system against the actual position values. The results of the tests were as follows:

Real Measurements [cm]		System Measurements [cm]		Time [s]
X	Y	X	Y	
20.00	35.00	20.31	34.36	84.782
40.00	3.00	38.48	3.04	143.368
-4.00	5.00	-4.95	6.93	50.442
15.00	-31.50	18.70	-36.23	141.694
34.00	-16.50	33.47	-17.21	43.432
-19.00	-25.20	-20.97	-25.07	92.844
-40.00	-1.00	-39.64	-0.67	119.512
15.00	3.50	14.66	2.48	57.232
-25.00	29.00	-24.98	29.16	82.899
-25.30	29.60	-25.87	30.21	82.889

Table 1. Real and System calculated positions.

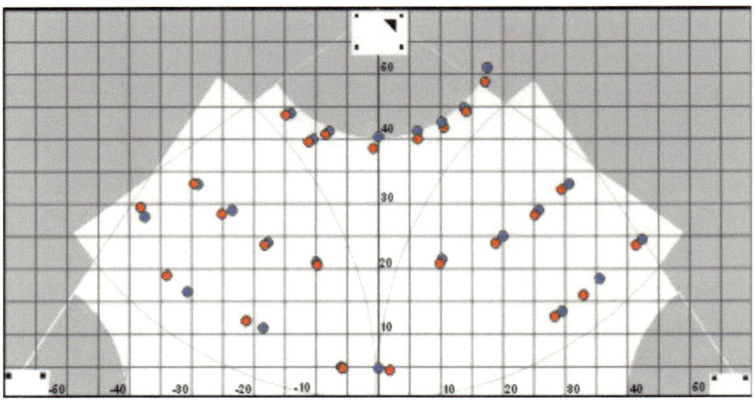

Fig. 14. Graphical representation of measured and real points for icon zone 1.

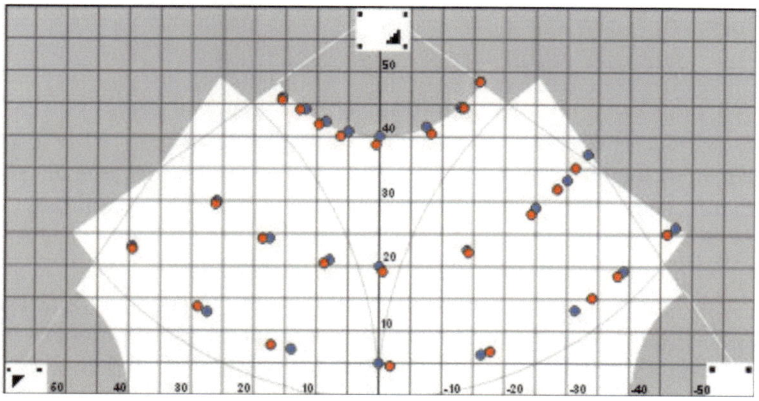

Fig. 15. Graphical representation of measured and real points for icon zone 2.

Experimental testing was repeated ten times and average measurements were registered. Figure 16, shows a graphic for the error in x axis for 3 zones of a working icon.

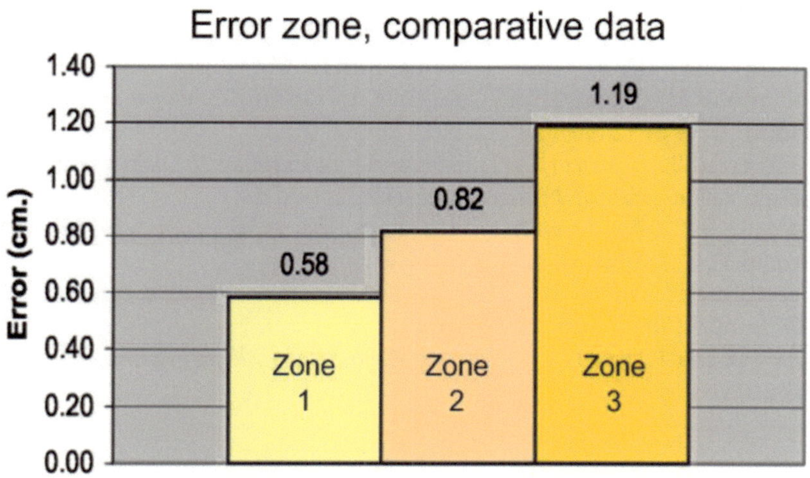

Fig. 16. Error for x coordinate

Average error in x and y can be established and for each of the measurements zones , in order to see in which of the three zones the system's behavior is more precise, resulting as follows:

Zone 1	Zone 2	Zone 3
x = **0.58 cm**	x= **0.82 cm**	x= **1.19 cm**
y = **0.79 cm**	y= **0.89 cm**	y= **1.47 cm**

Table 2. Average error for x and y for three different zones.

Previous data indicates through analysis and comparison of the obtained test results that: the precision of results that provides the system are directly proportional to the distance that the icon is captured, in addition also we can see from figure 12 that the greater the view angle, the greater error value too.

5. Conclusions

A system capable to obtain real time position of an object using a pan/tilt camera in hand as the sensor was developed. An iconographic symbol set was used to identify different working areas within an enclosed simulated working environment. Iconographic symbols projected or draw in the environment walls can be used to the purpose of get a calculated camera position. The camera has automated icon search capabilities, experimental measurements show feasible practical use in manufactured and assembly applications to find real-time positions in working tools for robot manipulators. Experimental test were carried out with some optimal laboratory conditions to get images such as good illumination, good contrast and specific sizes of experimental environment in order to assess the system. However, future work envisages an automated recalibration so for real applications in an arm robot manipulator with a camera mounted onto the arm in a hand-in-eye configuration. It is intended to preserve the use of basic geometric figures as it resulted very useful in this investigation and it can speed up the distance calculation in more complex scenarios.

6. References

Pajares M., Gonzalo., de la Cruz G, Jesús (2002). *Visión por computador* Ed. Ra-Ma. Colombia.R.M. Haralick and L.G. Shapiro (1993). *Computer and robot vision.* Ed. Addison-Wesley Publishing Co., New York.

J. de Lope, F. Serradilla, J. Zato (1997). Sistema de localización y posicionamiento de piezas usando visión artificial. *Inteligencia Artificial* 1(1):57-64

Pressman, Roger S. (2002). *Ingeniería del software. Un enfoque práctico.* Ed. McGraw-Hill. Madrid.Matlab Automation Server

http://www.mathworks.com/access/helpdesk/help/techdoc/matlab_external/f27470.html. visited Feb. 03, 2011.

Malvino Albert p, Leach Donald P. (1993). *Principios y aplicaciones digitales.* Ed. Marcombo Boixareu editores.

12

From Robot Arm to Intentional Agent: The Articulated Head

Christian Kroos, Damith C. Herath and Stelarc
MARCS Laboratories, University of Western Sydney
Australia

1. Introduction

1.1 Robots working together with humans

Robot arms have come a long way from the humble beginnings of the first Unimate robot at a General Motors plant installed to unload parts from a die-casting machine to the flexible and versatile tool ubiquitous and indispensable in many fields of industrial production nowadays. The other chapters of this book attest to the progress in the field and the plenitude of applications of robot arms. It is still fair, however, to say that currently industrial robot arms are primarily applied in continuously repeated manufacturing task for which they are pre-programmed. They are known for their precision and reliability but in general use only limited sensory input and the changes in the execution of their task due to varying environmental factors are minimal. If one was to compare a robot arm with an animal, even a very simple one, this property of robot arm applications would immediately stand out as one of the most striking differences. Living organisms must sense changes in the environment that are crucial to their survival and must have some flexibility to adjust their behaviour. In most robot arm contexts, such a comparison is currently at best of academic interest, though it might gain relevance very quickly in the future if robot arms are to be used to assist humans to a larger extend than at present. If robot arms will work in close proximity with and directly supporting humans in accomplishing a task, it becomes inevitable for the control system of the robot to have far reaching situational awareness and the capability to adjust its 'behaviour' according to the acquired situational information. In addition, robot perception and action have to conform a large degree to the expectations of the human co-worker.
Countless situations can be imagined (and are only a step away from current reality while fully autonomous mobile robots might still be far off):

- A robot arm lifting and turning a heavy workpiece such as a car engine for human inspection and repair;
- A robot arm acting as a 'third hand' for a human worker for all kinds of construction and manufacturing work that is yet too complex to be fully automated;
- A robot arm assisting a temporarily or permanently bedridden person and/or the nurses taking caring of the person. For the latter, one of the most important tasks would be again the careful lifting of the person;
- An intelligent robotic device assisting people with walking difficulties replacing the current clunky walkers;

- A robot arm assisting elderly people at home with all tasks that require considerable force (from opening a jar to lifting heavy items) or involve difficult to reach places (which might be simply the room floor).

To assess the social and economical impact that such a development would have, one might draw a parallel to the revolution that heavy machinery meant for construction and agriculture and with this for society at large. Within one generation, one might speculate, it could become inconceivable to imagine many workplaces and the average home in industrialised countries without assisting robot arms.

1.2 Joint action

Humans collaborate frequently with each other on all kinds of tasks, from jointly preparing a meal to build a shelter to write a book about robot arms. Even if the task is very simple such as carrying a load together, the underlying coordination mechanism are not. Collaborations with physical co-presence of the actors require a whole gamut of perceptive 'cues' to be observed and motor actions to be adjusted. This might be accomplished during execution or already during planning taking into account predictions of the co-workers' actions. In almost all situations so-called *joint attention* (to which we will return shortly) is an additional prerequisite. The emerging field of *joint action* research in psychology (Sebanz et al., 2006) tries to unravel the perceptive, cognitive and motor conditions and abilities that allow the seemingly effortless coordination of human action to accomplish a common goal. Sebanz et al. (2006) suggest an operational definition of joint action as 'any form of social interaction whereby two or more individuals coordinate their actions in space and time to bring about a change in the environment'. In this regard, the requirement for joint action builds on the concept of joint attention and extends it by requiring the prediction of actions of another. Joint action therefore depends on the abilities to (1) share representations, (2) predict actions, and (3) integrate predicted effects of one's own and the other's actions. These requirements do not change if the other is a machine or - narrowed down given the topic of this book - a robot arm. Admittedly, one could offload all the coordination work to the human co-worker by 'stereotyping' the action of the robot arm, i.e. reduce the movement vocabulary and make it easily predictable in all situations, but one would at the same time also severely limit the usefulness of the robot arm.

Arguably, we humans excel in joint actions because we perceive other humans as intentional agents similar to ourselves. Whether or not this would apply to robots is at the current state of research an unanswered question and, moreover, a question that poses difficulties to any investigation as there is no direct access to the states of the human mind. Some studies, though, provided partial evidence in favour of this using sophisticated experiment designs. Participants have been found to attribute animacy, agency, and intentionality to objects dependent on their motion pattern alone (Scholl & Tremoulet, 2000) and studies in Human-Robot Interaction (HRI) confirmed that robots are no exceptions (though clear differences remain if compared to the treatment of motor actions of other humans; see Castiello, 2003; Liepelt et al., 2010). Humans might also attribute emotions and moods to robots (e.g. Saerbeck & Bartneck, 2010). An important aspect of considering a robot as an intentional agent is the tacitly included assumption that the actions of the robot are neither random nor fully determined (as both would exclude agency), but a more or less appropriate and explainable response to the environment given the current agenda of the robotic agent.

Note that 'intentional agent' does not equate with human-like: animals are intentional agents as well, and there is long history of collaboration of humans with some of them, one of the most perspicuous examples being shepherds and their dogs. While high-level understanding of conspecifics as intentional beings like the self (so called 'theory of mind', see Carruthers & Smith (1996) for a theoretical review) might be a cognitive competency that is limited to humans and maybe (Tomasello, 1999) - or maybe not (Call & Tomasello, 2008) - other primates, understanding others as intentional beings similar to oneself is not a capability that emerged ex nihilo. Over the last two decades, research concerned with the development of this capacity has indicated that it is closely tied to what is now generally called *joint attention* (Tomasello et al., 2005).

1.3 Joint attention

The concept of joint attention refers to a triadic relationship between two beings and an outside entity (e.g. an object like an apple) whereby the two beings have a shared attentional focus on the object. Joint attention has been seen as a corner stone in the development of social cognition and failure to achieve it has been implicated in Autistic Spectrum Disorders (Charman, 2003). As pointed out by Tomasello (1999), for joint attention to be truly joint, the attentional focus of the two beings must not only converge on the same object but both participants must also monitor the other's attention to the object (secondary inter-subjectivity). This should be kept in mind when thinking about a robot arm collaborating with humans as it basically requires some kind of indicator that the control system is aware of the current human actions and - at least potentially - is able to infer the intention of the human co-worker. This indicator might be a virtual or mechatronic pair of eyes or full face. In previous research on joint attention, a variety of different definitions have been used, not all of them as strict as Tomasello's. This is because applying his definition poses substantial difficulties in verifying whether joint attention has occurred in an experimental set-up, in particular when investigating infants or non-humans, and by extension also makes modelling it in a machine more difficult.

Its link to understanding other people as intentional beings notwithstanding, joint attention is not uniquely human; it has been observed in monkeys (Emery et al., 1997) and apes (Carpenter et al., 1995). In the latter study, joint attention was heuristically defined in terms of episodes of alternating looks from an ape to the person and then to the object. This way of quantifying joint attention through gaze switching has become the one most frequently used, even though gaze alternation is not always a reliable indicator of joint attention as mentioned above. Furthermore, gaze alternation constitutes neither a sufficient nor a necessary condition for joint attention. On the one hand, it is very common among animals to use another animal's gaze direction as a clue to indicate important objects or events in the environment but the fact that the other animal paid attention to this event is of no consequence and not understood (Tomasello, 1999); on the other hand establishing joint attention, for instance, through the use of language is a much more powerful mechanism than just gaze following (since it includes the aspect of the object or event on which to focus). All of this will have an impact on designing a robot arm control system that is able to seamlessly and successfully cooperate with a human. Not surprisingly, joint attention in robotics poses challenges not to be underestimated (Kaplan & Hafner, 2004).

2. A virtual agent steps into the physical world

We went into some details with regard to joint action and attention to explain some of the basic motivations driving our use of a robot arm and shaping the realisation of the final system, the Articulated Head. Because of its genesis as a work of art, many of our aims and many of

Fig. 1. The Articulated Head.

the properties of the Articulated Head are probably far beyond the ordinary in robot arm research and development. On the hardware side, the Articulated Head consist of a Fanuc LR Mate 200iC robot arm with an LCD monitor as its end effector (see Figure 1). The Articulated Head represents the robotic embodiment of the *Prosthetic Head* (Stelarc, 2003) by Australian performance artist Stelarc, an Embodied Conversational Agent (ECA) residing only in virtual reality, and is one of the many faces of the *Thinking Head* developed in the Thinking Head Project (Burnham et al., 2008).

The Prosthetic Head (Figure 2) is a computer graphic animation based on a 3D laser scan of the head of the artist. Through deforming its underlying 3D mesh structure and blending the associated texture maps a set of emotional face expressions and facial speech movements are created. A text-to-speech engine produces the acoustic speech output to which the face motion are synchronised. Language input from the user is acquired through a conventional computer keyboard. Questions and statements from the user are sent to the A.L.I.C.E. chatbot (Wallace, 2009) which generates a response utterance. The Prosthetic Head has been presented at numerous art exhibitions, usually as a projection of several square meters in size.

The Articulated Head was born as a challenge to the traditional embodiment of ECAs in virtual reality. No matter how convincing the behavioural aspects and cognitive capabilities

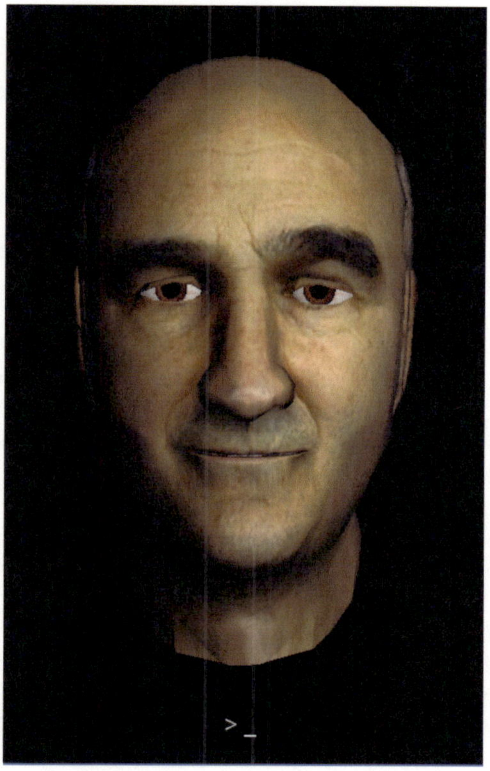

Fig. 2. The Prosthetic Head.

of a conventional ECA might be, it would always fall short of sharing the physical space with the interacting human. As physical co-presence is of great importance for humans (e.g. infants do not learn foreign language sounds from television; see Kuhl et al., 2003), transgressing the boundaries of virtual reality would enable a different quality of machine-human interaction. The robot arm enables the virtual agent to step out into the physical space shared with its human interlocutor. The sensory capabilities of the Articulated Head in the form of cameras, microphones, proximity sensors, etc. (Kroos et al., 2009) allow it to respond to the user's action in the physical world and thus engage the user on a categorically different level compared to interfacing only via written text and the 2D display of an animated face.

2.1 Problems of the physical world

With the benefits of the step into the physical world, however, come the difficulties of the physical world. Not only becomes perfect virtual perception noisy real world sensing, precise and almost delay-free visual animation imprecise and execution time-adherent physical activation, but also the stakes are set higher to achieve the ultimate goal of creating a believable interactive agent. The virtual world is (at least currently) much sparser than the physical world and thus offers substantially less cues to the observer. Less cues mean less opportunities to destroy the user's perception of agency which is fragile no matter how sophisticated the underlying control algorithms might be given the current state of art of artificial intelligence. In other words, compared to the virtual-only agent, many more aspects of the robotic agent must be modeled correctly, because failure to do so would immediately

expose the 'dumb' nature of the agent. This might not constitute a problem in some of the applications of human-robot collaboration we discussed above since the human co-worker might easily accommodate to shortcomings and peculiarities of the machine colleague but it can be assumed that in many other contexts the tender fabric of interactions will be torn apart, in particular, if the interactions are more complex. Statements in this regard are currently marred by their speculative nature as the appropriate research using psychological experiments has not been done yet. This is equally due to the lack of sufficiently advanced and interactive robots as to the difficulties to even simulate such a robot and systematically vary experiment conditions in so-called Wizard-of-Oz experiments where unknown to the participants a human operator steers the robot.
In our case of a robotic conversational agent, the overall goal of the art project was at stake: the ability to engage in a conversation, to take turns in a dialogue, to use language and speech more or less correctly, requires as a prerequisite an intentional agent. Thus, if the robot's actions had betrayed the goal of evoking and maintaining the impression of intentionality and agency, it would have compromised the agent as a whole: either by unmasking the integrated chatbot as a 'shallow' algorithm operating on a limited database with no deeper understanding of the content of the dialogue or by destroying the perception of embodiment by introducing a rift between the 'clever' language module and the failing robot.

2.2 Convincing behaviour

The cardinal problem encountered is the requirement to respond to a changing stimulus-rich environment with reasonable and appropriate behaviour as judged by the human observer. Overcoming this problem is not possible, we propose, without integration of the plenitude of incoming sensory information as far as possible and selection of the most relevant parts taking into account that (for our purposes) the sensory information is not a sufficiently complete description of the physical environment. Therefore, as a first step after low-level sensory processing, an attention mechanism is necessitated that prioritises information relevant to the current task of the agent over less important incoming data. An attention model not only takes care of the selection process, it also implicitly solves the problem of a vastly incomplete representation of the environment. For any control system that receives the output of the attention model, it is per se evident that it receives only a fragment of the available information and that, should this information not be sufficient for the current task, further data need to be actively acquired. In a second step then, the selected stimuli have to be responded to with appropriate behavior, which means in most cases with motor action though at other times only the settings of internal state variables of the system might be changed (e.g. an attention threshold).
There is another important issue here: when it comes to the movements of the robot not only the 'what' but also the 'how' gains significance. 'Natural' movements, i.e. movements that resemble biological movements, contribute crucially to the overall impression of agency as the Articulated Head has a realistic human face. Robot motion perceived as 'mechanical' or 'machine-like' would abet the separation of the robot and the virtual agent displayed on the LCD monitor, and thus create the impression of a humanoid agent being trapped in the machine. Again, if we allow a little bit of speculation, it can be hypothesised that robot arms engaging in joint action with humans will need to generate biological motions in order to make predictions of future actions of the robot arm easier and more intuitive for the collaborating human.

Joint	1	2	3	4	5	6
Motion range (deg)	340	200	388	380	240	720
Motion speed (deg/s)	350	350	400	450	450	720

Table 1. Robot joint motion range and speeds.

3. The robot arm

The robot arm employed is a Fanuc LR Mate 200iC used typically in industrial applications. It has six degrees of freedom. Table 1 shows the speed and the motion range of each individual joint of the robot arm with joint 1 being the closest to the mounting base. The robot is mounted on a custom made four-legged heavy steel structure which does not require fixing it to the floor for stable operation. The robot's work envelop is protected from inadvertent entry by human users through a series of glass structures and interlocks. The robot is controlled through an external control box using a proprietary handling application program language. In standard usage the robot is pre-programmed with the necessary movement instructions using a 'teaching pendant' with target points entered 'online' during the teaching phase prior to commissioning of the robot. However, in order to accommodate realtime interactive behavior desired by the Articulated Head, the robot interface was customised such that target points (i.e. motor goals, see section 7) could be created during the execution phase. This also meant that an additional layer of safety checks were needed to prevent the robot from trying to reach unreachable locations resulting in collisions and/or singularity conditions.

4. Inter-Component Communication and Sensing

4.1 Software Architecture

The communication framework (Herath et al., 2010) for our system combines approaches from open agent-oriented systems previously used for multimodal dialogue systems (e.g. Herzog & Reithinger, 2006) and frameworks for high-performance robotic platforms (e.g. Brooks et al., 2005; Gerkey et al., 2003). The driving motivation is to enable easy integration of components with different capabilities, written in different programming languages and potentially running on different platforms (including distributed platforms). A specific requirement for our application is realtime performance under massive data processing over streaming audio and video; this ruled out the existing multimodal dialogue platforms, and also led us to eschew standards-based APIs which incur overheads on message-passing to components. In common with other dialogue platforms, we use an event-driven framework, which has a number of desirable properties, such as: naturally modelling the non-linear nature of human interaction; providing the flexibility required for easy integration of components into a distributed architecture; dynamically prioritising software components and event types; and optimising the system via inter-component configuration commands for particular interaction states.

4.2 Sensors

We have adopted two commercially available camera systems for tracking people in 3D and faces in close proximity. A stereo camera mounted rigidly high on a wall opposite the Articulated Head looking downwards into the interaction space of the robot provides information about human movement. The commercial people tracking algorithm is based on an assumed depth profile of an average human and uses disparity images produced by a calibrated camera pair. It provides the localisation and height information of all people within the camera's field of view to the robot. The tracking system is capable of tracking multiple persons with considerable tolerance to occlusion and occasional disappearance from the field of view.

A monocular camera mounted above the top edge of the LCD screen provides fine grain information about humans directly interacting with the robot. Data from this camera feeds on to a face tracking algorithm that is capable of detecting and tracking a single face in the camera's field of view. The used algorithm has a high degree of accuracy withstanding considerable occlusion, scale variance and deformations.

Stereo microphones mounted on the back panel of the robot enclosure coupled to an auditory localiser provides accurate information of the instantaneous locations (azimuth) of a moving interlocutor in a noisy and reverberant environment. Localisation is limited to the half sphere in front of the robot and provides azimuth angle from about -90° to $+90^\circ$. The localisation is based on Faller & Merimaa (2004) which has been modified and adapted to the Articulated Head setup.

In addition to above components, various ancillary components such as proximity detectors, keyboard input device, gesture recognition system, text-to-speech system, dialogue management system, monitoring and a data logging systems are also implemented to support the various interactive aspects of the Articulated Head. Figure 3 shows the overall component topology.

5. Attention model

Human attention is a heavily researched area with thousands of scholarly articles. In general, attention is investigated in controlled psychological experiments focusing on specific aspects of the overall phenomenon of attention, say, visual attention activated by certain types of motion perceived in peripheral vision. A substantial amount of knowledge has been accumulated, though sometimes disparate or conflicting. One of the most important findings for attention systems in machines (Shic & Scassellati, 2007) is that attention is driven by two sources, saliency in the perceptual input (bottom-up or exogenous attention) and task-dependent attention direction (top-down or endogenous attention). The former is comparatively easier to handle and evaluate (e.g. data from human participants can be acquired using eye-tracking technology). Top-down attention mechanisms, on the other hand, pose severe difficulties as they typically involve high-level world knowledge and understanding. Unfortunately, however, top-down mechanisms appear to be more decisive: Even for a barn owl only 20% of attentional gaze control could be explained by low-level visual saliency (Ohayon et al., 2008).

Compared to human attention, modelling attention in artificial agents is less studied. Attention models have been primarily investigated and applied in virtual environments (Kim et al., 2005), avoiding the largely unsolved problem of real world object recognition. The identity of objects placed in a virtual environment can be made known directly to the attention model of the agent; an option that is clearly not available when dealing with a robot and real world sensing. In addition, the sensory input in real world sensing is always affected by considerable noise. The majority of the attention models for virtual agents is biologically inspired and thus complex (e.g. Bosse et al., 2006; Itti et al., 1998; Peters & Itti, 2006; Sun et al., 2008), though others amount to not much more than a fixed selection process of an input source based on the value of a single (or a few) parameter.

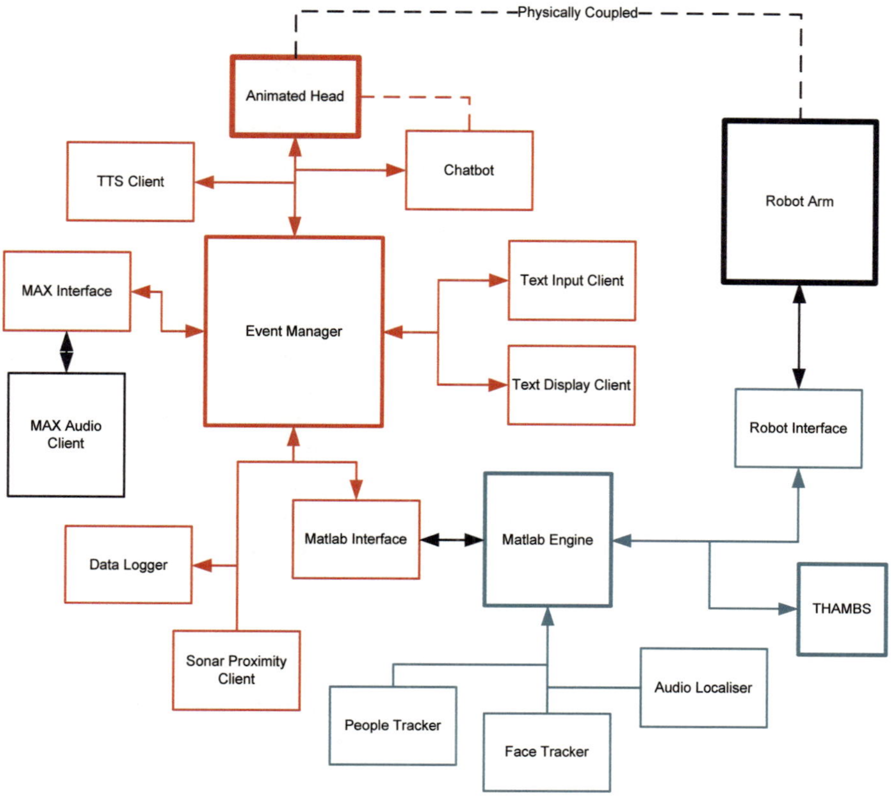

Fig. 3. Component topology. 'THAMBS' in the rightmost box stands for the Thinking Head Attention Model and Behavioural System and is described in section 5 to 8.

5.1 Attention models in robotics

A few attempts have been made to develop attention models for robots. One of the first implementations, named FeatureGate, used an artificial neural network that operated on 2D feature maps (Driscoll et al., 1998), specifically - in the tests presented in the paper - feature maps derived from synthetic images. In its handling of how the features were weighted, it allowed changes depending on the task, that is, top-down attention was partially established. However, despite its sophisticated algorithms, FeatureGate corresponds more to a target-detection system than an attention system as it does almost nothing other than find a given target among distractors in an efficient manner. This is in line with many of the experiments studying visual attention in humans, but these experiments use a simplified controlled experiment set-up to isolate aspects of the complex human attention system; they do not indicate that the human attention can be reduced to an efficient search method (e.g. Cavanagh, 2004). We would argue that when it comes to work with a robot, attention truly starts when there are several potential targets and the system has to make a choice: discard (temporarily) all but one target (the most relevant one given the current task) and focus on it.

Only bottom-up attention mechanisms were considered in the attention model developed in Metta (2001). The study focused on log-polar vision which simulates the distribution of the photoreceptors in the primate eye. They showed that this type of space variant vision is well suitable for implementing an attention system and controlling robot movements through it. It addresses implicitly the reduction of previous attention systems to target detection systems by having different sensory resolutions in the periphery and the foveal area. Thus, attention coincides with the fixation point, but events registered in the periphery could still attract attention and command fixation.

Like the model of Driscoll and colleagues, the visual attention system of Breazeal & Scassellati (1999) used with the robot Kismet was based on the guided search model of Cave & Wolfe (1990) and Wolfe (1994), but it went beyond it. The attention system did not only combine several different feature maps, but also modeled the influence of habituation effects and integrated the impact of the robot's motivational state on the generated attention activation map. In this way, their attention system became context-dependent and Kismet's behaviour emerged from the interaction of its own state and the state of the environment. For instance, the attentional gain for faces was increased during Kismet's *seek people* behaviour and decreased during its *avoid people* behaviour. Kopp & Gärdenfors (2001) postulated the imperative of an attention system for perceived intentionality of a robotic agent, but, unfortunately, they did not implement one. Their robot arm equipped with two cameras, one for peripheral vision (above arm) and one for central/focal vision (at arm), would have been, as they indicated, a very suitable platform for it.

A multimodal attention system to guide an interactive robot was proposed in Déniz et al. (2003). The researchers used feature and saliency maps to model bottom-up attention combining visual and acoustic features, but did not include top-down processes. They also did not treat visual and acoustical events equally: acoustic events could not change the focus of attention, they only reinforced the visual event closest to the acoustic event.

Attention models based on salience maps (the majority of those mentioned above) can be computationally very costly, particularly if an increasing number of features and larger feature and salience maps are used. Ude et al. (2005) demonstrated that with proper parallel processing in a distributed implementation, sufficient speeds were achieved to steer the visual system of the humanoid robot they used in realtime. The model of Ude et al. (2005) was further developed in (Morén et al., 2008) by strengthening the top-down aspects and exploring a new way of integrating bottom-up and top-down mechanisms. The authors combined the use of saliency maps from Itti and Koch's (Itti et al., 1998) model with a more flexible version of the feature-specific top-down mechanism of Cave's FeatureGate (Cave, 1999).

In this vein it appears as if attention models in robots have been recently recognised as a way to tackle problems with visual segmentation. As we mentioned earlier, this view seems at times to be more inspired by psychological experiments investigating visual attention than biological attention itself; they seem to model aspects of those experiments. As a consequence, robotic attention does not only fail to model the complexity of human attention - something which is expected and generally unavoidable given the current state of technology - but also reduces attention to an auxiliary function of the robot's perceptual system while it should be, if anything, its 'guide'. Nevertheless useful results can be obtained. Yu et al. (2007), for instance, devised an attention-based method to segment specified object contours from the image motion produced by the egomotion of a mobile robot. They employed a pre-attentive state for contour segmentation and competing motion-based bottom-up and contour-based top-down salience maps. Using Bayesian inference top down saliency biased the final probabilistic attention distribution toward the task-dependent object contour.

Developing and applying attention models is usually motivated with the proposed requirement that robots interacting with humans should possess an attention system similar to that of humans. More specifically, *joint* attention is argued to be a conditio sine qua non for cooperative human-robot action, machines learning from human instructors, 'theory of mind' in robots (the ability to predict what another can and cannot perceive), and similar high-level cognitive social capabilities. However, joint attention of robot and human is trivially not possible if the robot does not have the capacity of attention in the first place (at least in the form of being able to select specific elements of the input over others). In the literature reviewed above attention systems sometimes seem more to be a means to an auxiliary end than being an integral and essential part of the robots behavioural system. A different and more immediate motivation is brought forward in Bachiller et al. (2008) in their 'attention-based control model'. The authors view the attention system as an essential mediator between visual perception and action control that is needed to handle two important tasks: to select perceptual information relevant for action execution and to limit potential actions based on the perceived situational context. In the context of autonomous navigation of robots they employ bottom-up and top-down attention processes and also model overt and covert attention. Covert attention refers here to regions of interest that are pre-activated within the attention system through target selection, but are currently not the focus of attention (overt attention).

Finally, a method to switch autonomously between bottom-up and top-down attention in a mobile robot was introduced by Xu et al. (2010). The different attention modes are activated dependent on the state the robot is in (exploring, searching, or operating) which links attention back to behaviour - something that in our view is essential for attention: attention cannot be seen as a passive input information selection mechanism since it is tied to action and also actively changes what is perceived. The benefits of the latter was demonstrated in Xu et al. (2010) through steering the active stereo camera of the robot they used towards target area identified by the bottom attention system as relevant for the task and then apply the top-down attention to keep the target in the focus of attention.

5.2 The attention model of the Articulated Head

In the Articulated Head, the attention model is part of the Thinking Head Attention and Behavioural System (THAMBS) that manages all high-level aspects of the interaction including the generation of response behaviour (see next section) except for conversational matters that are taken care of by the chatbot. THAMBS goes beyond straightforward action selection insofar as it is also concerned with determining the specific characteristics of the motor behaviour associated with the response (see section 7) and in that behaviours can interact with each other and can change the way the sensory input is processed. It consists of four modular subsystems: (1) a perception system, (2) an attention system, (3) a central control system, and (4) a motor system. Figure 4 shows a diagram of THAMBS, with its subsystems and flow of information. THAMBS is currently implemented in Matlab (The MathWorks, Inc) and following object-oriented programming principles, its subsystems are represented as classes to ensure their strict modularity.

Despite the array of sensing devices, the robot's sensing of the world is relatively sparse since the sensing devices and their software are specialised on particular tasks (e.g. people tracking, face detection). It is multimodal, however, and complex enough to allow sophisticated interactions with human users. Nevertheless, the difference in the input compared to almost

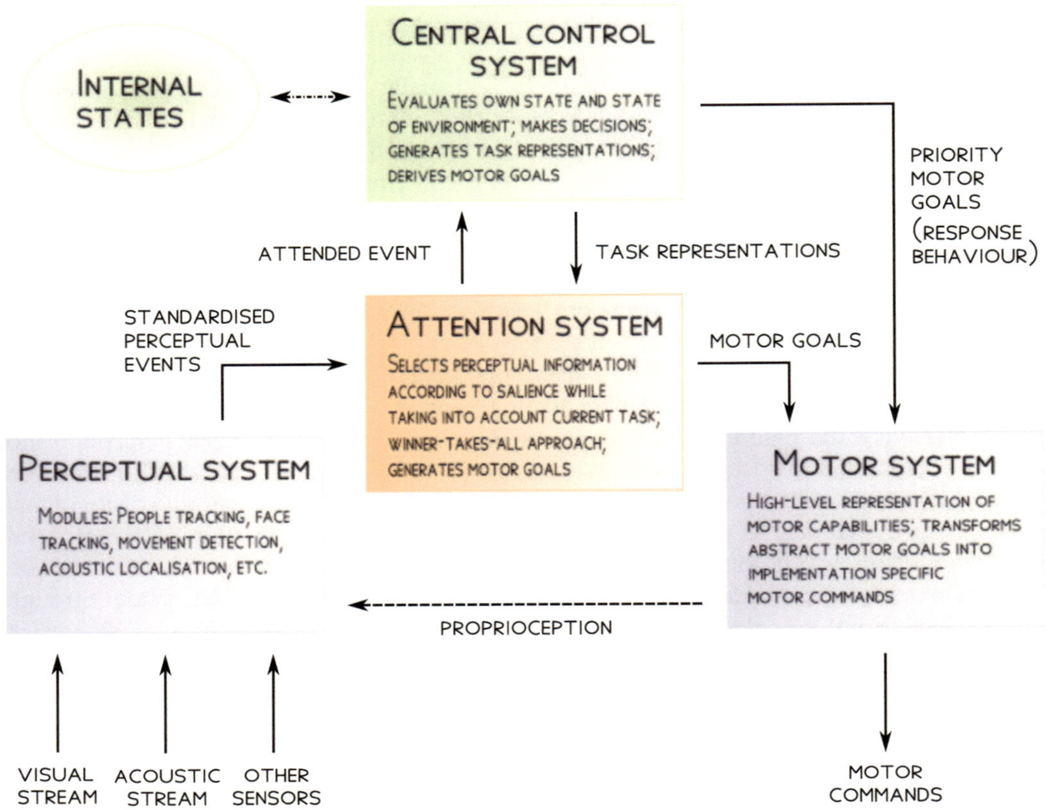

Fig. 4. Schematic of THAMBS and its subsystems.

all other attention models has implications on architecture and functionality of the attention system within THAMBS. First an upstream perception system transforms the information from the sensing modules into a standardised perception event making it possible to process very different events (e.g. a person being detected within the visual field of the Articulated Head as well as a character string being sent from the keyboard) with respect to their attentional importance not their detail characteristics. An attentional weight is assigned to the incoming perceptual event computed using a base weight assigned as a parameter value to the *type* of the perceptual event (e.g. acoustic localisation, people tracking) and an attention weight factor derived from the specific event *instance*, usually confidence values (the default value is 1):

$$w_p(i) = b_p(i)\; w_{base} \tag{1}$$

Note that both values might be changed during run time according to changes of the active task, for instance the base weight of face tracking might be increased to favour face-to-face interactions with a single person over 'distracting' other people in the area covered by the stereo camera. The resulting attention weight is checked against a threshold dependent again on the type of perceptional event. If the event passes, an attention focus is created (covered

attention) which is characterised by three properties: its weight, its decay function, and the spatial location in the real world it is referring to.
The weight is the attentional weight described above, however, for an already existing focus it is modified based on the duration of its existence.

$$w_p(i,t) = w_p(i)\,k(t) \tag{2}$$

where $k(t)$ is a decay function assigned to the attention focus that decides about its lifetime and its impact over time. It ensures that the attention focus outlives the potentially very short instance of the perceptual event that created it (e.g. a loud startling impulse-like noise) but at the same time that its strength is fading even if registration of the perceptual event is sustained (habituation). We found a generalisation of the simple exponential functions, the Kohlrausch function, preferable to the simple exponential functions employed in other attention models. The Kohlrausch function is often called a 'stretched exponential' and is known to be able to describe a wide range of physical and biological phenomena (Anderssen et al., 2004). It is given by:

$$k_{\tau,\beta}(t) = e^{-(\frac{\tau}{\beta})^{\beta}} \tag{3}$$

It is the additional parameter β that stretches or compresses the function. Thus with an appropriate setting a plateau at around zero is formed that guarantees in our implementation a high activation for a certain period immediately after the attention focus has been established ensuring that the focus can 'fend off' lower weight foci for some time. The decay function parameters are initialised dependent on the type of perceptual event, but again, they are modified dynamically during run time (in fact, the entire function can be replaced with a different one if e.g. a discontinuous function is needed. However, this is currently not used).
The last and most important defining property of an attention focus is the segment of 3D space it is referring to: the location of the event that attracted attention. Thus, the attention foci are spatially organised (compare space versus object-based attention in models of human attention; review in Heinke & Humphreys, 2004). This plays a decisive role in the identification of a new perceptual event as identical - per definition - to one of the already existing attention foci. Locations in spherical coordinates of the new event and all old foci are compared whereby underspecification always produces a positive value. If an incoming event is considered to be identical to one of the old attention foci, the old focus is kept. Its weight, however, is updated by combining of the new and old weights in supra-additive manner:

$$w_c = w_{old} + w_{new}\,a_p \qquad \text{with} \quad 0 \le a_p \le 1 \tag{4}$$

where a_p is a parameter specific to the perceptual event type of the new event. The decay function of the focus is not reset, which causes a slow but steady decline of the weight values even if new events are constantly reinforcing an old attention focus, for instance, a person standing still within the visual field of the Articulated Head. The procedure has a similar effect as the habituation modeled in Breazeal & Scassellati (1999).
A perceptual event might have several different features depending on its type. There are always the obtained sensory data values themselves, typically some form of tracking data (though in case of the keyboard 'sense', it is only a binary on/off signal and a character string) but in addition there might be velocities and, potentially, accelerations or other properties computed over the input values such as statistical and spectral moments or energy measures. Each feature on its own is able to invoke an attention focus: If one feature fails to create an attention focus because it can not pass the threshold, another feature might do so. For instance,

an avoidance behaviour might have blocked tracked people to become an attention focus, but very fast movements of a person might nevertheless 'break through' via the velocity feature. After all attention foci are generated and their weights computed, one of them is chosen as the single event that is attended by the system using a winner-takes-all-strategy based on the highest weight. Recently, we added an alternative, a persistence strategy. After a perceptual event providing an identification marker (currently only people tracking) has become the attended event through the default strategy, the attention system locks on the event based on its ID for a limited time - independent of its decaying weight - unless there is a very powerful distractor. The trigger for the persistence strategy is random at the time, but it was devised to be replaced with a trigger based on familiarity once face recognition is integrated in THAMBS. The attended event is passed on to the central control system (see next section) together with the information whether or not it is considered new by the attention system. To model pursuit movements based on a close attentional link between perception and action (Schneider & Deubel, 2002), the attention system is able to send a specific motor command, named `look_there`, directly to the motor system. It steers the robot arm to orient the normal to the monitor display plane (and with this the optical axis of the monovision camera) toward the spatial location of the attended event. This serves a two-fold purpose: to create the impression as if the virtual face displayed on the monitor is looking at the location of the event which attracted its attention and to provide the Articulated Head with more information about the source of the event via the monovision camera (see Figure 5).

(a) Idle (b) Idle (c) Focussing

Fig. 5. The Articulated Head being idle (a,b) and focussing on an interlocuter (c).

6. Behavioural system

The response behaviour of the Articulated Head is generated by the central control subsystem of THAMBS (except for verbal interactions). It is the highest-level processing stage for information about the environment that arrives from the sensors after being evaluated by the attention system. This information itself, however, is not independent from the behaviour (perception-action link) as the behaviour affects the sensing information either directly (e.g. the position and orientation of the monovision camera) or indirectly as the behaviour might cause a change in task priorities which in turn might trigger a modification in the attentional weights assigned to perceptual event types or single attention foci. The central control system is essentially still a stimulus-response system based on a set on conditional rules, but it is non-trivial since the rules are modified during run time and are at some points subject to probabilistic evaluation.

6.1 Behaviour triggers

In THAMBS the conditional rules are called *behaviour triggers* and realised as small decision trees. In most current triggers, however, only one branch leads to the activation of a behaviour while the remaining ones cause a termination of the trigger evaluation. This is bound to change with more complex behaviour options being implemented at future development stages. The trigger evaluation is implemented to be able to handle trees of arbitrary complexity fast and efficiently while requiring only a few lines of code in Matlab. The basic idea is to collect the test results as '0' or '1' characters in a single string while moving down the tree following the active branch and then treat the resulting string as representing a binary number and convert this number into an index into the possible actions associated with the terminal nodes. In pseudo-code:

```
(1) Collect all tests associated with the nodes of the decision tree
in a one-dimensional array of expressions (named 'conditions' here)
that evaluate to a Boolean value. Move from top to bottom and left to
right.
(2) Collect from left to right all possible actions associated with
the terminal nodes in a one-dimensional array of function handles
(named 'actions' here).
(3) Initialise an indicator variable 'indTest' with value 1,
another indicator variable 'indAction' with 0, and an empty string
array 'collectedTests'.
Note: It is assumed that array indexing starts with 1 not 0.

%%% loop through all (relevant) tests of the decision tree.
WHILE indTest <= SIZE(conditions)

  %%% evaluate the indicated condition
  isTrue = EVALUATE(conditions[indTest])

  %%% if condition evaluates to true, add a '1' else a '0' to the string
  %%% array that collects the test results
  IF isTrue
    add '1' to collectedTests
  ELSE
    add '0' to collectedTests
    %%% stop further testing in case of a branch type tree
    %%% (only one branch of the tree has a behaviour assigned)
    IF tree is of type branch
      BREAK
    ENDIF
  ENDIF

  %%% split according to type of tree: determine the index for the
  %%% next test
  IF tree is of type branch
     indTest = indTest + 1
  ELSE
     %%% treat the collected test string buffer as a binary number and
     %%% use it to update the index. In this way tests that belong to
     %%% branches, which have been already discarded, will be ignored
     indTest = 2 ^ (SIZE(collectedTests)) + BINARY_TO_DECIMAL(collectedTests)
  ENDIF
ENDWHILE
```

```
%%% treat the collected test string buffer as a binary number
%%% and convert it to a decimal number: the result is an index in the
%%% possible actions arranged according to the last level of the tree
indAction = BINARY_TO_DECIMAL(collectedTests) + 1;

%%% for type 'branch' only one branch has an action assigned: do we have
%%% it? If yes, set indAction to 1
IF tree is of type branch
  IF indAction == 2^SIZE(conditions)
    indAction = 1
  ELSE
    indAction = 0
  ENDIF
ENDIF

%%% Call the appropriate action function
IF indAction > 0
  EXECUTE(actions[indAction])
ENDIF
```

Behaviour triggers have a priority value assigned to them which decides about the order of their evaluation. A trigger with a higher priority is evaluated after a trigger with a lower priority since the associated behaviours of both might modify state variables of THAMBS and the changes made by the behaviour with the higher priority trigger should take precedence, that is, these changes should be the ones that persist and should not be overwritten. Typically behaviours specify a motor goal to be achieved. Motor goals are abstract representations of motor actions to be executed by the robot arm or the virtual head displayed on the monitor. They are context independent and thus no sensory information is required at this stage, though several attributes control their processing by the motor system later on. Other behaviours only modify values of state variables and with this cause a change in how future sensory information is processed or in how other motor goals are executed. The behaviours themselves are implemented as independent routines and their function handles are passed to the trigger evaluation routine.

6.2 Behaviour disposition

Two other important aspects of the central control system besides the generation of response behaviour need to be mentioned. First there is a set of subroutines that model endogenous processes. These are changes in THAMBS' state variables that are not activated - directly or indirectly - by stimuli from the environment. An example would be the spontaneous probability-driven awakening that happens sooner or later if the Articulated Head has fallen asleep (due to lack of stimuli in the environment; see section 9 for an overview of the behaviors of the Articulated Head). It contrasts with the awakening activated by a loud sound event via an ordinary behaviour trigger (Kroos et al., 2010). The endogenous processes would be more accurately assigned to a system other than the central control system as they emulate low-level functions of the mammalian brain located, for instance, in the brain stem. Future versions of THAMBS will parcel out these processes and subordinate them to a new subsystem.
Secondly, there is a preparatory phase. THAMBS currently employs a master execution loop running usually at 10 Hz through all the necessary tasks of its systems, starting with the endogenous processes, then handling perception, continuing with attention and so on.

However, we mentioned in section 5 that attention has a strong top-down component. This is specifically accounted for in the preparatory routine which is executed before the perception system becomes active in each evaluation cycle of the master loop. The routine can change thresholds of perception and attention, and in this way it can steer perception and attention toward stimuli relevant for its current task and its current inner state (active perception and active attention). Moreover, it is able to insert new behaviour triggers in the set of active behaviour triggers. For instance, the behaviour trigger `attend_close` activates a behaviour with the same name if a sizable number of people are in the visual field of the Articulated Head. The `attend_close` behaviour changes the weight of attention foci that are based on people-tracking to favour people closer to the Articulated Head over people further away. The trigger has limited lifetime and is currently inserted randomly from time to time. In future versions this will be replaced by an insertion based on the values of other state variables, e.g. the variable simulating anxiety. Note that the insertion of an behaviour trigger is not equivalent with activation of the associated behaviour. Indeed, taking the example above, the `attend_close` behaviour might never be activated during the lifetime of the trigger if there are no or only few people around. An Articulated Head made 'anxious' through the detection of a reduction in computational resources might insert the behaviour trigger fearing a crowd of people and dealing with this 'threatening' situation in advance.
The distinction between preemptive behavior disposition and actual response triggers is important because it constitutes an essential element in the differentiation of a simple context-independent stimulus-response system with the classical strict division of input and output from an adaptive system where the interaction with the environment is always bi-directional. Note also that the preparatory phase de-facto models expectations of the system about the future states of its environment and that contrary to the claims in Kopp & Gärdenfors (2001), this does not necessarily require full internal representations of the environment.

7. Motion generation

The motor subsystem of THAMBS is responsible for converting the abstract motor goals transmitted both from the attention system and the central control system into concrete motor primitives. At first, the motor system determines which one of the two motor goals - if both are in fact passed on - will be realised. In almost all cases the 'deliberate' action of the central control system takes precedence over the pursuit goal from the attention system. Only in the case of an event that attracts exceptional strong attention the priority is reversed. In humans, this could be compared with involuntary head and eye movements toward the source of a startling noise or toward substantial movement registered in peripheral vision. A motor goal that cannot currently be executed might be stored for later execution depending on a specific storage attribute that is part of the motor goal definition. For pursuit goals originating from the attention system the attribute is most of the time set to disallow storage as it makes only limited sense to move later toward a then outdated attention focus. On completion of the goal competition evaluation, the motor systems checks whether the robot is still in the process of executing motor commands from a previous motor goal and whether this can be interrupted. Each motor goal has an `InterruptStrength` and an `InterruptResistStrength` attribute and only if the value of the `InterruptStrength` attribute of the current motor goal is higher than the `InterruptResistStrength` of the ongoing motor goal, the latter can be terminated and the new motor goal realised. Again, if the motor goal cannot currently be executed it might be stored for later execution.

Motion generation in robot arms might be considered as a solved problem (short of a few problems due to singularities maybe) and as far as trajectory generation is concerned we would agree. The situation, however, changes quickly if requirements on the meta level of the motion beyond desired basic trajectory properties (e.g. achieving target position with the end effector or minimal jerk criteria) are imposed. In particular in our case, as mentioned in section 2.2, the requirement of the movements to resemble biological motion. Since there exists no biological model for joint system such as the Fanuc robot arm, an exploratory trial-and-error-based approach had to be followed. At this point a crucial problem was encountered: if the overall movement of the robot arm was repeated over and over again, the repetitive character would be quickly recognised by human users and perceived as 'machine-like' even if it would be indistinguishable from biological motion otherwise. Humans vary constantly albeit slightly when performing a repetitive or cyclical movement; they do not duplicate a movement cycle exactly even in highly practised tasks like walking, clapping or drumming (Riley & Turvey, 2002). In addition, the overall appearance of the Articulated Head does not and cannot deny its machine origin and is likely to bias peoples' expectations further. Making matters worse, the rhythmical tasks mentioned above still show a limited variance compared to the rich inventory of movement variation used in everyday idle behaviour or interactions with other people - the latter includes adaptation (entrainment) phenomena such as the adjustment of one's posture, gesture and speaking style to the interlocutor (e.g. Lakin et al., 2003; Pickering & Garrod, 2004) even if it is a robot (Breazeal, 2002). These situations constitute the task space of the Articulated Head while specialised repeated tasks are virtually non-existent in its role as a conversational sociable robot: one more time the primary difference between the usual application of a robot arm and the Articulated Head is encountered. Arguably, any perceivable movement repetition will diminish the impression of agency the robot is able to evoke as much as non-biological movements if not more.

To avoid repetitiveness we generated the joint angles for a subsets of joints from probability density function - most of the times normal distributions centred on the current or the target value - and used the remaining joints and the inherent redundancy of the six degrees of freedom robot arm to achieve the target configuration of the head (the monitor). Achieving a fixed motor goal with varying but compensating contributions of the participating effectors is known in biological motion research as *motor equivalence* (Bernstein, 1967; Gielen et al., 1995). The procedure we used not only resulted in movements which never exactly repeat but also increased the perceived fluency of the robot motion.

Idle movements, small random movements when there is no environmental stimulus to attract attention, are a special case. No constraint originating from a target configuration can be applied in the generation of these movements. However, completely random movements were considered to look awkward by the first author after testing them in the early programming stages. One might speculate that because true randomness is something that never occurs in biological motion, we consider it unnatural. As a remedial, we drew our joint angle values from a logarithmic normal (log normal) distribution with its mean at the current value of the joint. As can be seen in Figure 6, this biases the angle selection toward smaller values than the current one (due to a cut-off at larger values forced by the limited motion range of the joint; larger values are mapped to zero), but in general keeps it relatively close to the current value. At the same time in rare cases large movements in the opposite direction are possible.

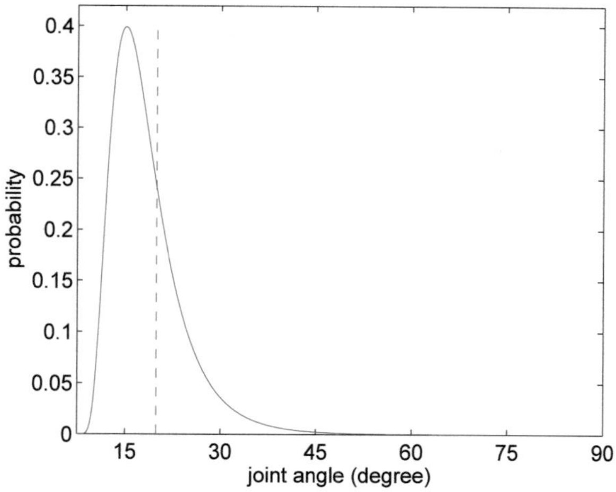

Fig. 6. Log normal probability distribution from which the new joint angle value is drawn. The parameters of the distribution are chosen so that the mean coincides with the current angle value of the robot joint. In this example it is at 24.7 degree indicated in the figure as dotted line and the cut-off is set to 90 degree.

The generation of the motor primitives realising an abstract motor goal is handled by specialised execution routines. The handles to these functions are stored as motor goal attributes and can be exchanged during runtime. The subroutines request sensory information if required such as the location of a person to be 'looked at' and transduce the motor goal in the case of the robot arm into target angle specifications for the six joints, and in case of the virtual head into high-level graphic commands controlling the face and eye motion of the avatar.
The joint angle values determined in this way are sent to the robot arm after they have passed safety checks preventing movements that could destroy the monitor by slamming it into one of the robot arm's limbs.

8. State variables and initial parameters

We described THAMBS from a procedural point of view which we deemed more appropriate with respect to the topic of evoking agency and more informative in general. However, this does not mean that there is not a host of state variables that provide the structure of THAMBS beyond the subsystems described in the previous section. In particular, the central control system has a rich inventory of them. They are organised roughly according to the time scale they operate on and their resemblance to human bodily and mental states. There are (admittedly badly named) 'somatic' states which constitute the fastest changing level, then 'emotional' states on the middle level and 'mood' states on the long term level. Except for the somatic states such as alertness and boredom those states are very sparsely used for the time being, but will play a greater role in further developments of THAMBS.
Although the behaviour of the Articulated Head emerges from the interplay of environmental stimuli, its own actions, and some pre-determined behaviour patterns (the behaviour triggers described in section 6.1), a host of initial parameter settings in THAMBS influences the overall behaviour of the Articulated Head. In fact, very often changing individual parameter settings creates patterns of behaviour that were described by exhibition visitors in terms of different

personalities or sometimes mental disorders. To investigate this further, however, a less heuristically driven approach is needed for modelling attention and behaviour control and rigorous psychological experiments. At the time of the writing both are underway.

9. Overview of most common behaviour patterns

If there is no environmental stimulus strong enough to attract the attention of THAMBS, the Articulated Head performs idle movements from time to time and the value of its boredom state variable increases. If it exceeds a threshold, the Articulated explores the environment with random scanning movements. While there is no input reaching the attention system, the value of the alertness state variable decreases slowly such that after prolonged time the Articulated Head falls asleep. In sleep, all visual senses are switched off and the threshold for an auditory event to become an attention focus is increased. The robot goes into a curled-up position (as far as this is possible with the monitor as its end effector). During sleep the probability of spontaneous awakening is very slowly increased starting from zero. If no acoustic event awakens the Articulated Head it wakes up spontaneously nevertheless sooner or later. If its attention system is not already directing it to a new attention focus, it performs two or three simulated stretching movements.

If there is only a single person in the visual field of the Articulated Head, it focuses in most instances on this person and pursues his or her movements. There might be, however, distractions from acoustic events if they are very clearly localised. If the person is standing still, the related attention focus gains for a short time a very high attentional weight, but if nothing else contributes, the weight fades, making it likely that the Articulated Head diverts its attention. Alternatively, the face detection software might register a face as the monovision camera is now pointing toward the head of the person and the person is not moving anymore. This would lead to a strong reinforcement of the attention focus and in addition the Articulated Head might either speak to the person (phrases like 'I am looking at you!', 'Did we meet before?', 'Are you happy?' or 'How does it look from your side?') or mimic the head posture. The latter concerns only rotations around the axis that is perpendicular to the monitor display plane in order to be able to maintain eye contact during mimicry.

If a visitor approaches the information kiosk (see Figure 7) containing the keyboard, the proximity sensor integrated into the information kiosk registers his or her presence. The Articulated Head turns toward the kiosk with a high probability because the proximity sensor creates an attention focus with a high weight. If the visitor loses the attention of THAMBS again due to inactivity or sustained typing without submitting the text, the Articulated Head would still return to the kiosk immediately before speaking the answer generated by the chatbot.

If there are several people in the vicinity of the Articulated Head, its behaviour becomes difficult to describe in general terms. It now depends on many factors which in turn depend on the behaviour of the people surrounding the installation. THAMBS will switch its attention from person to person depending on their movements, whether they speak or remain silent, how far they are from the enclosure, whether it can detect a face and so on. It might pick a person out of the crowd and follow him or her for a certain time interval, but this is not guaranteed when a visitor tries to actively invoke pursuit by waving his or her hands.

Fig. 7. The information kiosk with the keyboard for language-based interactions with the Articulated Head.

10. Validation

The Articulated Head is a work of art, it is an interactive robotic installation. It was designed to be engaging, to draw humans it encounters into an interaction with it, first through its motor behaviour, then by being able to have a reasonably coherent conversation with the interlocutor. Because of the shortcomings of current automatic speech recognition systems (low recognition rates in unconstrained topic domains, noisy backgrounds, with multiple speakers) a computer keyboard is still used for the language input to the machine but the Articulated Head answers acoustically with its own characteristic voice using speech synthesis. It can be very entertaining but entertainment is not its primary purpose but a consequence from its designation as a sociable interactive robot. In terms of measurable goals, interactivity and social engagement are difficult to measure, in particular in the unconstrained environment of a public exhibition.

So far the Articulated Head has been presented to the public at two exhibitions as part of arts and science conferences (Stelarc et al., 2010a;b) and hundred of interactions between the robotic agent and members of the audience have been recorded. At the time of the writing, a one year long exhibition in the Powerhouse Museum, Sydney, Australia, as part of the Engineering Excellence exhibition jointly organised by the Powerhouse Museum, Sydney, and the New South Wales section of Engineers Australia has just started (Stelarc et al., 2011). A custom-built glass enclosure was designed and built by museum staff (see Figure 8) and a lab area immediately behind the Articulated Head installed allowing research evaluating the interaction between the robot and members of the public over the time course of a full year.

Fig. 8. The triangular-shaped exhibition space in the Powerhouse Museum, Sydney.

This kind of systematic evaluation is in its earliest stages, preliminary observations point toward a rich inventory of interactive behaviour emerging from the dynamic interplay of the robot system and the users. The robot's situational awareness of the users' movements in space and its detection of face-to-face situations, its attention switching from one user and one sensory systems to the next according to task priorities that is visible in its expressive motor behaviour, all this entices changes in the users' behaviour which, of course, modify again the robots' behaviour. At several occasions, for instance, children played games similar to hide-and-seek with the robot. These games evolved spontaneously despite that they were never considered as an aim in the design of the system and nothing was directly implemented to support them.

11. Conclusion and outlook

Industrial robot arms are known for their precision and reliability in continuously repeating a pre-programmed manufacturing task using very limited sensory input, not for their ability to emulate the sensorimotor behaviour of living beings. In this chapter we have described our research and implementation work of transforming a Fanuc LR Mate 200iC robot arm with an LCD monitor as its end effector into a believable interactive agent within the context of a work of art, creating the Articulated Head. The requirements of interactivity and perceived agency imposed challenges with regard to the reliability of the sensing devices and software, selection and integration of the sensing information, realtime control of the robot arm and motion generation. Our approach was able to overcome some but certainly not all of these challenges. The corner stones of the research and development presented here are:

1. A flexible process communication system tying sensing devices, robot arm, software controlling the virtual avatar, and the integrated chatbot together;
2. Realtime online control of the robot arm;
3. An attention model selecting task-dependly relevant input information, influencing action and perception of the robot;
4. A behavioral system generating appropriate response behaviour given the sensory input and predefined behavioral dispositions ;
5. Robot motion generation inspired by biological motion avoiding repetitive patterns.

In many respects the entire research is still in its infancy, it is in progress as on the artistic side the Articulated Head is a work in progress, too. It will be continuously further developed: for instance, future work will include integrating a face recognition system and modelling memory processes allowing the Articulated Head to recall previous interactions. There are also already performances planned in which the Articulated Head will perform at different occasions with a singer, a dancer and its artistic creator. At all of these events the robot behaviour will be scripted as little as possible; the focus will be on interactivity and behaviour that instead of being fixated in few states *emerges* - emerges from the interplay of the robot's predispositions with the interactions themselves leading to a dynamical system that encompasses both machine and human. Thus, on the artistic side we will create - though only for the duration of the rehearsals and the performances - the situation we envisioned at the beginning of this chapter for a not too distant future: robots working together with humans.

12. References

Anderssen, R. S., Husain, S. A. & Loy, R. J. (2004). The Kohlrausch function: properties and applications, *in* J. Crawford & A. J. Roberts (eds), *Proceedings of 11th Computational Techniques and Applications Conference CTAC-2003*, Vol. 45, pp. C800–C816.

Bachiller, P., Bustos, P. & Manso, L. J. (2008). Attentional selection for action in mobile robots, *Advances in Robotics, Automation and Control*, InTech, pp. 111–136.

Bernstein, N. (1967). *The coordination and regulation of movements*, Pergamon, Oxford.

Bosse, T., van Maanen, P.-P. & Treur, J. (2006). A cognitive model for visual attention and its application, *in* T. Nishida (ed.), *2006 IEEE/WIC/ACM International Conference on Intelligent Agent Technology (IAT 2006)*, IEEE Computer Society Press, Hong Kong, pp. 255–262.

Breazeal, C. (2002). Regulation and entrainment in human-robot interaction, *The International Journal of Robotics Research* 21: 883–902.

Breazeal, C. & Scassellati, B. (1999). A context-dependent attention system for a social robot, *Proceedings of the 16th International Joint Conference on Artificial intelligence - Volume 2*, Morgan Kaufmann Publishers Inc., San Francisco, CA, USA, pp. 1146–1151.

Brooks, A., Kaupp, T., Makarenko, A., Williams, S. & Oreback, A. (2005). Towards component-based robotics, *International Conference on Intelligent Robots and Systems (IROS 2005)*, Edmonton, Canada, pp. 163–168.

Burnham, D., Abrahamyan, A., Cavedon, L., Davis, C., Hodgins, A., Kim, J., Kroos, C., Kuratate, T., Lewis, T., Luerssen, M., Paine, G., Powers, D., Riley, M., Stelarc, S. & Stevens, K. (2008). From talking to thinking heads: report 2008, *International Conference on Auditory-Visual Speech Processing 2008*, Moreton Island, Queensland, Australia, pp. 127–130.

Call, J. & Tomasello, M. (2008). Does the chimpanzee have a theory of mind? 30 years later, *Trends in Cognitive Sciences* 12(5): 187 – 192.

Carpenter, M., Tomasello, M. & Savage-Rumbaugh, S. (1995). Joint attention and imitative learning in children, chimpanzees, and enculturated chimpanzees, *Social Development* 4(3): 217–237.

Carruthers, P. & Smith, P. (1996). *Theories of Theories of Mind*, Cambridge University Press, Cambridge.

Castiello, U. (2003). Understanding other people's actions: Intention and attention., *Journal of Experimental Psychology: Human Perception and Performance* 29(2): 416 – 430.

Cavanagh, P. (2004). Attention routines and the architecture of selection, *in* M. I. Posner (ed.), *Cognitive Neuroscience of Attention*, Guilford Press, New York, pp. 13–18.

Cave, K. R. (1999). The FeatureGate model of visual selection, *Psychological Research* 62: 182–194.

Cave, K. R. & Wolfe, J. M. (1990). Modeling the role of parallel processing in visual search, *Cognitive Psychology* 22(2): 225 – 271.

Charman, T. (2003). Why is joint attention a pivotal skill in autism?, *Philosophical Transactions: Biological Sciences* 358: 315–324.

Déniz, O., Castrillión, M., Lorenzo, J., Hernández, M. & Méndez, J. (2003). Multimodal attention system for an interactive robot, *Pattern Recognition and Image Analysis*, Vol. 2652 of *Lecture Notes in Computer Science*, Springer Berlin / Heidelberg, pp. 212–220.

Driscoll, J., Peters, R. & Cave, K. (1998). A visual attention network for a humanoid robot, *Intelligent Robots and Systems, 1998. Proceedings., 1998 IEEE/RSJ International Conference on*, Vol. 3, pp. 1968–1974.

Emery, N. J., Lorincz, E. N., Perrett, D. I., Oram, M. W. & Baker, C. I. (1997). Gaze following and joint attention in rhesus monkeys (macaca mulatto), *Journal of Comparative Psychology* III(3): 286–293.

Faller, C. & Merimaa, J. (2004). Source localization in complex listening situations: Selection of binaural cues based on interaural coherence, *The Journal of the Acoustical Society of America* 116(5): 3075–3089.

Gerkey, B. P., Vaughan, R. T. & Howard, A. (2003). The player/stage project: Tools for multi-robot and distributed sensor systems, *International Conference on Advanced Robotics (ICAR 2003)*, Coimbra, Portugal, pp. 317–323.

Gielen, C. C. A. M., van Bolhuis, B. M. & Theeuwen, M. (1995). On the control of biologically and kinematically redundant manipulators, *Human Movement Science* 14(4-5): 487 – 509.

Heinke, D. & Humphreys, G. W. (2004). Computational models of visual selective attention: a review, *in* G. Houghton (ed.), *Connectionist Models in Psychology*, Psychology Press, Hobe, UK.

Herath, D. C., Kroos, C., Stevens, C. J., Cavedon, L. & Premaratne, P. (2010). Thinking Head: Towards human centred robotics, *Proceedings of 11th International Conference on Control, Automation, Robotics and Vision (ICARCV) 2010*, Singapore.

Herzog, G. & Reithinger, N. (2006). The SmartKom architecture: A framework for multimodal dialogue systems, *in* W. Wahlster (ed.), *SmartKom: Foundations of Multimodal Dialogue Systems*, Springer, Berlin, Germany, pp. 55–70.

Itti, L., Koch, C. & Niebur, E. (1998). A model of saliency-based visual attention for rapid scene analysis, *IEEE Transactions on Pattern Analysis and Machine Intelligence* 20(11): 1254–1259.

Kaplan, F. & Hafner, V. (2004). The challenges of joint attention, *in* L. Berthouze, H. Kozima, C. G. Prince, G. Sandini, G. Stojanov, G. Metta & C. Balkenius (eds), *Proceedings of the 4th International Workshop on Epigenetic Robotics*, Vol. 117, Lund University Cognitive Studies, pp. 67–74.

Kim, Y., Hill, R. W. & Traum, D. R. (2005). A computational model of dynamic perceptual attention for virtual humans, *14th Conference on Behavior Representation in Modeling and Simulation (brims)*, Universal City, CA., USA.

Kopp, L. & Gärdenfors, P. (2001). Attention as a minimal criterion of intentionality in robotics, *Lund University of Cognitive Studies* 89.

Kroos, C., Herath, D. C. & Stelarc (2009). The Articulated Head: An intelligent interactive agent as an artistic installation, *International Conference on Intelligent Robots and Systems (IROS 2009)*, St. Louis, MO, USA.

Kroos, C., Herath, D. C. & Stelarc (2010). The Articulated Head pays attention, *HRI '10: 5th ACM/IEEE International Conference on Human-Robot Interaction*, Osaka, Japan, pp. 357–358.

Kuhl, P. K., Tsao, F.-M. & Liu, H.-M. (2003). Foreign-language experience in infancy: effects of short-term exposure and social interaction on phonetic learning, *Proceedings of the National Academy of Sciences* 100: 9096–9101.

Lakin, J. L., Jefferis, V. E., Cheng, C. M. & Chartrand, T. L. (2003). The chameleon effect as social glue: Evidence for the evolutionary significance of nonconscious mimicry, *Journal of Nonverbal Behavior* 27: 145–162.

Liepelt, R., Prinz, W. & Brass, M. (2010). When do we simulate non-human agents? Dissociating communicative and non-communicative actions, *Cognition* 115(3): 426–434.

Metta, G. (2001). An attentional system for a humanoid robot exploiting space variant vision, *Proceedings of the International Conference on Humanoid Robots*, Tokyo, Japan, pp. 22–24.

Morén, J., Ude, A., Koene, A. & Cheng, G. (2008). Biologically based top-down attention modulation for humanoid interactions, *International Journal of Humanoid Robotics (IJHR)* 5(1): 3–24.

Ohayon, S., Harmening, W., Wagner, H. & Rivlin, E. (2008). Through a barn owl's eyes: interactions between scene content and visual attention, *Biological Cybernetics* 98: 115–132.

Peters, R. J. & Itti, L. (2006). Computational mechanisms for gaze direction in interactive visual environments, *ETRA '06: 2006 Symposium on Eye tracking research & applications*, San Diego, California, USA.

Pickering, M. & Garrod, S. (2004). Toward a mechanistic psychology of dialogue, *Behavioral and Brain Sciences* 27(2): 169–226.

Riley, M. A. & Turvey, M. T. (2002). Variability and determinism in motor behaviour, *Journal of Motor Behaviour* 34(2): 99–125.

Saerbeck, M. & Bartneck, C. (2010). Perception of affect elicited by robot motion, *Proceedings of the 5th ACM/IEEE International Conference on Human-Robot Interaction*, pp. 53–60.

Schneider, W. X. & Deubel, H. (2002). Selection-for-perception and selection-for-spatial-motor-action are coupled by visual attention: A review of recent findings and new evidence from stimulus-driven saccade control, *in* B. Hommel & W. Prinz (eds), *Attention and Performance XIX: Common mechanisms in perception and action*, Oxford University Press, Oxford.

Scholl, B. J. & Tremoulet, P. D. (2000). Perceptual causality and animacy, *Trends in Cognitive Sciences* 4(8): 299 – 309.

Sebanz, N., Bekkering, H. & Knoblich, G. (2006). Joint action: bodies and minds moving together, *Trends in Cognitive Sciences* 10(2): 70–76.

Shic, F. & Scassellati, B. (2007). A behavioral analysis of computational models of visual attention, *International Journal of Computer Vision* 73: 159–177.

Stelarc (2003). Prosthetic Head, New Territories, Glasgow, Interactive installation.

Stelarc, Herath, D., Kroos, C. & Zhang, Z. (2010a). The Articulated Head, NIME++ (New Interfaces for Musical Expression), University of Technology Sydney, Australia.

Stelarc, Herath, D., Kroos, C. & Zhang, Z. (2010b). The Articulated Head, SEAM: Agency & Action, Seymour Centre, University of Sydney, Australia.

Stelarc, Herath, D., Kroos, C. & Zhang, Z. (2011). The Articulated Head, Engineering Excellence Awards, Powerhouse Museum, Sydney, Australia.

Sun, Y., Fisher, R., Wang, F. & Gomes, H. M. (2008). A computer vision model for visual-object-based attention and eye movements, *Computer Vision and Image Understanding* 112(2): 126–142.

Tomasello, M. (1999). *The cultural origins of human cognition*, Harvard University Press, Cambridge, MA.

Tomasello, M., Carpenter, M., Call, J., Behne, T. & Moll, H. (2005). Understanding and sharing intentions: The origins of cultural cognition, *Behavioral and Brain Sciences* 28: 675–691.

Ude, A., Wyart, V., Lin, L.-H. & Cheng, G. (2005). Distributed visual attention on a humanoid robot, *Humanoid Robots, 2005 5th IEEE-RAS International Conference on*, pp. 381 – 386.

Wallace, R. S. (2009). The anatomy of A.L.I.C.E., *in* R. Epstein, G. Roberts & G. Beber (eds), *Parsing the Turing Test*, Springer Netherlands, pp. 181–210.

Wolfe, J. M. (1994). Guided search 2.0: a revised model of visual search, *Psychonomic Bulletin & Review* 1(2): 202–238.

Xu, T., Küandhnlenz, K. & Buss, M. (2010). Autonomous behavior-based switched top-down and bottom-up visual attention for mobile robots, *Robotics, IEEE Transactions on* 26(5): 947 –954.

Yu, Y., Mann, G. & Gosine, R. (2007). Task-driven moving object detection for robots using visual attention, *Proceedings of 7th IEEE-RAS International Conference on Humanoid Robots, 2007*, pp. 428–433.

13

Robot Arm-Child Interactions: A Novel Application Using Bio-Inspired Motion Control

Tanya N. Beran and Alejandro Ramirez-Serrano
University of Calgary
Canada

1. Introduction

Robot arms were originally designed in the 1960s for intended use in a wide variety of industrial and automation tasks such as fastening (e.g., welding and riveting), painting, grinding, assembly, palleting and object manipulation). In these tasks humans were not required to directly interact or cooperate with robot arms in any way. Robots, thus, did not require sophisticated means to perceive their environment as they interacted within it. As a result, machine type motions (e.g., fast, abrupt, rigid) were suitable with little consideration made of how these motions affect the environment or the users. The application fields of robot arms are now extended well beyond their traditional industrial use. These fields include physical interactions with humans (e.g., robot toys) and even emotional support (e.g., medical and elderly services).

In this chapter we begin by presenting a novel motion control approach to robotic design that was inspired by studies from the animal world. This approach combines the robot's manipulability aspects with its motion (e.g., in case of mobile robots such as humanoids or traditional mobile manipulators) to enable robots to physically interact with their users while adapting to changing conditions triggered by the user or the environment. These theoretical developments are then tested in robot-child interaction activities, which is the main focus of this chapter. Specifically, the children's relationships (e.g., friendship) with a robotic arm are studied. The chapter concludes with speculation about future use and application of robot arms while examining the needs for improved human-robot interactions in a social setting including physical and emotional interaction caused by human and robot motions.

2. Bio-inspired control for robot arms: simple and effective

2.1 Background: human robot interactive control

There are many different fields of human-robot interaction that have been developed within the last decade. The intelligent fusion scheme for human operator command and autonomous planner in a telerobotic system is based on the event based planning introduced in Chuanfan, 1995. This scheme integrates a human operator control command with an action planning and control for autonomous operation. Basically, a human operator passes his/her commands via the telerobotic system to the robot, which, in turn, executes the desired tasks. In many cases both an extender and material handling system are required during the implementation of tasks. To achieve proper control, force sensors have been used to measure the forces and moments provided by the human operator [e.g., Kim, 1998]. The sensed forces are then interpreted as the desired motion (translational and rotational) while the original compliant motion for the robot remains effective. To improve previous works, video and voice message has been employed, [e.g., Wikita, 1998], for information sharing during the human-robot cooperation. The projection function of the video projector is to project the images of the messages from the robot into an appropriate place. The voice message has the function to share the event information from the robot to the human. Fukuda et al. proposed a human-assisting manipulator teleoperated by electromyography [Fukuda, 2003]. The works described above simplify the many different applications in the field of human-robot interaction. The control mechanism presented herein allows robots to cooperate with humans where humans practically employ no effort during the cooperation task (i.e., minimal effort during command actions). Moreover, in contrast to previous work, where the human-robot cooperation takes place in a well structured engineered environment, the proposed mechanism allows cooperation in outdoor complex/rough terrains.

Human-robot arm manipulator coordination for load sharing

Several researchers have studied the load sharing problem in the dual manipulator coordination paradigm [e.g., Kim, 1991]. Unfortunately, these results cannot be applied in the scope of the human-arm-manipulator coordination. The reason is that in the dual manipulator coordination, the motions of the manipulators are assumed to be known. However, in the human-arm-manipulator coordination, the motion of the object may be unknown to the manipulator. A number of researchers have explored the coordination problem between a human arm and a robot manipulator using *compliant motion, predictive control* and *reflexive motion control* [Al-Jarrah, 1997; Al-Jarrah and Zheng, 1997; Iqbal, 1999]. In such scenarios the human-arm, by virtue of its intelligence, is assumed to lead the task while the manipulator is required to comply with the motion of the arm and support the object load. The intelligence of the arm helps perform complex functions such as task planning and obstacle avoidance, while the manipulator only performs the load sharing function. By coordinating the motions of the robotic arm with the user's arm, the uncertainty due to the environment can be reduced while load sharing can help reduce the physical strain in the human.

Complaint control

The basic ability for a robot to cooperate with a human is to respond to the human's intentions. Complaint motion control has been used to achieve both load sharing and trajectory tracking where the robot's motion along a specific direction is called complaint motion. This simple but effective technique can be used to guide the robot as it attempts to eliminate the forces sensed (i.e., precise human-robot interaction). However, diverse problems might occur that require different control approaches.

Predictive control

The problem in the framework of model-based predictive control for human-robot interaction has been addressed in numerous papers [e.g., Iqbal, 1999]. First, the transfer function from the manipulator position command to the wrist's sensor force output is defined. Then, the desired set point for the manipulator force is set to equal the gravitational force. Numerous results reported in the literature indicate that predictive control allows the manipulator to effectively take over the object load, and the human's forces (effort) stays close to zero. Moreover, manipulators have been shown to be highly responsive to the human's movement, and relatively small arm force can effectively initiate the manipulation task. However, difficulties still remain when sudden large forces are exerted to the robot to change the motion of the shared object (load) as the robot arm acts as another automated load to the human.

Reflexive motion control

Al-Jarrah [1997] proposed reflexive motion control for solving the loading problem, and an extended reflexive control was shown to improve the speed of the manipulator in response to the motion of the human. The results show that the controller anticipated the movements of the human and applied the required corrections in advance. Reflexive control, thus, has been shown to assist the robot in comprehending the intentions of the human while they shared a common load. Reflexive motion is an inspiration from biological systems; however, in reflexive motion control it is assumed that the human and the manipulator are both always in contact with an object. That is, there is an object which represents the only communication channel between the robot and the human. This is not always possible. Thus, mechanisms that allow human-robot cooperation without direct contact are needed.

In an attempt to enhance pure human-robot arm cooperation, human-mobile manipulator cooperation applications have been proposed [e.g., Jae, 2002; Yamanaka, 2002; Hirata, 2005; Hirata, 2007]. Here the workspace of the cooperation is increased at the expense of the added complexity introduced by the navigation aspects that need to be considered. Accordingly, humans cooperate with autonomous mobile manipulators through intention recognition [e.g., Fernandez, 2001]. Herein mobile-manipulators refer to ground vehicles with robot arms (Fig. 1a), humanoid robots, and aerial vehicles having grasping devices (Fig. 1b). In contrast to human-robot arm cooperation, here the cooperation problem increases as the mobile manipulator is not only required to comply with the human's intentions but simultaneously perceives the environment, avoids obstacles, coordinates the motion between the vehicle and the manipulator, and copes with terrain/environment irregularities/uncertainties, all of this while making cooperation decisions, not only between human and robot but also between the mobile-base and robot arm in real-time. This approach has been designated as active cooperation where diverse institutions are running research studies. Some work extends the traditional basic kinematic control schemes to master-slave mechanisms where the master role of the task is assigned to the actor (i.e., human) having better perception capabilities. In this way, the mobile manipulator not only is required to comply with the force exerted by the human while driving the task, but also contributes with its own motion and effort. The robot must respond to the master's intention to cooperate actively in the task execution. The contribution of this approach is that the recognition process is applied on the frequency spectrum of the force-torque signal measured at the robot's gripper. Previous works on intention recognition are mostly based on monitoring the human's motion [Yamada, 1999] and have neglected the selection of the optimal robot motion that would create a true human-robot interaction, reducing robot slavery and promoting human-robot friendship. Thus, robots will be required not only to help and collaborate, but to do so in a friendly and caring way. Accordingly, the following section presents a simple yet effective robot control approach to facilitate human-robot interaction.

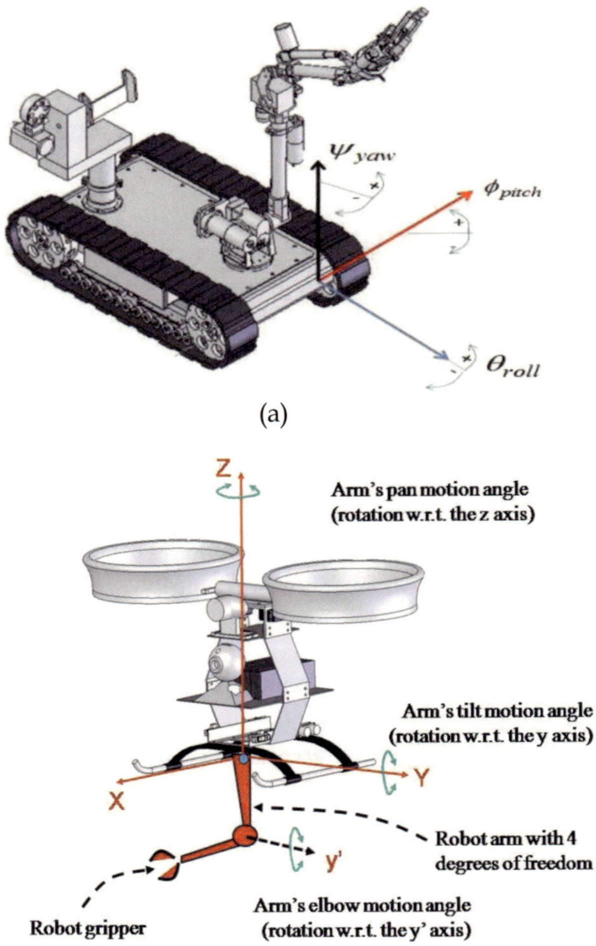

(b)

Fig. 1. Schematic diagrams of: a) Mobile manipulator, and b) Aerial robot with robotic arm.

2.2 Simple yet effective approach for friendly human-robot interaction

The objective of this section is to briefly present, without a detailed mathematical analysis, a simple yet effective human-robot cooperation control mechanism capable of achieving the following two objectives: *i*) Cooperation between a human and a robot arm in 3D dimensions, and *ii*) Cooperation between a human and a mobile-manipulator moving on rough terrain. Here the focus is placed on the former aspect as it is directly related to the experiments discussed in Section 3.

Many solutions have been developed for human-robot interaction; however, current techniques work primarily when cooperation occurs on simple engineered environments, which prevents robots from working in cooperation with humans in real human settings (e.g., playgrounds). Despite the fact that the control methodology presented in this section can be used in a number of mobile manipulators (e.g., ground and aerial) cooperating with humans, herein we focus on the cooperation between a human and a robot arm in 3D dimensions. This application requires a fuzzy logic force velocity feedback control to deal with unknown nonlinear terms that need to be resolved during the cooperation. The fuzzy force logic control and the robot's *manipulability* are used and applied to the control algorithm. The goal of using these combined techniques is to ensure that the design of the control system is stable, reliable, and applicable in a wide range of human cooperation areas. Herein, we specially consider those areas and settings where the associated complexities that humans and their environments impose on the system (robot arm) have a significant impact. When interaction occurs, the dynamic coupling between the *end-effector* (i.e., robot arm) and the environment becomes important. In a motion and force control scenario, interaction affects the controlled variables, introducing error upon which the controller must act. Even though it is usually possible to obtain a reasonably accurate dynamic model of the manipulator, the main difficulties occur from the dynamic coupling with the environment and similarly with the human. The latter is, in general, impossible to model due to time variation. Under such conditions a stable manipulator system could usually be destabilized by the environment/human coupling. Although a number of control approaches of robot interaction have been developed in the last three decades the *compliant motion control* can be categorized as the one performing well within the above described problems. This is due to the fact that compliant motion control uses *indirect* and *direct* force control. The main difference between these two approaches is that the former achieves force control via motion control without an explicit force feedback loop, while the latter can regulate the contact (cooperation) force to a desired value due to the explicit force feedback control loop. The indirect force control includes compliance (or stiffness) and impedance control with the regulation of the relation between position and force (related to the notion of impedance or admittance). The manipulator under impedance control is described by an equivalent mass-spring-damper system with the contact force as input. With the availability of a force sensor, the force signal can be used in the control law to achieve linear and decoupled impedance. Impedance control aims at the realization of a suitable relation between the forces and motion at the point of interaction between the robot and the environment. This relation describes the robot's velocity as a result of the imposed force(s). The actual motion and force is then a result of the imposed impedance, reference signals, and the environment admittance.

It has been found by a number of researchers that impedance control is superior over explicit force control methods (including hybrid control). However, impedance control pays the price of accurate force tracking, which is better achieved by explicit force control. It has also been shown that some particular formulations of hybrid control appear as special cases of impedance control and, hence, impedance control is perceived as the appropriate method for further investigation related to human-robot arm cooperation. Hybrid motion/force control is suitable if a detailed model of the environment (e.g., geometry) is available. As a result, the hybrid motion/force control has been a widely adopted strategy, which is aimed at explicit position control in the unconstrained task direction and force control in the constrained task direction. However, a number of problems still remain to be resolved due to the explicit force control in relation to the geometry.

Control architecture of human robot arm cooperation

To address the problems found in current human-robot cooperation mechanisms, a new control approach is described herein. The approach uses common known techniques and combines them to maximize their advantages while reducing their deficiencies. Figure 2 shows the proposed human-mobile robot cooperation architecture that is used in its simplified version in human-robot arm cooperation described in Section 3.

In this architecture the human interacting with the robot arm provides the external forces and moments to which the robot must follow. For this, the human and the robot arm are considered as a coupled system carrying a physical or virtual object in cooperation. When a virtual object is considered, virtual forces are used to represent the desired trajectory and velocities that guide the robot in its motion. In this control method the human (or virtual force) is considered as the master while the robot takes the role of the slave. To achieve cooperation, the changes in the force values, which can be measured via a force/torque (F/T) sensor, must be initialized before starting the cooperation. Subsequently, when the cooperation task starts, the measured forces will, in general, be different than the initialized values. As a result, the robot will attempt to reduce such differences to zero. According to the force changes, the robot determines its motion (trajectory and velocity) to compensate the changing in F/T values. Thus, the objective of the control approach is to eliminate (minimize) the human effort in the accomplishment of the task. When virtual forces are used instead of direct human contact with the robot the need to re-compute the virtual forces is eliminated.

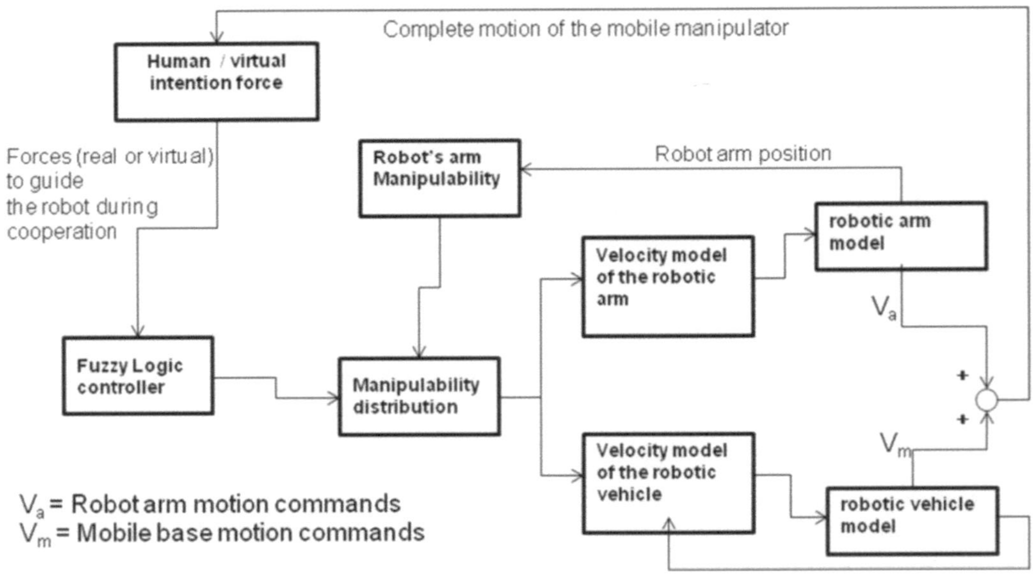

Fig. 2. Flow chart of the human-mobile robot cooperation.

Motion decomposition of the end-effector

The *manipulability* (w) of the robot arm captures the relation between the singular point and the gripper's end point. Here, the manipulability function of the robot arm (Fig. 2) is used to decompose the end-effector's desired motion based on the value of w. First the maximum w value of the arm has to be known before it can be used. If the manipulability is small, the end point of the robot's gripper is close to the singular point of the manipulator. That is, the capability of the robot arm to effectively react to the task while cooperating is reduced. On the other hand, if the value of w (manipulability) is large, the end point of the robot is far from the its singular point and the manipulator will find it easier to perform cooperating actions.

Thus the goal is to maintain the manipulability of the arm (and the *mobility* of the vehicle if working with a mobile manipulator) as large as possible, thus allowing the arm (and the vehicle when used) to effectively react to the unknown conditions of the environment and the cooperation tasks simultaneously. The fuzzy logic controller in Figure 2 is important in this case as the fuzzy rules can easily be tuned and used to distribute the robot arm's motion based on the manipulability value and the geometry of the environment (e.g., as the robot arm overcomes obstacles).

Control architecture of human-mobile manipulator cooperation

To finalize this section the cooperation between a human and a mobile manipulator is described for completeness. The motion of a mobile base is subject to *holonomic* or *nonholonomic* kinematics constraints, which renders the control of mobile manipulators very challenging, especially when robots work in non-engineered environments. To achieve the cooperation between the human and a mobile manipulator, a set of equations to represent the changes in forces and torques on the robot's arm caused by the interaction of the mobile manipulator on rough terrains is required. These equations can take different forms depending on the type or robot systems used (e.g., sensors). However, all forces and torques should be a function of the roll, pitch, and yaw angles of the vehicle as it moves. These formulations will indicate what portion of the actual sensed force must be considered for effective cooperation (i.e., human intention) and which portion is to be neglected (i.e., reaction forces due to the terrain or the disturbances encountered by the robot).

The control system of the manipulator for human-robot cooperation/interaction was designed considering the operational force by the human (operator) and the contact force between the manipulator and the mobile robot. The interacting force can be measured by a F/T sensor which can be located between the final link of the manipulator and the end-effector (i.e., the wrist of the manipulator). The human force and the operational force applied by the human operator denote the desired force for the end-effector to move while compensating the changing in the forces. The final motion of the manipulator is determined by the desired motion by the human force controller. To allow the arm to be more reactive to unknown changes (due to the human and the environment) the manipulability of the arm must be continuously computed. As the arm approaches the limits of its working environment the motion of the mobile manipulator relies more on the mobile base rather than the arm. In this way, the arm is able to reposition itself in a state where it is able to move reactively. In the experiments used in the next section the mobile base was removed. This facilitated the tests while simultaneously enhancing the cooperation.

The above control mechanism (Fig. 2) not only enhances human-robot cooperation but also enhances their interaction. This is due to the fact that the robot reacts not only to the human but also to the environmental conditions. This control mechanism was implemented in the studies presented in the following section.

3. Children's relationships with robots

We designed a series of experiments to explore children's cognitive, affective, and behavioral responses towards a robot arm under a controlled task. The robot is controlled using a virtual force representing a hypothetical human-robot interaction set a priori. The goal of using such control architecture was to enable the robot to appear dexterous, flexible while operating with smooth, yet firm biological type motions. The objective was to enhance and facilitate the human-robot cooperation/interaction with children.

3.1 Series of experiments

Experimental setup

A robot arm was presented as an exhibit in a large city Science Centre. This exhibit was used in all the experimental studies. The exhibit was enclosed with a curtain within a 20 by 7 foot space (including the computer area). A robot arm was situated on a platform with a chair placed .56 meters from its 3D workspace to ensure safety. Behind a short wall of the robot arm was one laptop used to run the commands to the robotic arm and a second laptop connected to a camera positioned towards the child to conduct observations of children's helping and general behaviors.
All three studies employed a common method. A researcher randomly selected visitors to invite them to an exhibit. The study was explained, and consent was obtained. Each child was accompanied behind a curtain where the robot arm was set up, with parents waiting nearby. Upon entering an enclosed space, the child was seated in front of a robot arm. Once the researcher left, the child then observed the robot arm conduct a block stacking task (using the bio-inspired motion control mechanisms described in Section 2). After stacking five blocks, it dropped the last block, as programmed.

Design and characteristics of the employed robot arm

The robot arm used in these experiments was a small industrial electric robot arm having 5 degrees of freedom where pre-programmed bio-inspired control mechanisms were implemented. To aesthetically enhance the bio-inspired motions of the robot the arm was "dressed" in wood, corrugated cardboard, craft foam, and metal to hide its wires and metal casing. It was given silver buttons for eyes, wooden cut-outs for ears, and the gripper served as the mouth. The face was mounted at the end of the arm, creating an appearance of the arm as the neck. Gender neutral colors (yellow, black, and white) were given to a non-specific gender. Overall, it was decorated to appear pleasant, without creating a likeness of an animal, person, or any familiar character yet having smooth natural type motions.
In addition to these physical characteristics, its behaviour was friendly and familiar to children. That is, it was programmed to pick up and stack small wooden blocks. Most children own and have played with blocks, and have created towers just as the robot arm did. This familiarity may have made the robot arm appear endearing and friendly to the children.
The third aspect of the scenario that was appealing to the children was that it was programmed to exhibit several social behaviours. Its face was in line with the child's face to give the appearance that it was looking at the child. Also, as it picked up each block with its grip (decorated as the mouth), it raised its head to appear to be looking at the child before it positioned the block in the stack. Such movement was executed by the robot by following a virtual pulling force simulating how a human would guide another person when collaborating in moving objects. Then, as it lifted the third block, the mouth opened slightly to drop the block and then opened wider as if to express surprise at dropping it. It then looked at the child, and then turned towards the platform. In a sweeping motion it looked

back and forth across the surface to find the block. After several seconds it then looked up at the child again, as if to ask for help and express the inability to find the block.

Fig. 3. Five degree of freedom robot arm on platform with blocks.

Measures

The child's reactions to the robot arm were observed and recorded. Then the researcher returned to the child to conduct a semi-structured interview regarding perceptions of the robot arm. In total, 60 to 184 boys and girls between the ages of 5 to 16 years (M = 8.18 years) participated in each study. We administered 15 open-ended questions. Three questions asked for general feedback about the arm's appearance, six questions referred to the robot's animistic characteristics, and six questions asked about friendship. These data formed the basis of three separate areas of study. First, we explored whether children would offer assistance to a robot arm in a block stacking task. Second, we examined children's perceptions of whether the arm was capable of various thoughts, feelings, and behaviours. Finally, the children's impressions about friendship with the robot arm were investigated.

3.2 Background

Only a generation ago, children spent much of their leisure time playing outdoors. These days, one of the favourite leisure activities for children is using some form of advanced technological device (York, Vandercook, & Stave, 1990). Indeed, children spend 2-4 hours each day engaged in these forms of play (Media Awareness Network, 2005). Robotics is a rapidly advancing field of technology that will likely result in mass production of robots to become as popular as the devices children today enjoy. With robotic toys such as Sony's AIBO on the market, and robots being developed with more advanced and sensitive responding capabilities, it is crucial to ask how children regard these devices. Would children act towards robots in a similar way as with humans? Would children prefer to play with a robot than with another child? Would they develop a bond with a robot? Would they think it was alive? Given that humans are likely to become more reliant upon robots in many aspects of daily life such as manufacturing, health care, and leisure, we must explore their psycho-social impact. The remainder of this chapter takes a glimpse on this potential impact on children by determining their reactions to a robot arm. Specifically, this section will explain whether children would offer assistance to a robot, perceive a robot as having humanistic qualities, and would consider having a robot as a friend.

Study 1: Assistance to a Robot Arm

Helping, or prosocial behaviours are actions intended to help or benefit another individual or group of individuals (Eisenberg & Mussen, 1989; Penner, Dovidio, Pilavin, & Schroeder, 2005). With no previous research to guide us, we tested several conditions in which we believed children would offer assistance (see Beran et al. 2011). The one reported here elicited the most helping behaviors.

Upon sitting in front of the robot arm the researcher stated the following:

- Are you enjoying the science centre? What's your favorite part?
- This is my robot (researcher touches platform near robot arm). What do you think?
- My robot stacks blocks (researcher runs fingers along blocks).
- I'll be right back.

The researcher then exited and observed the child's behaviors on the laptop. A similar number of children, who did not hear this introduction, formed the comparison group. As soon as children in each group were alone with the robot arm, it began stacking blocks. A significantly larger number of children in the introduction group (n = 17, 53.1%), than in the comparison group (n = 9, 28.1%), helped the robot stack the blocks, $X^2(1) = 4.15$, $p = 0.04$. Thus, children are more likely to offer assistance for a robot when they hear a friendly introduction than when they receive no introduction. We interpret these results to suggest that the adult's positive statements about the robot modeled to the child positive rapport regarding the robot arm, which may have created an expectation for the child to have a positive exchange with it. Having access to no other information about the robot, children may have relied on this cue to gauge how to act and feel in this novel experience. Interestingly, at the end of the experiment, the researcher noted anecdotally that many children were excited to share their experience with their parents, asked the parents to visit the robot, and explained that they felt proud to have helped the robot stack blocks. Other children told their parents that they did not help the robot because they believed that it was capable of finding the block itself. Overall, we speculate that the adult's display of positive regard towards the robot impacted children's offers of assistance towards it.

Study 2: Animistic impressions of a Robot Arm

Animism as a typical developmental stage in children has been studied for over 50 years, pioneered by Piaget (1930; 1951). It refers to the belief that inanimate objects are living. This belief, according to Piaget, occurs in children up to about 12 years of age. The disappearance of this belief system by this age has been supported by some studies (Bullock, 1985; Inagaki and Sugiyama, 1988) but not others (Golinkoff et al., 1984; Gelman and Gottfried, 1983). Nevertheless, the study of animism is relevant in exploring how children perceive an autonomous robot arm.

Animism can be divided and studied within several domains. These may include cognitive (thoughts), affective (feelings), and behavioural (actions) beliefs, known as schemata. In other words, people possess schemata, or awareness, that human beings have abilities for thinking, feeling, and acting. More specifically, thinking abilities may include memory and knowledge; feeling abilities include pleasant and unpleasant emotions; and behaviour abilities can refer to physical abilities and actions. Melson et al. (2009) provide some initial insights into several of these types of beliefs children hold towards a robotic pet (Sony's AIBO). Also, Melson et al. (2005) found that many children believed that such a robot was capable of the feelings of embarrassment and happiness, as well as recognition. Additional evidence of animism towards a robot was obtained by Bumby and Dautenhahn (1999) who reported that children may include human characteristics about robots in stories they create. The most recent study on animism presents surprising insights about animism. A team of researchers from the University of Washington's Institute for Learning and Brain Sciences [I-LABS, 2010] found that, "babies can be tricked into believing robots are sentient". The researchers used a remote-controlled robot in a skit to act in a friendly manner towards its human (i.e., adult) counterpart. When the baby was left alone with the robot, in 13 out of 16 cases the baby followed the robot's gaze, leaving researchers to conclude that the baby believed it was sentient. We extend these insightful findings of animism to children's cognitive, affective, and behavioural beliefs about a robot arm in the present study.

Responses to questions about the arm's appearance and animistic qualities were coded for this study. Two raters were used to determine the reliability of the coding, with Cohen's Kappa values ranging from 0.87 to 0.98, with a mean of 0.96 indicating very good inter-rater agreement. The majority of children identified the robot as male, and less than a quarter of the children identified it as female. One child stated the robot was neither, and about 10% did not know. The child's sex was not related to their response. About a third of the children assigned human names to the robot such as 'Charlie'. About a third gave names that refer to machines, such as 'The Block Stacker'. A pet name was rarely assigned, such as 'Spud', or a combined human-machine type name 'Mr. Robot'. When asked about their general impressions of the robot, a large majority gave a positive description, such as cool/awesome, good/neat, nice, likeable, interesting, smart, realistic, super, fascinating, and funny. Two children reported that the robot had a frightening appearance, and three children thought it looked like a dog. Another 17 did not provide a valid response.

Regarding its cognitive characteristics, more than half of the children stated the robot had recognition memory due to the ability to see their face, hair, and clothes; and that the robot was smart and had a brain (see Table 1). Other children provided a mechanical reason by stating it had a memory chip, camera, or sensors, or may have been programmed. Over a third of the children stated the robot could not remember them, for various reasons shown in the table. Children's perceptions about the robot's cognitive abilities in regards to knowledge are also shown in Table 1. About half of the children thought the robot did not have this capability, due to reasons such as not having a brain or interactions with them. Almost a third indicated that they believed the robot does know their feelings for various reasons such as from seeing the child and being programmed with this ability.

Regarding affective characteristics, the majority of children thought that the robot liked them, as shown in Table 2. A few children believed that the robot did not like them. Similarly, the majority of children reported that they thought the robot would feel left out if they played with a friend. Over a quarter of the children stated the robot would not feel left out, but provided explanations that would seemingly protect the robot from harm.

Robot can remember you		Robot knows your feelings	
Yes	97 (52.7%)	Yes	54 (29.3%)
Can see me	37	Can see me	18
Has memory chip, sensors	15	Has memory chip, sensors	5
Smart, has brain	3	Smart, has brain	3
If has a brain	6	Do not know why	17
If short duration	5	Not coded	11
If programmed	1		
Do not know why	24		
Not coded	6		
No	68 (37.0%)	No	103 (56.0%)
No brain, eyes, or memory	30	No brain, eyes, or memory	37
Too many people to remember	14	No interaction with me	19
Robot does not like me	3	If not programmed	8
If no brain	3	Do not know why	31
If long duration	2	Not coded	8
If not programmed	2		
Do not know why	11		
Not coded	3		
Do not know	19 (10.3%)	Do not know	27 (14.7%)

Table 1. Number and percentage of children reporting cognitive features of robot ($N = 184$).

Robot likes you		Robot feels left out	
Yes	118 (64.0%)	Yes	127 (69.0%)
Looks/smiles at me, friendly	38	No one to play with	62
I was nice/did something nice	20	Hurt feelings	36
Did not hurt me	13	I would include robot	9
It had positive intentions	9	Not fair	2
Do not know why	33	Do not know why	11
Not coded	5	Not coded	7
No	16 (8.7%)	No	53 (28.8%)
Ignored me/didn't let me help	10	No thoughts/feelings	29
No thoughts/feelings	4	Would include robot	16
Do not know why	2	Does not understand	3
Not coded	0	Do not know why	5
		Not coded	0
Do not know	50 (27.3%)	Do not know	4 (2.2%)

Table 2. Number and percentage of children reporting affective features of robot ($N = 184$).

In regards to its behavioral characteristics (Table 3), more than a third of the children stated the robot was able to see the blocks, with just over half of the children indicating that the robot could not see the blocks. A higher endorsement of the robot's ability to show action is evident in the table. That is, a large majority stated the robot could play with them, and even provided a variety of ideas for play. Examples include block building, and Lego®, catch with a ball, running games, and puzzles.

Robot sees blocks		Robot plays with you*	
Yes	77 (41.8%)	Yes	154 (83.7%)
Has eyes	32	Construction	103
Stacking	20	Ball game	26
Sensors, camera	13	Running game	12
Trained	5	Board game	12
Other	0	Other	17
Do not know why	7	Do not know why	5
Not coded	0	Not coded	5
No	94 (51.1%)	No	25 (13.6%)
Eyes not real	49	Physical limitation	11
Sensors, camera	19	Other	4
Missed a block	19	Do not know why	6
Guessed	1	Not coded	4
Do not know why	5		
Not coded	1		
Do not know	13 (7.1%)	Do not know	5 (2.7%)

Note* Many children provided more than one response.

Table 3. Number and percentage of children reporting behavioral features of robot ($N = 184$).

To further determine whether children considered the robot to be animate or inanimate, we analyzed the pronouns children used when talking about the robot arm. Almost a quarter of the children used the pronoun "it" in reference to the robot, another quarter stated "he", and half used both.

In summary, children seemed to adopt many animistic beliefs about the robot. Half thought that it would remember them, and almost a third thought it knew how they were feeling. Affective characteristics were highly endorsed. More than half thought that the robot liked them and that it would feel rejected if not played with. In their behavioral descriptions, more than a third thought it could see the blocks, and more than half thought the robot could play with them. It is evident that children assigned many animistic abilities to the robot, but were more likely to ascribe affective than cognitive or behavioral ones. There was additional evidence of human qualities according to the names children gave it, their descriptions of it, and the pronouns they used to reference it in their responses. These animistic responses, moreover, were more apparent in younger than older children.

Although some responses suggest that children believed the robot held human characteristics because of programming and machine design, the majority of statements referred to human anatomy (e.g., eyes, facial features, and brain), emotions, and intentions. We explain these findings in two ways. First, the robot arm presented many social cues.

That is, the eyes were at the same eye level as the children's, giving the impression of 'looking' at child, and it returned to this position many times while scanning for the block. Children may have interpreted this movement as expression of interest and closeness, which is one of the reactions to frequent eye contact among people (Kleinke, 1986; Marsh, 1988). Second, children may have projected their own feelings, thoughts, and experiences onto the robot arm, which Turkle (1995) has reported may occur with robots. This was particularly evident in the surprising finding that so many children believed that the robot would feel rejected and lonely if not included in play, as well as that the arm could engage in forms of play that clearly it could not (e.g., running). Third, children may have lacked knowledge of terms and principles to explain the robot's actions, thereby relying on terms that express human qualities such as 'remembering', 'knowing', and 'liking'. Fourth, because the arm moved autonomously, children may have developed the impression that it has intentions and goals, as is a typical reaction to any independently moving object (Gelman, 1990; Gelman and Gottfried, 1996; Poulin-Dubois and Shultz, 1990).

Study 3: Children's impressions of friendship towards a Robot Arm

Friendships are undoubtedly important for childhood development, and, as such, set the stage for the development of communication skills, emotional regulation, and emotional understanding (Salkind, 2008). In this study, and given the animistic responses obtained in the previous study, we set out to determine the extent to which children would hold a sense of positive affiliation, social support, shared activities, and communication towards a robot; all of which exemplify friendship. In addition, we questioned whether children would share a secret

Robot can cheer you up		Robot can be your friend	
Yes	145 (78.8%)	Yes	158 (85.9%)
Perform action for me	61	Conditional	31
Perform action with me	12	Being or doing things together	30
Cheerful appearance	20	Helpful	17
Connects with me	20	Knows me	12
Help me	7	Kind	11
Do not know why	17	Friendly	6
Not coded	8	Likeable	7
No	27 (14.7%)	Friend to robot	4
Limited abilities	16	Do not know why	28
Does not like me	1	Not coded	12
		No	19 (10.3%)
		Limited mobility	3
		Limited communication	2
		No familiarity	3
		No brain, feelings	4
Do not know why	8	Do not know why	4
Not coded	2	Not coded	3
Do not know	12 (6.5%)	Do not know	7 (3.8%)

Table 4. Number and percentage of children reporting positive affiliation ($N = 184$).

with a robot, as this behavior may also signify friendship (Finkenauer, Engels & Meeus, 2002). As shown in Table 4, more than three quarters of the children stated that the robot could improve their mood, with reasons varying from its actions to its appearance. Moreover, more than three quarters stated the robot could be their friend. Many reasons were given for this possibility. They included enjoying activities together, helping each other, kindness, likeability, and shared understanding.

According to Table 5, the majority of children stated they would talk to the robot and share secrets with it. Most children had difficulty explaining their reasons for their answers. Rather, they provided answers that described what they would talk about, such as what to play together. Interestingly, many children stated that they liked the robot and wanted to spend time becoming acquainted. This desire for a greater connection to the robot is also exemplified in their responses to sharing secrets. More than a third of the children stated they would tell the robot a secret. Some children (n = 24) stated that they thought it was wrong to tell secrets, suggesting that of those children who would generally tell secrets (n = 160), half of them (n = 84, 52.50%) would tell a robot. The most frequent reason given was because they believed that the robot would not share it - seemingly because the robot arm could not speak. Many of them also stated, however, that they considered the robot arm to be friendly.

Talk to robot		Tell robot secrets	
Yes	124 (67.4%)	Yes	84 (45.7%)
I like the robot	16	Robot will keep secret	30
To get to know each other	6	Friendship with robot	13
Robot has mouth	6	Positive response to secret	7
If robot could talk	22	Other	4
Gave examples	30		
Do not know why	37	Do not know why	22
Not coded	7	Not coded	8
No*	53 (28.8%)	No	92 (50.0%)
Robot cannot talk	20	Secrets are wrong	24
Robot cannot hear	6	Robot has limitations	18
Not human	5	Robot not trustworthy	24
Looks unfriendly	9	Robot is not alive	9
Do not know why	11	Do not know why	12
Not coded	4	Not coded	5
Do not know	7 (3.8%)	Do not know	8 (4.3%)

*Some children provided more than one reason

Table 5. Number and percentage of children reporting communication ($N = 184$).

The majority of children responded affirmatively to questions about affiliation, receiving support, communicating, and sharing secrets, which typically characterize friendship. Regarding affiliation, almost two thirds of the children thought the robot liked them, and many explained that it was because the robot appeared friendly. Children also attributed positive intentions to the robot, likely because it was moving independently and engaging in a child friendly task. More than three quarters of the children did believe that the robot could offer them support. The action of stacking blocks was often explained as a means of providing this support, perhaps to distract and entertain the child. A large majority of children stated that they would play with the robot in a variety of games. It is not surprising that many of them suggested building with blocks, considering that they had just observed this activity. Finally, about two thirds of the children stated they would talk to the robot and more than a third stated that they would share secrets. Again, these results suggest that children are willing to develop a bond with the robot.

Many children in our study stated that they would not engage in these friendship-behaviors with a robot and explained that the robot did not have the capabilities to do so. Reasons for these different perceptions of the robot have not been explored in the research but may plausibly include variation in children's knowledge of the mechanics of robots. In addition, a considerable proportion of children did not or could not provide an answer to the questions about friendship. It is possible that children were unable to differentiate human from robot characteristics, lacked sufficient understanding about the mechanics of robots, or were generally confused about the robot's abilities. Our use of terms in the interview, such as whether the robot would 'feel left out', describe human characteristics and may have mislead children into positively responding. Clearly, the results raise many questions for research, not the least of which is whether children actually *do* develop a friendship with a robot. Over time, and as a result of interactions with robots, children may develop a new system or schema of understanding, and subsequent vocabulary to articulate their sense of friendship with a robot, that is likely distinct from their friendships with children.

4. Implications for robot design

The fact that so many children ascribed life characteristics to the robot suggests that they have high expectations of them and are willing to invite them into their world. This presents a challenge to robot designers to match these expectations, if the purpose of the robot is to garner and maintain interest from children. Children may be primed for these interactions. In fact, children may become frustrated when a robot does not respond to their initiations and may actually persevere at eliciting a response (Weiss et al., 2009). Therefore, the robot may not need to be programmed to respond in an identical fashion to a specific initiation, as humans certainly do not, which may actually increase the child's engagement with a robot. This principle is well known as variable ratio reinforcement according to behaviourism learning theory (Skinner, 1969). Of course, children may become discouraged if the robot's response is erratic. Instead, we propose that a high, but not perfectly predictable response to the child's behaviours will lead to the longest and most interesting interactions.

In addition, our studies suggest that children can develop a collaborative relationship with a robot when playing a game together. This gives some suggestion of the nature of the relationship children may enjoy with a robot: one that allows give and take (Xin & Sharlin, 2007). This may enhance a child's sense of altruism and, hence, increase engagement with it. It is, thus, recommended that developers of such robots consider designing them to not only offer help, but be able to receive it.

4.1 Limitations

The studies and tests reported in this chapter have certain limitations that one must consider when interpreting the results. The tests and associated observations made during the study can be reproduced by using a variety of robot arms and even include mobile manipulators from where more detailed children-robot interaction studies can be made. Although the bio-inspired control mechanism used in this study worked well, tests using such control approaches should be performed on other robot types including humanoids and mobile robots. Such control architecture should also be tested in physical children-robot interaction to determine its suitability towards enabling seamless active engagement between children (humans) and robots.

Robot arms have indeed changed from their original industrial and automotive applications in the 1960s. Our studies show that children are ready to accept them as social objects for sharing personal information, offering mutual support and assistance, and regarding them as human in various ways. In the near future, we expect that humans will not only frequently and directly interact with and rely on robot arms and robots of diverse types for daily activities, but perhaps treat them and regard them as possibly human. Our studies cannot begin to address the numerous complex questions about the nature of the interactions people will have with robots. We offer a glimpse, however, of children's willingness to do so. Overall, the results are rather surprising given that the robot arm did not speak, performed only one task, and did not initiate physical interaction with the child. Are children merely responding to the robot arm as if it is a fancy puppet, and they are presenting their imagination in their responses? Perhaps, but regardless of the explanation, children in these studies demonstrated overwhelmingly their predisposition towards active engagement for bio-inspired motion control.

5. Acknowledgments

We give special thanks to the TELUS World of Science - Calgary for collaborating with us. This research would not have been possible without their support.

6. References

Abelson, R.P., 1981. Psychological status of a script concept. American Psychologist 36, 715-729.

Al-Jarrah, O.M. and Zheng, Y.F. *Arm-manipulator coordination for load sharing using reflexive motion control*, 1997, Proceedings IEEE International Conference on Robotics and Automation.

Al-Jarrah, O.M. and Zheng, Y.F. *Arm-manipulator coordination for load sharing using variable compliance control*, 1997, Proceedings IEEE Int. Conf. on Robotics and Automation.

Baron, R., Earhard, B., Ozier, M., 1995. In: Psychology: Canadian Edition. Allyn and Bacon, Scarborough.

Batki, A., Baron-Cohen, S., Wheelwright, S., Connellan, J., Ahluwalia, J., 2000. Is there an innate gaze module? Evidence from human neonates. Infant Behavior and Development 23, 223-229.

Beran, T. N., Ramirez-Serrano, A., Kuzyk, R., Nugent, S., & Fior, M. (2011). Would children help a robot in need? *International Journal of Social Robotics, 3*(1), 83-92.

BNET., 2008. Robot ideas to make life easier. Retrieved from http://findarticles.com/p/articles/mi_qn4176/is_20080928/ai_n28115686/

Bullock, M., 1985. Animism in childhood thinking: A new look at an old question. Developmental Psychology 21, 217-225.

Bumby, K.E., Dautenhahn, K., 1999. Investigating children's attitudes towards robots: A case study. In: Proceedings of the Third Cognitive Technology Conference. San Francisco, CA.

Chuanfan, G., Tzyh-Jong, T., Ning, X., and Bejczy, A.K. *Fusion of human and machine intelligence for telerobotic systems,* 1995. Proceedings of the IEEE Int. Conf. on Robotics and Automation.

Deutsch, J.M., 1943. The development of children's concepts of causal relations. In Barker, R., Kounin, J., Wright, H. (Eds.), Child behavior and development. McGraw-Hill, New York, pp. 129-145.

Eisenberg N, Mussen PH (1989) The roots of prosocial behavior in children. Cambridge University Press, Cambridge

Farroni, T., Csibra, G., Simion, F., Johnson, M.H., 2002. Eye contact detection in humans from birth. In: Proceedings of the National Academy of Sciences 99, pp. 9602-9605.

Fernandez, V., Balaguer, C., Blanco, D., and Salichs, M.A. *Active human-mobile manipulator cooperation through intention recognition,* 2001, IEEE Int. Conf. on Robotics and Automation.

Fukuda, O., Tsuji, T., Kaneko, M., and Otsuka, A., *A human-assisting manipulator teleoperated by EMG signals and arm motions,* 2003, IEEE Transactions on Robotics and Automation, 19(2): p. 210-222.

Gelman, R., Spelke, E.S., Meck, E., 1983. What preschools know about animate and inanimate objects. In Rogers, D., Sloboda, J. (Eds.), The acquisition of symbolic skills. Plenum, New York, pp. 297-327.

Gelman, R., 1990. First principles organize attention to and learning about relevant data: Number and the animate-inanimate distinction. Cognitive Science 14, 79-106.

Gelman, S.A., Gottfried, G.M., 1996. Children's causal explanation of animate and inanimate motion. Child Development 67, 1970-1987.

Golinkoff, R.M., Harding, C., Carlson, V., Sexton, M.E., 1984. The infant's perception of causal events: The distinction between animate and inaminate objects. In Lipsitt, L.P., Rovee-Collier, C. (Eds.), Advances in infancy research. Vol. 3. Ablex, Norwood, pp. 145-151.

Heerink, M., Kröse, B. J.A., Evers, V., Wielinga, B., 2008. The influence of perceived adaptiveness of a social agent on acceptance by elderly users. In: Proceedings of the 6th International Conference of the International Society for Gerontechnology, pp. 57-61.

Hirata, Y., Kume, Y., Wang, Z.-D., and Kosuge, K., *Handling of a Single Object by Multiple Mobile Manipulators in Cooperation with Human Based on Virtual 3-D Caster Dynamics.*

JSME Int. J. Series C Mechanical Systems, Machine Elements and Manufacturing, 2005. 48(4): p. 613-619.
Hirata, Y., Matsuda, Y., and Kosuge, K. *Handling of an object in 3-D space by multiple mobile manipulators based on intentional force/moment applied by human*, 2007, IEEE/ASME International Conference on Advanced intelligent Mechatronics.
Holland, V.M., Rohrman, N.L., 1979. Distribution of the feature [+animate] in the lexicon of the child. Journal of Psycholinguistic Research 8, 367-378.
Huang, I., Lee, H.W., 1945. Experimental analysis of child animism. Journal of Genetic Psychology 66, 69-74.
Hymel, S., Vaillancourt, T., McDougall, P., Renshaw, P.D., 2002. Peer acceptance and rejection in childhood. In Smith, P.K., Hart, C.H. (Eds.), Blackwell handbook of childhood social development. Blackwell Publishers, Oxford, pp. 265-284.
Iqbal, K. and Zheng, Y.F. *Arm-manipulator coordination for load sharing using predictive control*, 1999, Proceedings IEEE International Conference on Robotics and Automation.
Inagaki, K., Sugiyama, K., 1988. Attributing human characteristics: Developmental changes in over- and underattribution. Cognitive Development 3, 55-70.
I-LABS, 2010, Institute for Learning & Brain Sciences http://ilabs.washington.edu/research/research_themes.html
Itakura, S., Ishida, H., Kanda, T., Shimada, Y., Ishiguro, H., Lee, K., 2008. How to build an intentional android: Infants' imitation of a robot's goal-directed actions. Infancy 13, 519-532.
Jae, H.C., *Interactive force control of an operator-mobile manipulator coordination syste,* 2002, Journal of Robotic Systems,. 19(4): p. 189-198.
Jipson, J., Gelman, S., 2007. Robots and rodents: Children's inferences about living and nonliving kinds. Child Development 78, 1675-1688.
Kahn, P. H., Jr., Friedman, B., Perez-Granados, D. R., Freier, N. G., 2006. Robotic pets in the lives of preschool children. Interaction Studies: Social Behavior and Communication in Biological and Artificial Systems 7, 405-436.
Keil, F., 1979. In: Semantic and conceptual development: An ontological perspective. Harvard University Press, Cambridge.
Kim, K.I. and Zheng, Y.F. *Unknown load distribution of two industrial robots*, 1991, Proceedings IEEE International Conference on Robotics and Automation.
Kim, K.I. and Zheng, Y.F. *Human-robot coordination with rotational motion*, 1998, Proceedings Robotics and Automation IEEE International Conference.
Kleinke, C.L., 1986. Gaze and eye contact: A research review. Psychological Bulletin 100, 78-100.
Liu, C., Conn, K., Sarkar, N., Stone, W., 2008. Online affect detection and robot behavior adaptation for intervention of children with autism. IEEE Transactions on Robotics 24, 883-896.
Loring-Meier, S., Halpern, D.F., 1999. Sex differences in visuospatial working memory: Components of cognitive processing. Psychonomic Bulletin and Review 6, 464-471.
Marett, R.R., 1914. In: The threshold of religion. Methuen and Co, London.

Marsh, P.E., 1988. In: Eye to eye: How people interact. Salem House, Topsfield.

Melson, G.F., Kahn, P.H., Beck, A.M., Friedman, B., Roberts, T., Garrett, E. 2005. Robots as dogs? Children's interactions with the robotic dog AIBO and a live Australian shepherd. CHI, ACM Press, 1649-1652.

Melson, G.F., Kahn, Jr., P.H., Beck, A., Friedman, B., Roberts, T., Garrett, E., Gill, B.T., 2009. Children's behavior toward and understanding of robotic and living dogs. Journal of Applied Developmental Psychology 30, 92-102.

Michotte, A., 1963. In: The perception of causality. Basic Books, New York.

Muran, J.C., 1991. A reformulation of the ABC model in cognitive psychotherapies: Implications for assessment and treatment. Clinical Psychology Review 11, 399-418.

Nyborg, H., 1983. Spatial ability in men and women: Review and new theory. Advances in Behavior Research and Therapy 5, 89-140.

Oakes, M.E., 1947. In: Children's explanations of natural phenomena. Bureau of Publications Teachers College, New York.

Parker, J.G., Asher, S.R., 1987. Peer relations and later personal adjustment: Are low-accepted children at risk? Psychological Bulletin 102, 357-389.

Parker, J.G., Rubin, K.H., Price, J.M., DeRosier, M.E., 1995. Peer relationships, child development, and adjustment: A developmental psychopathology perspective. In Cicchetti, D., Cohen, D.J. (Eds.), Developmental psychopathology: Risk, disorder, and adaptation. Vol. 2. Wiley, New York, pp. 96-161.

Pellegrini, A.D., Smith, P.K., 1998. The development of play during childhood: Forms and possible functions. Child Psychology and Psychiatry Review 3, 51-57.

Penner LA, Dovidio JF, Pilavin JA, Schroeder DA (2005) Prosocial behavior: multilevel perspectives. Annu Rev Psychol 56: 365-392

Piaget, J., 1929. In: The child's conception of the world. Harcourt Brace, New York.

Piaget, J., 1930. In: The child's conception of physical causality. Routledge and Kegan Paul, London.

Piaget, J., 1951. In: The child's conception of the world. Routledge and Kegan Paul, London.

Pitsch, K., Koch, B. 2010. How infants perceive the toy robot Pleo. An exploratory case study on infant-robot-interaction, Second Int. Symp. on New Frontiers in Human-Robot-Interaction, 80-87.

Polanyi, M., 1968. Logic and psychology. American Psychologist 23, 27-43.

Poulin-Dubois, D., Schultz, T.F., 1990. The infant's concept of agency: The distinction between social and non-social objects. Journal of Genetic Psychology 151, 77-90.

Richards, L., 2005. Handling qualitative data: A practical guide. Thousand Oaks: Sage.

Saarni, C., 1999. In: The development of emotional competence. Guilford, New York.

Safran, J.D., 1990. Towards a refinement of cognitive therapy in light of interpersonal therapy: I. Theory. Clinical Psychology Review, 10, 87-105.

Scaife, M., van, Duuren, M., 1996. "Because a robot's brain hasn't got a brain, it just controls itself"-Children's attributions of brain related behaviour to intelligent artifacts. European Journal of Psychology of Education 11, 365-76.

Shonkoff, J.P., Phillips, D.A., (Eds.). 2000. From neurons to neighborhoods: The science of early childhood development. Committee on Integrating the Science of Early Childhood Development. National Academy Press, Washington, DC.

Siegel, L., 1993. Amazing new discovery: Piaget was wrong. Canadian Psychology 34, 239-245.

Simmons, A.J., Goss, A.E., 1957. Animistic responses as a function of sentence contexts and instructions. Journal of Genetic Psychology 91, 181-189.

Skinner, B.F. (1969). Contingencies of Reinforcement: A Theoretical Analysis. Englewood Cliffs, NJ: Prentice-Hall.

Sodian, B., Bullock, M., 2008. Scientific reasoning - Where are we now? Cognitive Development 23, 431-434.

Syrdal, D.S., Koay, K.L., Walters, M.L., Dautenhahn, K. 2009. The boy-robot should bark! Children's impressions of agent migration into diverse embodiments. Proceedings of the New Frontiers in HRI Symposium.

Tamis-LeMonda, C., Shannon, J.D., Cabrera, N.J., Lamb, M.E., 2004. Fathers and mothers at play with their 2- and 3-year-olds: Contributions to language and cognitive development. Child Development 75, 1806-1820.

Tatsuya Nomura, Takayuki Kanda, Tomohiro Suzuki: Experimental Investigation into Influence of Negative Attitudes toward Robots on Human-Robot Interaction, AI&Society, Vol.20, No.2, Mar.2006

Thomas, R.M., 2005. Comparing theories of child development (6th ed.) Belmont: Thomson.

Tremoulet, P.D., Feldman, J., 2000. Perception of animacy from the motion of a single object. Perception 29, 943-951.

Turkle, S., 1995. In: Life on the screen: Identity in the age of the Internet. Simon and Schuster, New York.

Tylor, E.B., 1871. In: Religion in primitive culture. Primitive culture: Vol. 2. John Murray, London. (Reprinted 1958, New York: Harper Torchbooks).

Wakita, Y., Hirai, S., Hori, T., Takada, R., and Kakikura, M. *Realization of safety in a coexistent robotic system by information sharing*, 1998, Proc. Robotics and Automation, IEEE Int. Conference.

Weiss, A., Wurhofer, D., Tscheligi, M. 2009. "I love this dog" - Children's emotional attachment to the robotic dog AIBO. International Journal of Social Robotics 1, 243-248.

Weiss, A., Wurhofer, D., Bernhaupt, R., Beck, E., Tscheligi, M. 2008. This is a flying shopping trolley: A case study of participatory design with children in a shopping context. Proceedings of the Tenth Anniversary Conference on Participatory Design, 254-257.

Xin M, Sharlin E (2007) Playing games with robots - a method for evaluation human-robot interaction. N Sarkar (Ed) Human-robot interaction. Itech Education and Publishing, Vienna, Austria, pp. 469-480

Yamada, Y., Umetani, Y., Daitoh, H., and Sakai, T. *Construction of a human/robot coexistence system based on a model of human will-intention and desire*, 1999, Proc. IEEE Int. Conf. on Robotics and Automation.

Yamanaka, E., Murakami, T., and Ohnishi, K. *Cooperative motion control by human and mobile manipulator*, 2002, 7th International Workshop on Advanced Motion Control. 2002.

York, J., & Vandercook, T., & Stave, K. (1990). Determining favorite recreation/leisure activities. *Teaching Exceptional Children, 22*(4), 10-13.